The *Geographical Guide* of Ptole...,
of Alexandria

This volume offers a detailed study of Ptolemy of Alexandria's *Geographical Guide*, whose eight books contain a wealth of geographical information unavailable elsewhere and represent the culmination of the Greco-Roman discipline of geography.

Written near the middle of the second century AD, the *Geographical Guide* is the most anomalous of the surviving works of ancient geographical scholarship but offers a vivid record of the expansion of geographical knowledge in antiquity. Roller examines this peculiar text, which offers unique data about explorations in the far reaches of the inhabited world, from Thoule and Hibernia in the northwest to Kattigara in the southeast, and from Serike in northeastern Asia southwest into central Africa. He positions the *Guide* within the tradition of ancient geography and gives close attention to the reason why Ptolemy wrote the guide and how it contributes to the genre of geographical scholarship. There is also an emphasis on the topographic and ethnic material within the *Guide* that is new or unique, especially explorations in sub-Saharan Africa and knowledge of the world beyond India. Because the *Guide* was written over half a century after the previous extant geographical work—the first books of Pliny's *Natural History*—the book also assesses how knowledge of geography changed during this period.

This work is an essential text for students and scholars of ancient geography, and is also of interest to anyone working on the cultural history of the Roman Empire during this period.

Duane W. Roller is professor emeritus of classics at the Ohio State University. He received his PhD from Harvard University, and is the author of numerous academic books, including *Cleopatra: A Biography* and *Empire of the Black Sea*. He spent 34 years in archaeological field work in the eastern Mediterranean and was the recipient of four Fulbright awards.

Routledge Monographs in Classical Studies

The War Cry in the Graeco-Roman World
James Gersbach

Religion and Apuleius' *Golden Ass*
The Sacred Ass
Warren S. Smith

Studies in Ancient Greek Philosophy
In Honor of Professor Anthony Preus
Edited by D. M. Spitzer

Personal Experience and Materiality in Greek Religion
K.A. Rask

A Cognitive Analysis of the Main Apolline Divinatory Practices
Decoding Divination
Giulia Frigerio

Processions and the Construction of Communities in Antiquity
History and Comparative Perspectives
Edited by Elena Muñiz-Grijalvo and Alberto del Campo Tejedor

Didactic Literature in the Roman World
Edited by T. H. M. Gellar-Goad and Christopher B. Polt

Atheism at the Agora
A History of Unbelief in Ancient Greek Polytheism
James C Ford

For more information on this series, visit: www.routledge.com/Routledge-Monographs-in-Classical-Studies/book-series/RMCS

The *Geographical Guide* of Ptolemy of Alexandria

An Analysis

Duane W. Roller

Routledge
Taylor & Francis Group

LONDON AND NEW YORK

First published 2024
by Routledge
4 Park Square, Milton Park, Abingdon, Oxon OX14 4RN

and by Routledge
605 Third Avenue, New York, NY 10158

Routledge is an imprint of the Taylor & Francis Group, an informa business

British Library Cataloguing-in-Publication Data
A catalogue record for this book is available from the British Library

ISBN: 978-1-032-16441-0 (hbk)
ISBN: 978-1-032-16442-7 (pbk)
ISBN: 978-1-003-24859-0 (ebk)

DOI: 10.4324/9781003248590

Typeset in Times New Roman
by Apex CoVantage, LLC

Contents

List of Maps   *vi*
Preface   *vii*
Acknowledgements   *x*

Introduction   1

1   Ptolemy and the *Geographical Guide*   7

2   Ptolemy's Introduction to the *Guide*   27

3   Northern, Central, and Western Europe   46

4   Italia and Eastern Europe   70

5   Libya   95

6   Asia   119

7   The Final Portion of the *Guide*   178

Epilogue   184

Appendix 1: The Text of the Ivernia Section of the Guide   *190*
Appendix 2: Toponyms Mentioned in Book 1   *193*
Appendix 3: The Caption for Map 1 of Europe   *201*
Abbreviations   *203*
Bibliography   *205*
List of Passages Cited   *211*
Index   *216*

Maps

1.1	The Inhabited World as Known to Ptolemy	17
3.1	Map 1 of Europe	47
3.2	Map 2 of Europe	52
3.3	Map 3 of Europe	55
3.4	Map 4 of Europe	58
3.5	Map 5 of Europe	63
4.1	Map 6 of Europe	71
4.2	Map 7 of Europe	73
4.3	Map 8 of Europe	76
4.4	Map 9 of Europe	82
4.5	Map 10 of Europe	90
5.1	Map 1 of Libya	96
5.2	Map 2 of Libya	101
5.3	Map 3 of Libya	105
5.4	Map 4 of Libya	111
6.1	Map 1 of Asia	120
6.2	Map 2 of Asia	128
6.3	Map 3 of Asia	132
6.4	Map 4 of Asia	137
6.5	Map 5 of Asia	144
6.6	Map 6 of Asia	149
6.7	Map 7 of Asia	153
6.8	Map 8 of Asia	158
6.9	Map 9 of Asia	161
6.10	Map 10 of Asia	164
6.11	Map 11 of Asia	167
6.12	Map 12 of Asia	172

Preface

In the second century AD, the Roman empire encompassed what Romans believed was the best part of the earth and the most civilized portions of humanity. During the reign of Trajan (AD 98–117) it reached its widest extent, from the British Isles to the Persian Gulf. In the north the Rhine-Danube line was the effective boundary, with some territories across the lower Danube, and in the south the limit was the Sahara, although Roman control extended well up the Nile.

It was in this environment that the last great work of ancient geographical scholarship was produced. The eight books of the *Geographike Hyphegesis* (*Geographical Guide*) of Klaudios Ptolemaios (Claudius Ptolemy) of Alexandria, written near the middle of the second century AD, are the culmination of the Greco-Roman discipline of geography. After its publication, only a few limited works and the idiosyncratic *Christian Topography* of Kosmas Indikopleustes appeared.

Yet the *Geographical Guide* itself is a peculiar work. Ptolemy was primarily a mathematician and astronomer, and even though geography and astronomy had intersected since the time of Hipparchos of Nikaia in the second century BC, and Ptolemy in his *Tetrabiblios* had considered ethnographic issues, the *Guide* is not so much a geographical treatise as an instruction manual for creating a map of the known world. But buried within it is a vast amount of unique data about the ancient Mediterranean world at its most expansive, with over 6000 toponyms, many of which had not been cited previously. Before Ptolemy, the last extensive geographical treatise was the geographical chapters of the *Natural History* of Pliny the Elder, completed in the AD 70s, and in the half century or more between that time and the *Geographical Guide* the Roman world had continued to spread, especially to the east, and Roman traders and merchants had gone far beyond the Empire into central Africa and eastern Asia. The geographical coverage of the *Guide* incorporates their experiences and thus moves beyond Pliny's treatise.[1]

The eight books of the *Guide* make it the second-longest geographical work surviving from antiquity. Much of it is a list of toponyms and their coordinates, bracketed by cartographic instructions in Books 1, 7, and 8. The current volume focuses on the geographical information, the toponyms and ethnyms, their location, and, most important, the unique nuggets on ancient exploration, especially in East and Central Africa and the farthest regions of southern Asia, that appear nowhere else in extant ancient literature.

Units of Measurement

The creation of any geographical text in antiquity has the pervasive problem of units of measurement. Until the establishment of the Roman mile, probably in the second century BC, there was no standard unit of long-distance measurement in the ancient world. Greek stadia, Persian parasangs, and Egyptian *schoinoi* all competed, often in the same locality. All of these units were variable: in the 20s BC Strabo noted that when he was traveling up the Nile and reached the customs post at the Hermopolitic guard station, the *schoinos* changed from 120 to 60 stadia.[2] Elsewhere there were other lengths for it. Modern efforts to find exact equivalents for these units are an exercise in futility and are doomed to failure, although very roughly there are about eight stadia to a mile, or five to a kilometer.[3] Obviously any attempt to determine the length of a *schoinos* is problematic; Ptolemy used a conversion factor of 30 stadia (*GG* 1.12.3). The parasang ranged from 30 to 60 stadia, which means a variable unit is calculated in terms of another variable one. The only distance from antiquity that can be exactly expressed is the Roman mile, at 1.48 km.

Ptolemy mostly used stadia—although whose stadion remains uncertain—with occasional references to Roman miles and *schoinoi*. But all these distances may have been converted from other units without explanation, or go back to the most ancient unit of distance, travel days, whether by foot, camel, or ship. Any conversion provided in the present text remains problematic.

The Maps

Accompanying this volume are 27 maps which represent the landforms and some features of the inhabited world as Ptolemy represented them. These are designed to show how Ptolemy saw the world, and are based on his coordinates as well as the medieval maps—which represent no visual tradition from antiquity—that accompany many of the manuscripts.

Transliteration

Translation of ancient toponyms and ethnyms is another difficult issue. The form presented in an ancient text, whether in Greek or Latin, may not be that used by the local inhabitants, and may have been preserved through a complex derivative chain of multiple earlier languages. Moreover, by the Roman period, there was the additional problem of expressing toponyms in the Greek language that were originally Latin (or had come into Latin from indigenous versions). In this work, names are generally presented as Ptolemy had them, except for some that are obviously Greek versions of Latin forms (e.g., Noviomagus and Camulodunum, not Noiomagos and Kamoulodounion), as well as others that have a common English rendering (e.g., Rome or Athens). Yet no scheme of transliteration is completely infallible and there will always be inconsistencies; one suspects that in the multilingual world of classical antiquity such concerns were less important than they appear now.

Other Considerations

To the modern reader, the *Geographical Guide* seems an amalgamation of two separate works: the instructions for map making that occupy portions of Books 1 and 7 and all of Book 8, and the geographical treatise that is the remainder. Because of this it falls into the trap of all interdisciplinary works, covering as it does issues of astronomy, mathematics, and geography; rarely is the entire work of equal interest to a particular scholar. This volume is intended to focus on the geographical material; the astronomy and mathematics have been well treated by others.[4] Citations to the *Geographical Guide* (*GG*) in the following pages are within the text; other references are in the notes.

Notes

1 The *Circuit of the Inhabited World*, by a certain Dionysios, who lived in Alexandria during the reign of Hadrian (AD 117–138), while of interest, is brief (1,186 hexameters) and derivative. Yet one wonders if Ptolemy and Dionysios crossed paths.
2 Strabo, *Geography* 17.1.41.
3 Aubrey Diller, "The Ancient Measurements of the Earth," *Isis* 40 (1949) 6–9.
4 See, especially, J. Lennart Berggren and Alexander Jones, *Ptolemy's Geography: An Annotated Translation of the Theoretical Chapters* (Princeton 2000).

Acknowledgements

The author would like to extend his gratitude to Stanley M. Burstein, Georgia L. Irby, cartographer Esther Rodríguez-Gonzáles, Letitia K. Roller, Richard Stoneman, Lisbet Thoresen, and Amy Davis-Poynter, Marcia Adams, and many others at Routledge. Thanks are also due to the Emeritus Academy of the Ohio State University for grant support that assisted in the completion of this volume. As usual, most of the writing was done in his study in Santa Fe.

Introduction

Geography as a scholarly discipline began in the late third century BC with the publication of the *Geographika* of Eratosthenes of Kyrene. He was the first to lay out a comprehensive grid of the known world, and in the process located several hundred places.[1] In an earlier treatise, *The Measurement of the Earth*, he had determined the size of the earth and the position of the inhabited portion—essentially the Mediterranean basin and its surroundings—on its surface. Yet long before Eratosthenes, individuals had been interested in places and peoples, the routes between them, and the nature of the land and sea. This was necessary practical knowledge for seamen, traders, and armies. Such data were collected randomly without any scholarly analysis: the world of the *Odyssey* and the tribulations of Odysseus are the best demonstration of this early environment.

From the beginnings of seamanship it had been known that the world was curved: ships sank below the horizon and the land rose up as one approached it. This eventually led to the idea of the spherical earth, a belief that had developed among the Pythagoreans of southern Italy by the early fifth century BC and which was an essential component of Eratosthenes' calculations. Parmenides of Elea, a Pythagorean by training, may have been the first to suggest such a theory.[2] This was not an idea easily accepted: even Eratosthenes, hundreds of years later, had to remind his readers that his theories were based on such a concept.[3]

But Eratosthenes faced another problem in establishing the discipline of geography: he functioned in a world where long-distance measurements were hardly necessary and often unavailable. All that needed to be known regarding the relationship between two places was the length of time it took to travel from one to the other. That it took 65 days for camel caravans to bring frankincense and other aromatics from southern Arabia to the exporting seaport of Gaza was the essential piece of information.[4] The actual distance was irrelevant.

Moreover, had someone wanted to calculate the distance to Gaza, it was almost impossible to do so. The measurements of length in Greek antiquity were variable and could not easily be converted from one to the other. Additionally, as a means of comprehending long distances they became increasingly unreliable: the 65 days across the Arabian Peninsula could not be accurately determined in any units of length. Eratosthenes, who used stadia in both of his geographical treatises, made

DOI: 10.4324/9781003248590-1

many conversions from travel days into stadia, which gave his material an appearance of greater precision than actually existed.

Approximately a century after the publication of Eratosthenes' works, these flaws were pointed out by the mathematician and astronomer Hipparchos of Nikaia, in his *Against the Geography of Eratosthenes*, produced by the 120s BC. He noted that Eratosthenes' distances and positions, even if expressed in stadia, were originally from travelers' reports, and thus many of his figures were inaccurate.[5] Hipparchos realized that correct positioning could only be achieved through astronomical calculation, although he had no good way of systematically putting his theory into practice and therefore could not revise much of Eratosthenes' material.[6] With a rudimentary understanding of trigonometry (not available to Eratosthenes), and also the better knowledge of the western Mediterranean that had become available to the Greco-Roman world since the fall of Carthage in 146 BC, Hipparchos was able to make some corrections and additions to the earlier scholar's data. He provided as many astronomically calculated measurements as possible, but he was also well aware that his inability to determine longitude hampered his efforts (he knew the process, using eclipse data, but had no reliable way of implementing it).[7]

Hipparchos was most adept at providing the theoretical structure for obtaining accurate geographical information; his application of the methodology was less successful. But his great achievement was understanding that geographical precision required use of mathematics and astronomy. In this he anticipated Ptolemy nearly 300 years later, who was familiar with Hipparchos' work.

Thus by the second century BC the abilities and limitations of geographical scholarship had been deftly set out by Eratosthenes and Hipparchos. Yet determining locations and distances on the surface of the earth remained a problematic endeavor, due to the lack of fixed units of distance measurement, uncertain conversion factors, and the inability to apply astronomical information systematically. Nevertheless Eratosthenes had, in theory, positioned about 400 places, extending from the British Isles and the northern coast of the Black Sea east and south to India and the edge of the Sahara Desert and west to the Atlantic. However uncertainly these may have been located, it was an astounding feat given the limitations that he faced.

Another important element of geographical scholarship was mapping, a tool that had existed since the earliest times. Ptolemy made it clear that the primary purpose of his treatise was to provide instructions for a map:

> What is before us at present is to lay out the inhabited world as properly and accurately as possible.
>
> *GG* 1.1.2

Although the word that he used for the process, *katagraphein*, could mean a literary description of the inhabited world (known as the *oikoumene* in Greek), what follows in the *Guide* makes it implicit that he was referring to a map.

Maps had existed in the Greek world since at least the sixth century BC, and would have been more useful than written descriptions in a culture with limited

literacy. Some ancient maps have survived, but most of the evidence is from literature: maps may often have been ephemeral documents created for a single purpose,
such as for soldiers in the field. The origin of Greek mapmaking is attributed to the
Ionian monist Anaximandros of Miletos around 600 BC.[8] The most famous early
incident involving a map in the Greek world was when Aristagoras, the tyrant of
Miletos, used one to attempt to persuade (unsuccessfully) the king of Sparta, Kleomenes, to join the Greek resistance to Persian expansionism.[9]

By Hellenistic times there were public maps, such as the one in the Lyceum
in Athens commissioned by Theophrastos, perhaps for teaching purposes.[10] Later
geographical scholars may not have created maps but nevertheless used them.[11] An
evolution in mapping technique was when Krates of Mallos, the Pergamene ambas-
sador to Rome in the 160s BC, constructed the first globe, ten feet in diameter.[12] He
was primarily a Homeric scholar, and his intent was to illustrate points of Homeric
topography, but it was the first attempt to show the entire earth.

By the second century BC the Romans had begun to have a major international presence. The Roman Republic now controlled territory from the Greek
world to the Iberian Peninsula, and Romans were developing an interest in geographical issues. Previously they had added little to the discipline, since Roman
territory was largely limited to regions already known to the Greeks. But after
the fall of Carthage in 146 BC this began to change. Polybios of Megalopolis, in
the service of Publius Cornelius Scipio Aemilianus, was present with his patron
at the destruction of Carthage and was sent to inspect the former Carthaginian
territories on the Atlantic coast of Africa, going as far as the equator.[13] Roman
entanglement with Mithridates VI of Pontos (ruled from around 120 to 63 BC)
led to an awareness of the Black Sea coasts and the Caucasus Mountains. In the
middle of the first century BC, Julius Caesar, while stationed in Gaul, learned
about that region and the adjacent portion of the British Isles and later went
far up the Nile in the company of Kleopatra VII of Egypt.[14] Juba II of Mauretania (ruled 25 BC–AD 23) explored northwest Africa and the Arabian Peninsula.
Roman-inspired traders in the Augustan period perfected the route to India and
went beyond the Himalayas into central Asia;[15] at the same time others went
into the Baltic, adding the toponym Scatinavia or Scadinavia (Scandinavia) to
the known places.[16]

Although the Romans contributed significantly to the range of topographical
and geographical knowledge, these penetrations beyond the previously known
world were for pragmatic rather than scholarly reasons. Men like Cn. Pompeius
(Pompey the Great, the first Roman to explore the Caucasus), Julius Caesar, and
even Polybios visited these distant regions for military purposes, not exploration.
Any knowledge received about the landscape was secondary. Traders and merchants were perhaps the primary source of information—and always had been—
going far beyond the world known to most Mediterranean peoples. A document
known as the *Periplous of the Erythraian Sea*, produced in the middle of the first
century AD, is a trading and sailing guide to the Indian Ocean, with emphasis on
the routes to India.[17] It is a rare surviving example of a genre that flourished in
the expanding Roman world and which informed interested parties how to locate

overseas ports, what to take to them, and what to bring home. Needless to say it is replete with rare geographical information.

The practical knowledge thus obtained from reports such as the *Periplous of the Erythraian Sea*, Roman military exploration, and many other sources, often found its way into new scholarly treatises. Over 200 geographical authors are known by name from the Greco-Roman world, but only a handful survive.[18] Artemidoros of Ephesos wrote 11 books on geography at the end of the second century BC, including what may be the first examination of western Europe.[19] The existing fragments demonstrate how geographical writing began to flourish in late Hellenistic times. The earliest extensive extant geographical work is the 17-book *Geography* of Strabo of Amaseia, completed in the AD 20s. As one of the longest surviving works from classical antiquity, quoting nearly 200 predecessors, from Homer to scholars of Strabo's own era, it is the primary source today for the history of ancient geography.

Romans were more interested in maps than the Greeks, because of the ideological purpose of showing the various parts of the ever-expanding Roman empire and how it dominated the inhabited world. Public maps flourished, with the prime example, the one created by Marcus Vipsanius Agrippa in the late first century BC, set up in the Porticus Vipsania in Rome.[20] The fragments of the *Forma Urbis Romae*, showing the city of Rome from around AD 200, and the late-antique Peutinger Map, based on earlier material, are the best surviving evidence for the genre.[21] Yet the extant maps in the manuscripts of the *Geographical Guide* are no earlier than AD 1300, and are based on Ptolemy's written data rather than any visual tradition from antiquity.[22]

Geography had always been a Greek discipline—even such romanized authors as Juba II and Strabo wrote in Greek—but by the first century AD there were geographical treatises in Latin, such as the three-book *Chorographia* of Pomponius Mela and the geographical books of the *Natural History* of Pliny. Pomponius Mela's work was written in the AD 40s. Although brief, it has information unavailable elsewhere and shows a surprisingly wide knowledge of the geographical tradition. More extensive are the five geographical books of Pliny the Elder, completed in the AD 70s. They were largely designed to apply Roman imperial ideology to geographical scholarship, and to emphasize the central position of the Roman world, both culturally and geographically. It remains the most thorough study of geography in the Latin language.[23] But Ptolemy, whose knowledge of Latin was limited, showed no direct awareness of either of these treatises.

Yet the Romans, despite their more pragmatic interest in geographical data, were able to mitigate the vagaries of uncertain units of measurement by establishing the first standardized length of distance, their mile (1.48 km.). Originally a military need, it came to be applied to the first Roman roads. Beginning in the fourth century BC, constructed roads started to radiate out from Rome, such as the Via Latina into the interior of Latium, and the Via Salaria to the northeast. The first long road was the Appia, begun in 312 BC, extending across the Pomentine Plain to Tarracina on the coast in southern Latium, and in time as far as Tarentum and Brundisium on the Adriatic. From its original segment comes the oldest known milestone, 53 miles from Rome. By the following century word *milia*, or miles, a contraction from the military term *milia passuum*, or "thousand paces," was in regular use.[24] The Roman mile, originally dependent on soldiers' marches, would at first have been somewhat variable, but with the establishment of milestones it became standardized. The thousands of examples surviving throughout Roman territory testify to the importance of the first regularized unit of long-distance measurement in the Mediterranean world.

Soon the mile made its way into Greek literature, as *milion*. Polybios, in the second century BC, reported the distances along Roman roads in miles, and thereby created a Greek neologism, *miliazo*, or "to measure in miles." He also provided the first known conversion factor from stadia, 8 1/3 to a mile.[25] Unfortunately this became anomalous, with exactly eight stadia to a mile more common.[26] Thus, while useful, Polybios' data also compounded the uncertainty of understanding measurements since it is often difficult to determine whether distances published as miles passed through Polybios' conversion factor or not. Nevertheless the increasing use of Roman miles, even by Greek writers, gave geographical calculations a precision that they did not previously have, although the fact still remained that the only accurate way of locating places was through astronomical positioning.

How much Ptolemy knew about these issues in the history of geographical scholarship remains uncertain. His sparse citation of sources means that it can be difficult to determine whom he relied upon. He mentioned Hipparchos twice (*GG* 1.4.2, 1.7.4), but failed to cite any of his major geographical predecessors. Eratosthenes, Strabo, Pomponius Mela, and Pliny are all absent (at least by name) in his text. But in his time the Roman mile was in all probability the most widely used long-distance measurement, even in the eastern parts of the empire which historically had depended on other units. Yet Ptolemy cited it on only two occasions, once in Britannia and again in territory southeast of the Black Sea (*GG* 1.15.6, 9); nevertheless these two widely separated locales demonstrate that miles were probably used extensively in preparing his treatise, but converted to stadia.

Thus Ptolemy functioned in a world with several hundred years of geographical scholarship, but which he hardly acknowledged. Yet his ability to collect thousands of toponyms demonstrates an astute knowledge of the contemporary state of the Roman world, even if his sources were unconventional.

Notes

1 Duane W. Roller, *Eratosthenes' Geography* (Princeton 2010).
2 Diogenes Laertios 8.48, 9.21.
3 Eratosthenes, *Geography* F25 (= Strabo, *Geography* 1.4.1).
4 Pliny, *Natural History* 12.64.
5 Hipparchos, *Against the Geography of Eratosthenes* F11–15.
6 D. R. Dicks, *The Geographical Fragments of Hipparchus* (London 1960) 32–3.
7 Hipparchos, *Against the Geography of Eratosthenes* F11 (= Strabo, *Geography* 1.1.12).
8 Agathemeros 1.1; Diogenes Laertios 2.2.
9 Herodotos 5.49–50.
10 Diogenes Laertios 5.51.

11 Eratosthenes, *Geography* F51; Hipparchos, *Against the Geography of Eratosthenes* F12.
12 Strabo, *Geography* 2.5.10. It is unlikely that Anaximandros of Miletos produced a globe, although one was attributed to him (Diogenes Laertios 2.2).
13 Polybios 34.15.7; Marijean H. Eichel and Joan Markley Todd, "A Note on Polybius' Voyage to Africa in 146 BC," *CP* 71 (1976) 237–43.
14 Lucan 10.188–92, 268–331; Suetonius, *Divine Julius* 52.1; Appian, *Civil War* 2.90; Leandro Polverini, "Cesare e la geografia," *Semanas de estudios romanos* 14 (2005) 59–72.
15 Duane W. Roller, *Ancient Geography* (London 2015) 164–6.
16 Pomponius Mela 3.54; Pliny, *Natural History* 4.96.
17 Lionel Casson, *The Periplus Maris Erythraei* (Princeton 1989).
18 *EANS* 999–1002.
19 R. Stiehle, "Der Geograph Artemidoros von Ephesos," *Philologus* 11 (1856) 193–244.
20 Pascal Arnaud, "Texte et carte de Marcus Agrippa: historiographie et données textuelles," *GA* 16–17 (2007–2008) 73–126; Richard J. A. Talbert, "Urbs Roma to Orbis Romanus: Roman Mapping on the Grand Scale," in *Ancient Perspectives: Maps and Their Place in Mesopotamia, Egypt, Greece and Rome* (ed. Richard J. A. Talbert, Chicago 2012) 167–70.
21 Tina Najbjerg and Jennifer Trimble, "The Severan Marble Plan Since 1960," in *Formae Urbis Romae* (ed. Robert Meneghini and Riccardo Santangeli Valenzani, Rome 2006) 75–101; Talbert, "Urbs Roma" 163–91.
22 Berggren and Jones, *Ptolemy's Geography* 41–50.
23 Duane W. Roller, *A Guide to the Geography of Pliny the Elder* (Cambridge 2022).
24 *CIL* 1.2.21.
25 Polybios 12.2a.4; 34.11.8; 34.29.4.
26 Strabo, *Geography* 7.7.4.

1 Ptolemy and the *Geographical Guide*

The Life and Works of Ptolemy

Despite Ptolemy's extensive professional output in the fields of astronomy, mathematics, geography, and other disciplines, his life and career are hardly known.[1] The sparse data are largely from late sources, and may be as much supposition as actual information. His name, Klaudios Ptolemaios, shows that his family had Roman citizenship, presumably given to an ancestor during the reign of the emperors Tiberius, Claudius, or Nero, all members of the Claudian family of Rome and who sequentially ruled from AD 14 to 68. This means that two or more generations before Ptolemy the family was one of distinction.

Nothing more is known about his family or his personal life other than the fact that he is consistently associated with Alexandria in Egypt and its environs. The name Ptolemaios is recorded from the earliest period of Greek history: the first known bearer of it was the father of Agamemnon's charioteer.[2] It occurred occasionally thereafter, especially in Macedonia, but became common after Ptolemy the son of Lagos accompanied Alexander the Great on his eastern expedition and then established himself as king in the new city of Alexandria in 305 BC. This created a dynasty that lasted until the death of Kleopatra VII in 30 BC, and which produced over a dozen homonymous successors to the first Ptolemy. The name spread through the royal families of the Hellenistic world, and also became common outside royalty during the rest of classical antiquity. But there is no way of connecting Klaudios Ptolemaios with other holders of the name.

There are enough known dates of his life and career to show that they spanned much of the second century AD. His *Mathematical Syntaxis* (popularly known today as the *Almagest*, a name derived through Arabic and Latin) contains dates ranging from AD 127 to 141.[3] Since by all accounts it was his earliest work—it was mentioned in several later treatises, including the *Geographical Guide* (*GG* 8.2.3)—one might assume that Ptolemy was born around the beginning of the second century AD.

There is evidence that he erected an inscription at Kanobos (Canopus, at modern Abukir, about 15 km. northeast of Alexandria). It was an ancient city, documented from the sixth century BC, and a religious center that by the late Hellenistic period had become a festive resort town.[4] Ptolemy's inscription is not extant, but was

DOI: 10.4324/9781003248590-2

mentioned by Olympiodoros of Alexandria, a Neo-Platonist philosopher of the sixth century AD, with the implication that it was still visible.[5] He also noted that Ptolemy lived for 40 years at a locale called the Wings of Kanobos, an enigmatic name that may either be part of building (unlikely) or a neighborhood in or near Kanobos.

Olympiodoros may have recorded the inscription, and its text appears in various medieval manuscripts of Ptolemy's works. It is a lengthy list of astronomical parameters. But of particular interest in understanding Ptolemy's biography is the explicit statement at the end, "erected at Kanobos in the tenth year of Antoninus," or AD 147/148. The 40 years that Ptolemy lived in Kanobos may be an accurate statement, or just the common generalization that this was the length of a person's professional life.

The final datum concerning the life of Ptolemy is from the *Suda*, the encyclopedia of the tenth century AD, whose entry for Ptolemy placed him in the time of Marcus Aurelius (reigned AD 161–180), a statement perhaps more approximate than precise.[6] Nevertheless the collected biographical material about Ptolemy is consistent. Born at the beginning of the second century AD, his professional life began in the AD 120s, and he spent many years working at the Wings of Kanobos. He was certainly there in AD 147/148, and may have lived into the AD 160s. Nothing is known about his education, students, or other associates.

Like most scholars in Alexandria during the Hellenistic and Roman periods, Ptolemy wrote in a variety of genres, although the emphasis was inevitably toward mathematics. The *Mathematical Syntaxis* in 13 books is his most famous work, setting forth the "Ptolemaic System" whose geocentric orientation was the basic of astronomical understanding until the heliocentric revolution of the sixteenth century. His subsequent efforts, some of which are only partially extant, or exist merely in Arabic or Latin versions, include ones on the fixed stars, astronomical calculations, planetary motion, geometry, astrology, music, and optics. One work which is of particular interest is the *Tetrabiblios*, a discussion of the relationship between astrology and astronomy. Yet it also has a geographical component, considering at length the characteristics of people living in different places on the earth.[7] This was not a new idea: the treatise *Airs, Waters, and Places*, of unknown authorship but attached to the Hippokratic corpus and probably from the fifth century BC, had explored this concept. Ptolemy's work examined the personal qualities of a number of ethnic groups, reflecting where they lived: reporting, for example, that the Hispanics love freedom, the Hellenes love learning, and the Parthians, Medians, and Persians were known for their cleanliness. There are many other examples: over a hundred regions of the world and their inhabitants are defined by personal characteristics. These researches may have helped Ptolemy conceive of a broader work on geography. Moreover, in the *Mathematical Syntaxis* he not only listed various terrestrial parallels—organized according to maximum length of day—but looked ahead to "a separate and geographical treatise."[8] The *Geographical Guide* may have been his last major work, probably begun after the Kanobic inscription was erected; a suggested date in the AD 150s or even 160s is most probable.[9]

The Sources of the *Geographical Guide*

Unlike his predecessors in the ancient geographical tradition, Ptolemy mentioned an astonishingly limited number of sources, in fact, only a dozen. By contrast, Strabo had nearly 200, and Pliny about 150. But Ptolemy emphasized that his major source was travel reports (*historia periodike, GG* 1.2.2), which he seems to have decided not to attribute to their authors.[10] Needless to say, one cannot determine all the sources that Ptolemy used, and as with any geographical work there would have been oral information, probably gathered in the waterfront tavernas of Alexandria, as well as various Roman official documents, including legionary reports and those relating to *coloniae*.[11]

Yet most of his named sources are not known elsewhere, and only five are geographical authors, with a particular emphasis on a certain Marinos of Tyre. The remaining seven are merchants, traders, and seamen, some of whose information may have been transmitted orally to Ptolemy or Marinos. The sources are cited only in Book 1 of the *Geographical Guide*. Ptolemy showed no awareness of the canonical Greek authors such as Homer, Herodotos, or the dramatists; presumably they had nothing significant to offer toward his research.[12] Yet his limited number of citations is misleading; as a resident near Alexandria he had access to the finest library in the world and thus essentially all human knowledge written in Greek and a number of other languages, and there are vague references to additional sources (e.g., *GG* 1.17, 7.7.4).[13] These are generally later than Marinos, and usually represent instances when Ptolemy had serious disagreements with his data. Ptolemy seems to have the greatest criticism of the earlier scholar's material with his information about East Africa. There is an extensive amount of detail about this region in the *Geographical Guide*, with several references to people who had sailed along its coast, as well as to India and beyond. Particularly singled out were unnamed merchants who had gone from the frankincense territory of Arabia south as far as Rhapta (near modern Dar es Salaam).

Moreover, Ptolemy had access to Tacitus' *Annals*, since a phrase in it, "ad sua tutanda" ("to protect their possessions"), was turned into a non-existent toponym, Siatoutanda.[14] This would also demonstrate that Ptolemy's Latin was limited; as noted in the following, Marinos would not have made this mistake. In addition, Ptolemy was probably aware of the Map of Agrippa in Rome, or perhaps a publication emanating from it.[15]

The earliest source named is Timosthenes of Rhodes, who was the naval chief of staff for Ptolemy II in the first half of the third century BC, writing *On Harbors*, a nautical guide for Ptolemaic seamen.[16] Only a few fragments of the treatise survive, and he was cited twice by Ptolemy in a passage directly quoting Marinos, perhaps an indication that Marinos may have been more connected to the history of geography than Ptolemy. The two references are in close succession.[17] The first is a distance (incorrectly preserved) up the Nile from Kanobos and the second a sailing distance along the northern African coast.

Chronologically the next source is the mathematician and astronomer Hipparchos of Nikaia, active during the second century BC. He was noted for a catalogue of stars, which is lost but was used frequently by Ptolemy in his *Mathematical Syntaxis*, yet his only known work on geography was his *Against the Geography of Eratosthenes*, most of whose extant fragments appear in the *Geography* of Strabo.[18] It is probable that Ptolemy used Hipparchos' material extensively in the *Geographical Guide*, but he was only mentioned twice. Hipparchos was said have reported the elevation of the north celestial pole at a number of cities, although no further details were provided (*GG* 1.4.2). The other citation is regarding the location of the southern star of the Little Bear (Ursa Minor), known today as Polaris (α UMi), which was 12 2/5 degrees from the pole.[19] This is hardly a geographical note, but it is in a direct quotation from Marinos' discussion of the celestial poles and the Little Bear. Like the case of Timosthenes, Ptolemy may have not accessed Hipparchos directly.

Two other barely known geographical authors are named in the *Geographical Guide*, and thus were probably cited by Marinos. A certain Philemon wrote about northern Europe: his surviving material is on the characteristics of amber, the Baltic, and Ivernia (Ireland), with only the mention of Ivernia found in the *Geographical Guide*.[20] Strabo, whose *Geography* was completed by AD 23, did not mention him, but Pliny did, writing half a century later, so he probably lived toward the middle of the first century AD. There is no further information about him.

Diodoros of Samos wrote at least three books on the sailing route to India, including details of astronomical phenomena (*GG* 1.7.6). The citation in the *Geographical Guide* is the only mention of him. Since he was not known to Pliny, he probably lived near the end of the first century AD. The single extant notice mentions the situation when one sails "from Indike to Limyrike." Limyrike is actually part of Indike (India), a region in the southwest on the Malabar Coast.[21] Diodoros may have been confused about the topography of India, or there may have been some error in transmission. It is also possible that his report reflects an earlier period when Greeks limited the toponym "Indike" to the region around the Indos River.

In addition to Marinos there are seven other sources cited in the *Geographical Guide*, all acquired by Ptolemy through Marinos. Four have Greek names and three Roman, and all are otherwise unknown. A certain Alexandros reported on the world east of India as far as the Malay Peninsula, and even beyond, perhaps an account of his own journey or part of a general examination of the extreme eastern portion of the known world (*GG* 1.14.1–3).

The remaining six were all merchants, traders, or (in the case of the Romans) military officials. What kind of reports they produced cannot be determined. None of the six, as well as Alexandros, was known to Pliny, so it can be assumed that they lived late in the first century AD or early in the following century. Diogenes seems to have been a regular on the Red Sea-India run, but claimed (as so many have from ancient to modern times) to have been blown off course. He ended up on the East African coast, as far south as the region of Zanzibar (*GG* 1.9.1). Dioskoros provided distances along the same coast (*GG* 1.9.4; 1.14.3), perhaps building on Diogenes' data. Theophilos was also in the area (*GG* 1.9.1; 1.14.4). Diogenes, at least, may have gone inland to the lakes of central Africa that were the source of the

Nile, perhaps naming the Selene Mountains, known in early modern times as the Mountains of the Moon (*GG* 4.8.3). These reports of three explorers in roughly the same region may suggest that Marinos or Ptolemy had access to a work on recent travels in East Africa.

Two Romans, Septimius Flaccus and Julius Maternus, made expeditions into central Africa (*GG* 1.8.5, 1.10.2). Flaccus went south of the famous oasis and cultural center of Garama on a journey lasting three months. Maternus, perhaps slightly later, also set out from Garama and went as far as the rhinoceros territory. Although the animal had been known in the Mediterranean world since at least the third century BC, it was still rare, and specific mention of it as part of Maternus' expedition suggests that acquisition was a primary objective.[22]

In addition, there was the personality recorded as "Maes, the one who is also Titianos," said to be a Macedonian from a family of merchants, and who commissioned an expedition to the far east. Four fragments of a work by him have been theoretically identified, but it is by no means certain what he may have published, or how he did so. Moreover, there is no evidence how his material came to Marinos. All the citations are from the *Geographical Guide*.[23] Maes would have been a rough contemporary of Gan (or Kan) Ying, who was sent west from China in AD 97, commissioned to reach Daqin, the westernmost part of the world, and its capital Andu, perhaps Antioch-on-the-Orontes. He never made it west of Parthia, but the chronological conflation of him and Maes is an interesting point, and the period around AD 100 seems to have been one of attempted Mediterranean-Chinese contacts.[24]

Maes sent his people to the land of the Seres, or Silk People, far to the east. They recorded the distances that they traveled, providing the first report of such data on the route to the Seres, especially its eastern portions. The work may have been in the same genre as the extant *Parthian Stations* of Isidoros of Charax, written in the Augustan period and reporting the distances from the Euphrates to Alexandria in Arachosia.[25]

Maes Titianos is an interesting personality, one of several merchants who sent their people to far-flung areas. A similar example is Annius Plocamus, who in the middle of the first century AD commissioned a mercantile expedition to Taprobane (modern Sri Lanka) and established trade between that locality and the Roman world.[26] Maes Titianos is less well known, recorded only through the single entry in the *Geographical Guide*.[27] Yet his identification as a Macedonian does not mean that he was from Macedonia, although this is possible, but the ethnym became common in Hellenistic times referring to those descended from the Macedonians who were spread across the Seleukid empire, remnants of military settlements or other ethnic movements in the years after Alexander the Great.[28] Thus it is more probable that Maes Titianos was from Asia Minor or the Levant, regions more suitable to his area of interest. His double name—seemingly both an indigenous one and a Greco-Roman one—follows a practice common in the diverse world of the Late Hellenistic period: the best-known example is Paul of Tarsos, styled (as popularly translated) "Saul who is also called Paul."[29] But any details about Maes' origin or career must remain speculative beyond the bare details reported by Ptolemy.

Although Marinos—Ptolemy's primary source—said that he did not trust the reports of merchants, these accounts of far-flung travelers are among the most interesting material preserved in the record of ancient geography, and are demonstrative that merchants and traders generally went farther than anyone else. But none of them are mentioned beyond the specific incidents (no more than two each) with which they are connected, and so their material, however fascinating, becomes isolated points of data.

Finally, there is Marinos himself.[30] He was mentioned over 20 times in the *Geographical Guide*, more than all other sources put together. He was introduced as the most recent author on geography and cartography (*GG* 1.6.1). After praising Marinos' diligence and skills, and his astute evaluation of his predecessors and correction of their information, Ptolemy then began a lengthy criticism of his material, noting that his treatise was not without numerous flaws and offering many corrections. There are also hints that Marinos' treatise was difficult to read. It was a lengthy work, although no book beyond the third is mentioned, and appeared in more than one edition, with any map lacking from the latest one. There were inconsistencies between the various versions (*GG* 1.7, 1.17.1), and both Marinos' latitudes (north-south) and longitudes (east-west) for the inhabited world were said to be in error (*GG* 1.7–9, 1.12.13). There were also issues with many distances, such as that to Kattigara (in east Asia) (*GG* 1.14), and a large number of problems of detail (*GG* 1.15–16). Moreover, Marinos' technique of map making was at fault (*GG* 1.19–20). Ptolemy saw making Marinos' work more accessible as his main goal:

> We have suitably decided to contribute as much as is necessary to the man's treatise in order to make it more reasonable and serviceable.
>
> *GG* 1.6.2

Thus, his intent was essentially editorial: Marinos' treatise was a difficult work to process, but Ptolemy would make it easier to read. This implies he did not see the *Geographical Guide* as a replacement to Marinos' work, but as a supplement.[31]

Many of the assumed errors by Marinos are inscrutable to the modern reader, although some of the minor details are quite obvious. Ravenna and Tergeste (modern Trieste) are not on the same latitude (the latter is over a degree to the north), a surprising error in this well-known region (*GG* 1.15.3). Marinos could be inconsistent: he reported that London was 59 miles north of Noviomagus (modern Chichester), which is essentially correct, but elsewhere placed Noviomagus to the north of London, probably a failure to reconcile sources (*GG* 1.15.6, 2.3.38).

The most obvious major error apparent to the modern reader concerns the journey of Julius Maternus to the rhinoceros country of Agisymba (*GG* 1.8.1). In his first edition, Marinos placed Agisymba 24,680 stadia south of the equator. Even given the difficulties in determining the length of the stadion,[32] one can consider this distance to be about 5000 km., which would place Agisymba well beyond of the southern end of Africa. It was calculated by Marinos in terms of the daily marches of Maternus' expedition and sailing days along the coast (presumably obtained from Diogenes). Marinos eventually realized that there was a problem, and in a later edition reduced the figure to 12,000 stadia (perhaps 2400 km.), essentially an arbitrary correction but still far too much (to about the latitude of Zimbabwe). This shows the perils of long-distance calculations based on travel times, and how so often there was no other way of obtaining a figure.

The question remains who Marinos was. As with Maes Titianos, his identity has been the subject of much speculation, and an presumed edition of his work was allegedly known to the Arab explorer and scholar al-Mas'udi in the tenth century AD, but this was probably merely a compilation from the citations in the *Geographical Guide*.[33] Ptolemy provided little assistance in identifying his predecessor, only that he was said to be from Tyros (Tyre) (*GG* 1.6.1). This is universally assumed to be the famous Phoenician city, known for its purple dye, seamanship, and expansionism. But there was also a Tyros (often Tylos) on the Persian Gulf (modern Bahrain), a Phoenician outpost as well as an obscure locality with the same name on the east coast of the Peloponnesos.[34] But it is most likely that Marinos came from Tyre in Phoenicia.

The single characterization of Marinos by Ptolemy is that he was dismissive of the ability of merchants to provide accurate information. Ptolemy was quite explicit about Marinos' feelings:

It seems that he [Marinos] was distrustful of the reports of merchants. At least he did not agree with the account of Philemon, who recorded the east-west length of the island of Ivernia as 20 days, which he said he had learned from the reports of merchants. He says that they are not interested in extracting the truth, because they are engaged in business, and they often increase the distance somewhat because of boastfulness.

GG 1.11.8

The essential truth of this statement does not hide the inevitable fact that merchants' reports were often the only information available; moreover this strong and definitive statement precludes any assumption that Marinos was a merchant himself.

Nevertheless Marinos did have access to Roman and Latin sources, a language that Ptolemy did not seem to know. The *Geographical Guide* recorded 20 legionary camps, as well as the expeditions of Julius Maternus and Septimius Flaccus, which suggests use of official Roman documents and records.[35] Although there is little reason to doubt that Marinos was from Phoenician Tyre, which was within the province of Syria in his day, his use of Roman military information makes it unlikely that he was merely a Greek geographer, and his personal association with Tyre may have been minimal, although the city had a long intellectual tradition, which, however, seems to have diminished by the first century AD.[36]

In Ptolemy's Greek text, the name is Marinos, but its Latin form (Marinus) is a Roman cognomen, especially that of a prominent Roman family of the late first and early second centuries AD, including L. Julius Marinus, the suffect consul of AD 93, and his son L. Julius Marinus Caecilius Simplex, suffect consul in AD 101.[37] The younger Marinus had a wide-ranging career, including command of the Legio XI

Claudia[38] and postings in Syria, Lykia, and Achaia. He is the most likely candidate for Ptolemy's Marinos.[39]

For a provincial to achieve the highest levels of the Roman administration, yet retain an affiliation with his city of birth, travel widely, and then devote his later years to a scholarly career reflecting his experiences was quite possible in the Roman world of the early second century AD. The outstanding example is L. Flavius Arrianus of Nikomedeia in Bithynia, who had a career much like that of the younger Marinos: a consulship (around AD 130), extensive travels, and then a later life of scholarship. Moreover, he retained enough affinity for his origins to write in Greek, a long tradition among highly placed scholarly Romans, including Juba II of Mauretania and the emperor Claudius.

Whoever Marinos was—and there seems little reason to doubt the identification with L. Julius Marinus Caecilius Simplex—his geographical work, which was lengthy and produced in several editions, had its final publication around AD 110 or slightly later.[40] One might presume that like Arrian, he retired from politics and began work on his treatise, which was completed before the emperor Trajan embarked on his Parthian campaign in AD 114, of which there is no mention in the *Geographical Guide*.

The Structure of the *Guide*

As a guide to making maps, Ptolemy's emphasis was on the visual aspect of geography, so that one would not be limited to a literary description in understanding the world (as was the case with readers of Pomponius Mela and Pliny, and to some extent Polybios), but the world can now be **seen**, and thus a visual comprehension of its full extent can be obtained.[41] Book 1 of the *Guide* lays out the issues regarding the drawing of maps of the inhabited world,[42] both in spherical and plane forms. But of particular interest to students of geography are Ptolemy's critiques of Marinos (known only through these passages) as well as numerous reports, presented in the context of correcting Marinos, about expeditions to previously unknown parts of the world. These expeditions are not documented elsewhere.

Book 2 is the first one devoted to the actual toponymic list; after some introductory material, it begins the catalogue of over 6000 place names (more than recorded in any other ancient text), with their coordinates.[43] These were based on a determination of latitudes north or south of the equator, and longitude east from a zero meridian just west of the Makarioi Nesoi, the ancient Blessed Islands, almost certainly the Canaries. Topographically the places located in Book 2 extend from the north Atlantic through northern and western Europe and into the northern Balkans.

Books 3 through 7 continue the toponymic survey. Book 3 is about Italy and its environs, the Danube, western and northern Black Sea regions, the remainder of the Balkans, and the Greek peninsula. Book 4 is about Libya (Africa), Egypt, and Aithiopia. Book 5 examines Asia Minor, the eastern Black Sea region, the Levant and Mesopotamia, and Book 6 continues east through Asia as far as it was known. Book 7 completes the survey with Indike (India) and places to its east, as well as Taprobane (Sri Lanka). In these remote regions there is a greater emphasis—although still limited—on ethnography, more than elsewhere in the *Geographical Guide*. Then, after a summary of the inhabited world, the process begins of how to make a map, with the actual regional maps outlined in Book 8. If physical maps were included in Ptolemy's text, they had vanished by medieval times.

Ptolemy's Grid of the Known World

In order to locate his places on the surface of the inhabited world, Ptolemy used a refinement of a grid system that had had its origins in the fourth century BC, if not earlier. This fixed the position of places in terms of their latitude (parallels) and longitude (meridians). Although these are the modern terms, derived from Latin *latitudo* and *longitudinis*, or width and length, the words were not used geographically in antiquity. The Greek for latitude was *klima*, or "slope," based on the belief that the earth sloped toward the poles, and thus different latitudes would reflect this.[44] The word for longitude was *mesembria*, or "midday," since it would be noon simultaneously at all points on the same longitude. This concept developed in the fifth century BC and was known to Herodotos.[45] Combining the *klimai* and the *mesembriai* (in their geographical sense) created a grid that ideally could locate every known point on the surface of the earth.

The first person to create a list of places on the same *klima*, or latitude, was Dikaiarchos of Messana. A talented student of Aristotle's, he created a base parallel across the inhabited world. It had long been believed that it was oblong, with its east-west dimension significantly longer than the north-south one, a point of view probably based on the reality of the shape of the Mediterranean.[46] Dikaiarchos created an east-west line from the Pillars of Herakles to the Imaos (Himalaya) Mountains.[47] It was astonishingly straight except in its western portion, which includes Sardinia between the Pillars and Sicily, an indication that the western Mediterranean was still little known. This line, although modified over time, became a standard element of geographical research.[48] In addition, Dikaiarchos made some rudimentary attempts to determine distances along his parallel; these, although generally erroneous, are the first attempt to make such calculations based on linear measurements, in this case using stadia.[49]

Early attempts to create a meridian, the *mesembria* line, were also more theoretical than empirical: Herodotos noted that Egypt and the mouth of the Istros (Danube) River were on the same longitude, quite reasonable.[50] With the various speculations about latitude and longitude lines, and the refinements over the centuries, it was theoretically possible to position any point on the earth, although the received data for the places, and their relationship to one another, were still based on overland distances. This was a flawed methodology that led to inaccuracies, as was realized by Hipparchos in the second century BC.[51] Nevertheless, as more places came to be known within the inhabited world, the grid system became more complex and thorough, and thus by the second century AD Marinos and Ptolemy were able to plot thousands of places.

Ptolemy relied on the Babylonian system of dividing the zodiac into 360 parts, which since at least the second century BC had also been used by Greek scholars to apply to terrestrial circles.[52] He, or Marinos, presumably converted existing distances in stadia or other units—such as those preserved by Strabo and probably derived (in part) from Polybios[53]—into a system of degrees. The Greek word normally translated "degree" is *moira*, common since the beginnings of Greek literature to mean "a portion,"[54] and which by Hellenistic times came to have its more familiar mathematical meaning; it was perhaps Hipparchos who was the first to use it in that sense.[55] Yet, as was so often the case in Greek geographical practice, the adoption of the concept of degrees gave a sense of greater precision than actually existed, since most of the degrees were converted from stadia or other units, resulting in a high amount of uncertainty in any modern determination of Ptolemy's degree measurements.

However innovative this system was, it was heavily flawed, since the repeated accumulation of overland distances—rarely if ever in straight lines—resulted in an east-west length to the inhabited world that was far greater that its actuality. This was a problem limited to the longitudes; for the latitudes, the length of the longest day at various points on a line provided accurate information.[56] But the east-west distances were almost impossible to determine accurately, and Ptolemy believed that the inhabited world extended over half the circumference of the earth.[57]

The easternmost documented points (Sarata and the land of the Sinai) were placed a full 180 degrees east of the zero meridian, which was just west of the Blessed Islands (Canaries) off the coast of West Africa (*GG* 7.3.5, 4.6.34). In actuality the distance from the Canaries to the far eastern points is no more than 130 degrees, which meant that the *Geographical Guide* created a distorted sense of the distances around the world, something that misled Renaissance explorers heading west from Spain. Aristotle had been the first to suggest vaguely that one could reach India by sailing west from the Pillars of Herakles,[58] and, using Ptolemy's data, Columbus and others could believe that it was only halfway around the world to the easternmost part of Asia, when in fact it was two thirds. Yet this belief that it was shorter to Asia from the Pillars than it actually was became a major facilitator in Renaissance exploration.

Although a relatively minor point, the exact placement of Ptolemy's zero meridian is difficult to determine. There is no doubt that the Blessed Islands formed the basis of the meridian, and Ptolemy identified six of them, placing them all in a north-south line extending through 5 1/2 degrees of latitude. The names he used (Aprositos, Hera, Pluvalia, Kasperia, Kanaria, and Kentouria) correspond, with some orthographical variants, to those established by the explorers of Juba II of Mauretania, the effective discoverers of the islands. He published his findings in his *Libyka*, and provided several figures for their location;[59] the most meaningful today, recorded by Pliny, is that they were 750 miles from Gades (modern Cádiz) and the islands themselves extended over 750 miles, but the numbers and orientation in Pliny's text are confused and contradictory. What is certain, however, is that the islands extend east-west over several hundred kilometers, which does not reconcile with Ptolemy's north-south positioning.

Ptolemy placed the Blessed Islands off the coast of Inner Libya and opposite the coastal toponyms of this region that lay between 7 and 14 degrees east longitude,

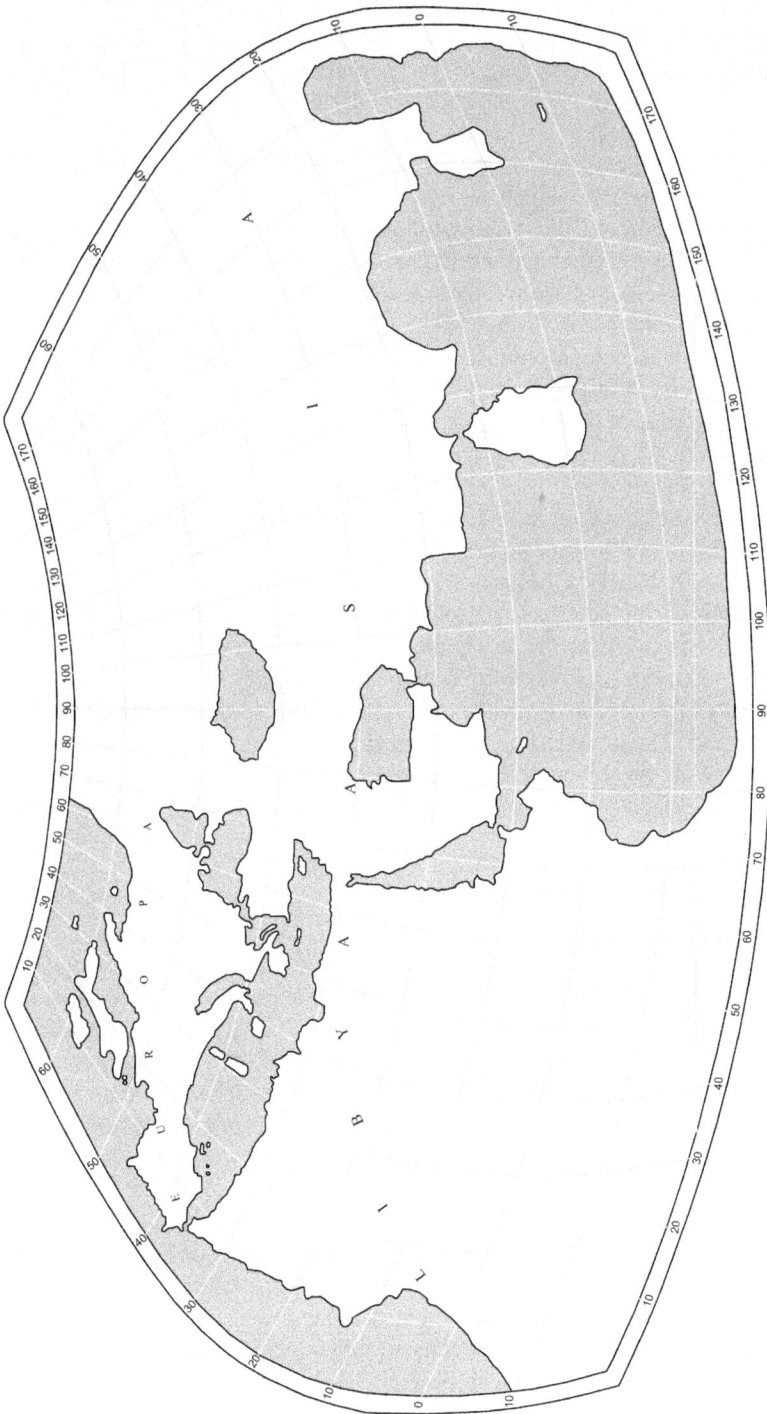

Map 1.1 The Inhabited World as Known to Ptolemy

but this is of little help in correcting their orientation error. Other than these locations, the westernmost point plotted on Ptolemy's maps is the Sacred Cape of Lusitania (the southwestern point of the Iberian Peninsula), at 2 1/2 degrees east longitude, but the longitude of the cape actually lies several degrees to the east of any point in the islands.

All these variables make it impossible to determine exactly where Ptolemy placed his zero meridian, which was one degree west of the Blessed Islands, but from what point in those islands remains inscrutable. No place on the maps is at the zero meridian; all the inhabited world is to its east. The best interpretation is that the meridian was west of the westernmost island: Ptolemy took the farthest west point of land that he knew about, and then placed his meridian one degree farther west. If he had complete information on the islands, this would have been modern La Palma, but it is not possible to reconcile this with Ptolemy's names, and his north-south alignment of a series of east-west trending islands in a further adds to the confusion.[60]

The north-south extent of the inhabited world was less problematic, given the easier means of calculating latitude. Moreover all of it was in the northern hemisphere or slightly south of the equator: the southernmost known point, Cape Prason on the east coast of Africa, lies at 15 degrees south (*GG* 4.8.2). To the south of the cape was unknown land extending to the south pole (*GG* 4.8.7). To the north the known land only went as far as Thoule (Thule), at "the end of the known sea," considered to be at 63 degrees north latitude (*GG* 2.3.32). Thus the perceived east-west extent of the inhabited world extended through 180 degrees of longitude but the north-south distance was only a matter of 78 degrees, confirming the east-west orientation of the inhabited world. It was within these limits that Ptolemy, following Marinos, located thousands of toponyms.

The Toponymic Lists

The toponymic lists of the *Guide* start near the beginning of Book 2 and continue part way through Book 7. In Book 2, Ptolemy first set forth his methodology for the actual guide itself (*GG* 2.1.1–11), noting that the positioning by degrees (*moirographia*, perhaps a neologism on his part) of well-known places could be considered nearly accurate, but less-visited sites were located only approximately and in terms of their relationship to the former.[61] He created a series of columns in his text, with toponyms followed by longitude (length) and latitude (width), thereby leaving space for later corrections. In theory, using Eratosthenes' circumference of the earth at 252,000 stadia, a degree of longitude at the equator would be 700 stadia (somewhat over 100 km.), but Ptolemy used a conversion of 500 stadia to a degree (*GG* 7.5). It was thus necessary to divide the degrees into fractions, as little as a twelfth of a degree, or 42 stadia (about eight km.), the smallest interval in the *Guide*.

Moreover, Ptolemy told his readers that his maps would be oriented with north toward the top and right toward the east, and that the entering of locations was best done from north to south and west to east, generally using political boundaries that would allow the world map to be divided into regional ones, a process outlined in Book 8.

This arrangement of the inhabited world reflected the lengthy east-west nature of the Mediterranean, and has persisted since the fourth century BC, with some exceptions in late antiquity and medieval times.[62] As soon as it was realized that almost the entire inhabited world lay north of the equator, it was easy to assume that north should be at the top of any visual representation. Throughout the *Guide* Ptolemy often used "above" and "below" for "north" and "south," an orientation used by his predecessors, such as Pliny the Elder, who regularly referred in this way to the map of Agrippa in Rome.[63] Yet Ptolemy felt it necessary to point out that this was indeed the scheme that he followed, suggesting that it was not inevitably taken for granted even in the second century AD.

Each region of the inhabited world consisted of a series of toponyms of varying types (towns, rivers and their mouths, and mountains are the most common). Places on the coasts would usually be presented first, and then those in the interior, followed by islands. Plotting places on the coasts inevitably created a coastline, which was not actually positioned but determined by connecting coastal points with one another, hardly the most accurate method, since twists and turns between plotted points would not be shown. Yet this would be the only reasonable one available: coasts tend to become straight lines from a perspective of out to sea, and offshore islands often cannot be distinguished from the mainland.[64]

At times there are brief explanatory comments regarding what follows, or the positioning of localities (especially islands) by the seas surrounding them. Mainland toponyms are also located by rivers and mountains, and the boundaries of other regions. Lists of local ethnic groups, which often do not have any coordinates, are interspersed within the toponymic catalogues, yet ethnographic comments are limited, although more common in the remote extremities of the world.

A final point stressed by Ptolemy is that he would make the meridians parallel. This would cause distortions at the extremities of the earth, an issue he dismissed, probably since the inhabited world did not extend beyond 63 degrees north or 15 degrees south, and thus any failure to show the meridians contracting was not particularly relevant. Yet he was certainly familiar with this phenomenon. The globe of Krates of Mallos would have shown the convergence of the meridians, and later, whether quoting Krates or not, Strabo had laid out the issues of depicting the earth on a globe.[65] Ptolemy certainly knew about globes, and discussed how to make a map on one (*GG* 1.22.1–6), but also chose to create flat maps, while insuring that his readers understood the difference between these two types.

Having set out his basic rules for the guide that follows, Ptolemy moved immediately to his catalogue of toponyms, beginning at the northwestern corner of the inhabited world, "a description of the northern side, beyond which lies the Hyperboreian Ocean" (*GG* 2.2.1). Each of the catalogue books of the *Guide* follows roughly the same pattern, with introductory passages summarizing the region to be discussed, followed by the actual list of toponyms with their longitudes and latitudes. The introductions can either be terse or highly descriptive, and often provide ethnyms of the peoples living in the region of the toponyms that follow. Yet rarely is there any geographical or cultural material of the sort found in Book 1, although when such information does appear, as with the products of Taprobane (*GG* 7.4.1), it is generally unique and valuable.

The toponyms can be of several types.[66] Especially common, particularly in more remote areas, are river mouths and coastal features, thus demonstrating Ptolemy's dependence on the coastal sailing manuals (*periploi*) that had pervaded Greek geographical thought since earliest times. As an example, of the 30 toponyms from Ivernia (Ireland), only seven are in the interior; the remainder are mouths of rivers (15), coastal features (five, marking the extremities of the island), and coastal towns (three) (*GG* 2.2.1–10). To be sure, the situation is different in the better-known parts of the inhabited world such as Italy and Asia Minor, but even there the mouth (and often the source) of virtually every river is plotted, essential information for shipping and trade. Mountains often seem to exist merely to provide a location for the sources of rivers, and are frequently speculative or non-existent.

From the first part of the toponymic catalogue in Book 2, describing Ivernia, to the final portion, a remarkably detailed listing of over 60 places on Taprobane (*GG* 7.4.1–13), there appear over 6000 toponyms, many of which are cited nowhere else. By contrast, Strabo, writing 150 years before Ptolemy, had "nothing distinct" to say about Ireland and reported no local toponyms. Pliny, half a century later, was equally uninformative.[67] For Taprobane, Strabo had few comments, although Pliny was somewhat more detailed.[68] But neither provided anything close to the richness of toponymic information in the *Geographical Guide*, again demonstrative of the vast amount of data Ptolemy (and Marinos) could access.

The Conclusion of the *Guide*: How to Make the Maps

The toponymic catalogues terminate in the middle of Book 7, and the remainder of the *Guide*, after some summary material, is concerned with the process of mapping, with instructions for making regional maps of the inhabited world, a process laid out in Book 8. There are a total of 26 such maps, laid out in the same manner as the toponymic lists, or from Ivernia to Taprobane. Each map has a caption, which lists various key locations, defined by the length of the longest day at various points on the maps and their distances (expressed in hours) east or west of Alexandria in Egypt. Some of this is inconsistent with that previously presented, leading to questions about the compositional history of the work. There is no literary conclusion to the *Guide*; when the data in Book 8 are provided, the work comes to an end.

Understanding the *Guide*

The *Geographical Guide* is a difficult work to assess. It casts a wide net, both intellectually and geographically, from mathematics and astronomy to topography and cultural history, with its various aspects not always integrated. The manuscript tradition is complex and confusing; the thousands of toponyms and their associated numbers were easily subject to scribal error or emendation. Ptolemy's literary style in the analytical chapters is difficult and often awkward, although the geographical material, with its catalogic nature, is more straight-forward.[69] The material from the middle of Book 7 to the end of the treatise is often contradictory and repetitive when compared to that in the previous geographical catalogues, and it has even been suggested that this portion of the work—and indeed the entire treatise—is a Byzantine compilation from several sources.[70] But this remains an unlikely hypothesis that merely reflects the modern frustration at comprehending the text. Nevertheless, as is often the case with ancient literature, the subheadings and chapter divisions may have been added after the time of Ptolemy, yet his use of the word *geographike* in the title (assuming this is his diction) shows that he saw the treatise as a work of geography, despite his practiced role as a mathematician and astronomer. In this he was indebted to his predecessor Hipparchos.

The Geographical Extent of the *Guide*

Greco-Roman civilization included an almost constant expansion of topographical knowledge, from the early information gathered by Bronze Age seafarers to the broad extent known to Ptolemy. In earliest times the Greek horizon was limited to the eastern Mediterranean and adjacent regions, eventually adding southern Italy, Sicily, and parts of the Black Sea coasts: this was the environment of Greek mythology and the Homeric poems. Expansionism began in the eighth century BC: Greeks founded over a hundred settlements east and west of their heartland. These included cities and trading posts around the perimeter of the Black Sea to its farthest point, Tanais, at its extreme northeast. By the end of the following century the western Mediterranean was also known, largely through the efforts of the Ionian city of Phokaia and its major outpost, Massalia (modern Marseille). Many of these settlements were at the mouth of major rivers: Massalia was near the mouth of the Rhodanos (Rhone), and Tanais on the river of the same name (the modern Don). Such a location facilitated trade far into the interior. Across southern Asia, contact with the Persian empire by eastern Greeks led to an understanding of routes into their territory, as well as to Mesopotamia, Arabia, and even India.

In the fifth century BC traders had begun to penetrate the Alps and the great forests of northern Europe, and exploited the ancient amber route to the Baltic. To the south, they had crossed the Sahara into tropical Africa, bringing indigenous wares and topographical knowledge to the Mediterranean world. Greeks also learned about Carthaginians who had explored the Atlantic coasts of Europe and Africa, and in some cases they attempted to replicate these voyages. The Carthaginians had gone as far north as the British Isles and perhaps Ireland, and south to the equator.

The eastern expedition of Alexander the Great in the second half of the fourth century BC allowed closer scrutiny of India and its environs, the Arabian Peninsula, and Taprobane (Sri Lanka). At the same time Pytheas of Massalia penetrated the Arctic, going as far as the mysterious island of Thoule (perhaps Iceland) and touching Scandinavia. He may also have entered the Baltic and have come to understand the riverine trade routes from it to the Black Sea. Trade developed with the elusive Silk People in far eastern Asia through a route first perfected from northern India, and then directly from the Mediterranean.

The Romans, beginning in the first century BC, added much knowledge about northern Europe, the Baltic, and Scandinavia. There was also additional information about sub-Saharan Africa and the upper Nile. And wherever official expeditions went, traders and merchants went farther: Greek and Roman trade objects have been discovered in central Scandinavia, southern Africa, and remote eastern Asia. There were even rumors of undiscovered continents, across the Atlantic and east of the Indian Ocean.[71]

Ptolemy inherited all this knowledge. He knew about practically the entire eastern hemisphere, from Scandinavia and Thoule in the north to the Selene Mountains and Cape Prason (probably modern Cape Delgado in Mozambique) in the south. To the east his knowledge went as far as the land of the Sinai (in some way associated with the Chinese), and into southeast Asia and perhaps to Java, although the positioning and identification of these places is far from certain. The only part of the continent of Asia that was totally unknown was the extreme northeast, beyond a line roughly from the Baltic to the Caspian Sea, and then east to the assumed location of the External Ocean. This was a region candidly described by Ptolemy as "unknown land"; the same phrase was used for extreme southern Africa (*GG* 8.24.2, 4.8.1). Ptolemy thus reflected ancient topographical knowledge at its widest extent, and all later ancient scholarship on the topic tended to be derivative. It was only in the Renaissance that Europeans gained additional knowledge about the inhabited world.

The Later History of the *Guide*

As the most extensive topographical survey written, the *Guide* was regularly consulted throughout later antiquity and into Byzantine and medieval times. Pappos of Alexandria, active around AD 300, was, like Hipparchos and Ptolemy, a mathematician who also had an interest in geography.[72] His *Chorographia of the Oikoumene* is only known through an Armenian edition, probably of the seventh century AD.[73] Although its textual tradition is thus complex, he seems to have quoted the *Guide* extensively.

Later in the fourth century AD, the historian Ammianus Marcellinus specifically mentioned that he had consulted Ptolemy for material on the Black Sea and probably made use of the *Guide* elsewhere in his treatise.[74] Sometime between the second and the sixth century AD, a certain Markianos, from Herakleia Pontika on the Black Sea, made a collation of Greek geographical writings (a collection known since the nineteenth century as the *Minor Greek Geographers*), which included such important texts as the *Periplous* of the Carthaginian explorer Hanno, and the *Periodos Dedicated to King Nikomedes* (often erroneously called Pseudo-Skymnos), as well as Markianos' own work, the *Periplous of the External Sea*. In it he often converted distances between points implicit in Ptolemy's coordinates into stadia. But he may not have consulted Ptolemy directly, since he credited a little-known Protagoras as his source.[75] Markianos, who cannot be dated precisely, mentioned Ptolemy and Protagoras in tandem as "ancient" authors, suggesting that they might have been roughly contemporary.[76]

Cassiodorus, a political leader and later a monk from southern Italy during the sixth century AD, wrote eulogistically about the *Guide*, noting that it was well researched and suitable for Christian monks.[77] His contemporary Kosmas Indikopleustes, in his *Christian Topography*, was aware of Ptolemy's research, but only cited him by name in connection with his astronomical efforts.[78]

Later Byzantine geographical authors had their own knowledge of the *Guide*, especially Johannes Tzetzes in the twelfth century, whose *History of Books*, or *Chiliades*, adapted into verse some portions of the *Guide*. Moreover, there was also an Arabic tradition of the treatise from as early as the ninth century, with a translation surviving from 1453. There may have also been versions in Syriac.[79] The Arabic translations can be useful for correcting Ptolemy's text.

The *Guide* itself was allegedly rediscovered around 1300, probably by Maximos Planoudes (ca. 1255–1305), who was from Nikomedeia in Bithynia but professionally active in Constantinople, where he may have found a copy of the treatise.[80] He used Ptolemy's instructions to create the maps that were to accompany the text, which are attached to a number of the extant manuscripts (and printed editions) but represent no visual tradition from antiquity.[81] There are about 50 manuscripts of the *Guide*, with the earliest from the end of the thirteenth century, but several originating with Planoudes and containing his maps.[82] The oldest one is only a few years earlier than those of Planoudes and is in poor shape.[83] The entire manuscript tradition is complex and contradictory, and is one of the major reasons that there has been a school of thought suggesting that the treatise as it exists today is an amalgamation, not an organic whole.[84]

Latin editions appeared in the fifteenth century, but the first printed Greek edition was by Erasmus, produced at Basel in 1533. The Teubner edition by C. F. A. Nobbe appeared in 1845.[85] The only English edition is by Edward Luther Stevenson in 1932, but it is so deeply flawed as to be useless.[86] Modern editions of importance include that of Berggren and Jones, with translations, analysis, and commentary on the theoretical chapters (portions of Book 1, 7, and 8), and that edited by Alfred Stückelberger and Gerd Graßhoff, with complete text and German translation, but limited notes and erratic handling of the toponyms.[87] The current volume builds on these previous editions and provides an analytical discussion of the *Geographical Guide*, with an emphasis on its geographical data.

Notes

1 G. J. Toomer, "Ptolemy," *DSB* 11 (1975) 186–206; Alexander Jones, "Ptolemy ('Claudius Ptolemaeus')," *EANS* 706–9.
2 Homer, *Iliad* 4.228.
3 Ptolemy, *Mathematical Syntaxis* 9.7, 11.5. These are dates when Ptolemy made his own observations.
4 Herodotos 2.15; Strabo, *Geography* 17.1.17; Plutarch, *Solon* 26.1.
5 Olympiodoros, *Commentary on Plato's Phaidon* 17.72b; Alexander Jones, "Ptolemy's *Canobic Inscription* and Heliodorus' Observation Reports," *SCIAMVS* 6 (2005) 53–97, with text and translation.
6 *Suda*, "Ptolemaios [2]".
7 Ptolemy, *Tetrabiblios* 2.2–3. See the diagram in Germaine Aujac, *Claude Ptolémée, Astronome, Astrologue, Géographe: Connaissance et Représentation du Monde habité* (Paris 1993) 300–1.
8 Ptolemy, *Mathematical Syntaxis* 2.13; Germaine Aujac, "The 'Revolution' of Ptolemy," in *Brill's Companion to Ancient Geography: The Inhabited World in Greek and Roman Tradition* (ed. Serena Biachetti et al., Leiden 2016) 313–34.
9 Erich Polaschek, "Ptolemaios: Das geographische Werk," *RE Supp* 10 (1965) 680–833.

10 Katherine Clarke, *Between Geography and History: Hellenistic Constructions of the Roman World* (Oxford 1999) 10.

11 Alfred Stückelberger, "Zu den Quellen der *Geographie*," in *Klaudios Ptolemaios: Handbuch der Geographie, Ergänzungsband* (ed. Alfred Stückelberger and Florian Mittenhuber, Basel 2009) 122–33.

12 Whether Ptolemy was aware of the attempts of Heron of Alexandria, active in the mid-first century AD, to determine astronomically the distance between Rome and Alexandria, remains speculative (Otto Neugebauer, *A History of Ancient Mathematical Astronomy* [Berlin 1975] 845–8; Nathan Sidoli, "Heron's *Dioptra* 35 and Analemma Methods: An Astronomical Determination of the Distance Between Two Cities," *Centaurus* 47 [2005] 236–58).

13 Mark T. Riley, "Ptolemy's Use of His Predecessors' Data," *TAPA* 125 (1995) 235–6.

14 Tacitus, *Annals* 4.73; *GG* 2.11.27.

15 Jochen Bertheau, "Die mitteleuropäischen Ortsnamen in der Geographie des Klaudios Ptolemaios," *OT* 8 (2002) 14–16.

16 Duane W. Roller, "Timosthenes of Rhodes," in *New Directions in the Study of Ancient Geography* (ed. Duane W. Roller, University Park 2019) 56–79.

17 *GG* 1.15.2, 1.15.4 = Timosthenes F29, 11.

18 Dicks, *Geographical Fragments* 56–103.

19 *GG* 1.7.4 = Hipparchos, *Against the Geography of Eratosthenes* F45. Since antiquity the star has steadily moved closer to the pole and is now less than a degree away; hence its current name (Berggren and Jones, *Ptolemy's Geography* 65).

20 Pliny, *Natural History* 4.95; 37.33, 36; *GG* 1.11.8.

21 *Periplous of the Erythraian Sea* 56; Casson, *Periplus* 213–14.

22 Athenaios 5.201c; Kenneth F. Kitchell, Jr., *Animals in the Ancient World From A to Z* (London 2014) 161–3.

23 Maes, *FGrHist* #2213.

24 Krisztina Hoppál, "The Roman Empire According to the Ancient Chinese Sources," *AAntHung* 51 (2011) 263–306.

25 Isidoros, *FGrHist* #781.

26 Pliny, *Natural History* 6.84–5.

27 Mattäus Heil and Raimund Schulz, "Who Was Maes Titianus?" *JAC* 30 (2015) 72–84.

28 Getzel M. Cohen, "*Katoikiai, katoikoi* and Macedonians in Asia Minor," *AncSoc* 22 (1991) 41–50.

29 Acts 13:9; M. Cary, "Maës, qui et Titianus," *CQ* 6 (1956) 130–4. For further examples of this style of nomenclature, see Heil and Schulz, "Who Was Maes Titianus," 75–6.

30 N. C. Photinos, "Marinos von Tyros," *RE Supplement* 12 (1970) 791–838.

31 Klaus Geus, "The Problem of Practical Applicability in Ptolemy's *Geography*," in *Knowledge, Text and Practice in Ancient Technical Writing* (ed. Marco Formisano and Philip van der Eijk, Cambridge 2017) 186–99.

32 Diller, "Ancient Measurements" 6–9.

33 Reinhard Wieber, "Marinos von Tyros in der Arabischen Überlieferung," in *Historische Interpretationen: Gerold Walser zum 75. Geburtstag* (ed. Marlis Weinmann-Walser, Stuttgart 1995) 161–90.

34 Herodotos 7.89; Strabo, *Geography* 16.3.4; *SIG* 407.

35 Klaus Geus, "Wer ist Marinos von Tyros? Zur Hauptquelle des Ptolemaios in seiner *Geographie*," *GA* 26 (2017) 13–22.

36 Pliny, *Natural History* 5.76; Roger M. Batty, "A Tale of Two Tyrians," *Classics Ireland* 9 (2002) 1–18. For a contrasting view regarding Marinus' role as geographer, see Pascal Arnaud, "Marin de Tyr," in *Sources de l'histoire de Tyr* (ed. Pierre-Louis Gatier et al., Beirut 2017) 87–100.

37 *PIR* I273, 274.

38 The legion is mentioned in the *Geographical Guide* (3.10.10).

39 Geus, "Wer ist Marinos," 18–20.

40 Berggren and Jones, *Ptolemy's Geography* 23.

41 Clarke, *Between Geography* 128.
42 The Greek world *oikoumene* is what is translated in this volume as "inhabited world"; this is a very specific and indeed technical term for the portion of the world that people live in.
43 An exact count is impossible to determine because of repetitions and variant forms, but the total number of unique names is between 6,000 and 7,000.
44 Eratosthenes, *Geography* F68 = Strabo, *Geography* 2.1.20.
45 Herodotos 2.34.
46 Eudoxos of Knidos F276a = Agathemeros 2.
47 Dikaiarchos F122–5; Agathemeros 2, 5; Paul T. Keyser, "The Geographical Work of Dikaiarchos," in *Dicaearchus of Messana: Text, Translation, and Discussion* (ed. William W. Fortenbaugh and Eckart Schütrumpf, New Brunswick 2001) 353–72.
48 Eratosthenes, *Geography* F47 = Strabo, *Geography* 2.1.1–3.
49 Dikaiarchos F124 = Strabo, *Geography* 2.4.1–3.
50 Herodotos 2.34.
51 Hipparchos, *Against the Geography of Eratosthenes* F12–15.
52 Berggren and Jones, *Ptolemy's Geography* 14.
53 Polybios 34.5.7–6.14 = Strabo, *Geography* 2.4.2–3.
54 Homer, *Iliad* 10.252–3, etc.
55 Hipparchos, *Commentary on the Phenomena of Eudoxos and Aratos* 1.7.11.
56 See, for example, the lists by Eratosthenes (*Geography* F60), Strabo (*Geography* 2.5.38–41), and Pliny (*Natural History* 6.212–20).
57 S. I. Shcheglov, "The Accuracy of Ancient Cartography Reassessed: The Longitude Error in Ptolemy's Map," *Isis* 107 (2016) 687–706.
58 Aristotle, *On the Heavens* 2.14.298a.
59 Juba F43–4 = Pliny, *Natural History* 6.201–5.
60 Duane W. Roller, *Scholarly Kings: The Writings of Juba II of Mauretania, Archelaos of Kappadokia, Herod the Great and the Emperor Claudius* (Chicago 2004) 57–8; Christian Marx, "The Western Coast of Africa in Ptolemy's Geography and the Location of His Prime Meridian," *HGSS* 7 (2016) 27–52.
61 Klaus Geus, "Ptolemaios über die Schulter geschaut—zu seiner Arbeitsweise in der *Geographike Hyphegesis*," in *Wahrnehmung und Erfassung geographischer Raume in der Antike* (ed. Michael Rathmann, Mainz 2007) 159–66.
62 Agathemeros 2; Aristotle, *Meterologika* 2.5.362b; G. W. Bowersock, "The East-West Orientation of Mediterranean Studies and the Meaning of North and South in Antiquity," in *Rethinking the Mediterranean* (ed. W. V. Harris, Oxford 2005) 167–78.
63 See, for example, *Natural History* 5.73, where Engada is said to lie below (*infra*) the Essenes, a reference to their relative positions on the map.
64 S. I. Shcheglov, "The Length of Coastlines in Ptolemy's *Geography* and in Ancient *Periploi*," *HGSS* 9 (2018) 9–24.
65 Strabo, *Geography* 2.5.10; J. Oliver Thomson, *History of Ancient Geography* (Cambridge 1948) 203.
66 As an illustration, a translation of *GG* 2.2.1–12 appears in Appendix 1.
67 Strabo, *Geography* 4.5.4; Pliny, *Natural History* 4.103.
68 Strabo, *Geography* 15.1.14–15; Pliny, *Natural History* 6.81–91.
69 Alfred Stückelberger, "Zu Sprache und Stil der *Geographie* der Ptolemaios," in *Klaudios Ptolemaios: Handbuch der Geographie, Ergänzungsband* (ed. Alfred Stückelberger and Florian Mittenhuber, Basel 2009) 432–9.
70 See, for example, Leo Bagrow, "The Origin of Ptolemy's Geographia," *GeogAnn* 27 (1945) 318–87; in contrast, see Berggren and Jones, *Ptolemy's Geography* 5.
71 Roller, *Through the Pillars* 50–2.
72 Alain Bernard, "Pappos of Alexandria," *EANS* 611–12.
73 Robert H. Hewsen, "The *Geography* of Pappus of Alexandria: A Translation of the Armenian Fragments," *Isis* 62 (1971) 186–207.

74 Ammianus Marcellinus 22.8.10.
75 Andreas Kuelzer, "Protagoras," *EANS* 700–1.
76 Markianos 2.2; Aubrey Diller, *The Tradition of the Minor Greek Geographers* (New York 1952) 45.
77 Cassiodorus, *Institutiones* 1.25.
78 Kosmas 177, 182.
79 Berggren and Jones, *Ptolemy's Geography* 51–2.
80 David Pingree, "Maximus Planudes," *DSB* 11 (1975) 18.
81 Aubrey Diller, "The Oldest Manuscripts of Ptolemaic Maps," *TAPA* 71 (1940) 62–7.
82 Renate Burri, "Übersicht über die griechischen Handschriften der ptolemäischen *Geographie*," in *Klaudios Ptolemaios: Handbuch der Geographie, Ergänzungsband* (ed. Alfred Stückelberger and Florian Mittenhuber, Basel 2009) 10–20.
83 See the list in Burri, "Übersicht" 12–20.
84 The various arguments are summarized by Polaschek, "Ptolemaios" 680–833. On the difficulties in interpreting the manuscripts, see Diller, "The Oldest Manuscripts" 62–7; also Berggren and Jones, *Ptolemy's Geography* 41–5.
85 C. F. A. Nobbe, *Claudii Ptolemaei Geographia* (Leipzig 1843–1845).
86 Edward Luther Stevenson, *Claudius Ptolemy, the Geography* (New York 1932); The translation has been characterized as "a complete failure" and "of very little use to scholarship": Aubrey Diller, "Review of Stevenson, *Geography of Claudius Ptolemy*," *Isis* 22 (1935) 533–9.
87 Berggren and Jones, *Ptolemy's Geography*; Alfred Stückelberger and G. Graßhoff, *Klaudios Ptolemaios: Handbuch der Geographie* (Basel 2006). See also the detailed discussion in Germaine Aujac, *Claude Ptolémée* 305–408.

2 Ptolemy's Introduction to the *Guide*

Introduction

Ptolemy opened the *Geographical Guide* with a number of definitions and the methodology of collecting the data relevant to his study. A significant part of the book is his critique of Marinos of Tyre. There is also some discussion of the means of making a map—both on a globe and a flat surface—material expanded upon in Books 7 and 8. The book is divided into 23 sections, each of which is titled and discusses a particular issue, and whose pattern is followed in this chapter.

1 On the Difference Between Geography and Chorography

Ptolemy's first concern was to define two technical terms essential to his presentation, *geographia* and *chorographia*, since a distinction between the two was an essential part of his thesis (*GG* 1.1.1–9). The word *geographia* was an invention of Eratosthenes in the second half of the third century BC, derived from surveyors' terminology and thereafter used to describe a discipline that continues to be called "geography."[1] But Ptolemy chose to define the word in the sense of making an imitation through drawing of the entire known portion of the world, along with what is generally connected with it.

This is a perfectly reasonable definition, but lacking in specific details of presentation, for which he used *chorographia*, a term perhaps developed by Polybios and originally meaning the study of places and distances.[2] Ptolemy defined what the word meant to him, reporting that *chorographia* records places individually, each one by itself, and entering almost everything and encompassing the smallest localities such as harbors, towns, districts, the branches of the major rivers, and so forth.

In other words, he saw it as a more specific and detailed version of *geographia*, which was lacking the detail of the former. Yet the definitions of both *geographia* and *chorographia* are innovative, since Ptolemy emphasized their relevance to the making of maps, something not obviously apparent in previous usages of the word.[3]

To explain his definition further, Ptolemy used an artistic analogy of the difference between drawing a small part of the human body as opposed to a larger anatomical unit.[4] Moreover, the level of detail should reflect the ability of someone viewing it to comprehend it; this seems so obvious that Ptolemy may have

DOI: 10.4324/9781003248590-3

been arguing against an unnamed predecessor. In addition, depicting *chorographia* required what Ptolemy called *topographia*, a word not documented before the first century BC but perhaps in existence earlier, and meaning the description of a particular region.[5] *Topographia* required a skilled draftsman to be depicted properly, but *geographia* as he defined it did not need such a skill.

Ptolemy also emphasized that *chorographia* was not mathematically oriented, but *geographia* had to take into account the size and shape of the earth and its position in the cosmos and relationship to the fixed stars, issues that were more astronomical. But the two disciplines were in fact intertwined with one another, since it was impossible for anyone to view the earth as a whole. According to Ptolemy, mathematics and astronomy were "the highest and most beautiful disciplines"—betraying his primary interests—and by applying them to geography and chorography he, without naming him, was reflecting Hipparchos and his application of them to geography.[6]

2   What Must Be Assumed for Geography

Having made the distinction between geography and chorography, Ptolemy then described the necessary knowledge for anyone who wished to be a geographer, and how this was different from learning to be a chorographer (*GG* 1.2.1–8). Not unexpectedly, the potential geographer had to begin with systematic research, using travelers' reports, but taking care that these sources had the proper training in order to provide accurate data, especially surveying and astronomical knowledge.

Surveying was necessary to determine distances: although not named by Ptolemy, one might think of Baiton, who did the surveying for Alexander the Great, recording the distances along his route.[7] But astronomy was less subject to error, and data could be obtained through certain instruments. This included the *astrolabos* (or *astrolabon*), which could determine the position of a heavenly body, and the *skiotheres*, a shadow-casting instrument, essentially a sundial.[8] Surveying was said to be less accurate than astronomy, in part because in order to calculate the distances between two points it was necessary to know their direction from one another, although Ptolemy did not explain why this was needed. More importantly, he emphasized that distances between known points (measured in stadia) were rarely in a straight line. This was obvious in overland journeys, which tended to follow the twists and turns of existing roads. For sea travel, there could be variances due to the winds and sailing speeds. The best that Ptolemy could offer was that one needed to make estimates that subtracted the excess, since all reported distances would be longer than any direct route. But even when an exact distance between two points was obtained, the points could not be plotted properly on the surface of the earth with only this information. Thus one had to resort to astronomical calculations, basically the method that Eratosthenes had used to determine the circumference of the earth, although somewhat restricted by the fact that the size of the earth prevented any empirical measurement of its circumference, or any ability to relate a specific calculated distance to places on its surface.

3   How a Measurement in Stadia of the Earth's Circumference Can Be Obtained from the Measurement of Stadia in a Straight Interval, and the Reverse, Even if it is Not on a Single Meridian

Ptolemy described how the technique of measuring the distance between two points on a single meridian could be used to determine the circumference of the earth (*GG* 1.4.1–2). This is a summary account of the methodology of Eratosthenes, who used Alexandria and Syene as his endpoints.[9] Ptolemy, however, offered an alternative that he believed was better, using an instrument of his own invention, a *meteoroskopion*, which was a refined version of his *astrolabos* and which did not require measurement on a single meridian or parallel.

4   It is Necessary to Give Preference to Phenomena Rather Than Records of Travel

Having precisely laid out the superiority of astronomical data (merely "phenomena" in the title), Ptolemy reemphasized the inevitable systemic failure of relying merely on travel data (which had been a necessity for Eratosthenes). At this point he introduced Hipparchos by name into the text, and it can be assumed that his material lay behind much of the previous argumentation in Book 1. Noting that if there were astronomically determined observations for every point recorded in the inhabited world, the map would be free of errors, Ptolemy emphasized that such data are rare, and those that existed were primarily the work of Hipparchos. About 30 places are known where he determined latitudes, usually by the length of the longest day, and Ptolemy tended to replicate his observations, sometimes with slight variants.[10]

Longitude was of course the problem: the basic method was relating the times of eclipses in two different places to one another, but this was virtually impossible even though the technique was well understood by Hipparchos and Ptolemy; the latter was correct in noting that such data were almost non-existent.[11] In fact there seems to have been only one case known to him, the lunar eclipse of 20 September 331 BC, documented at Arbela in Assyria and at Carthage three hours apart, which gave a distance of 45 degrees of longitude between the two places, which in fact was 11 degrees in error.[12] Another instance, on 30 April AD 59 and seemingly not known to Ptolemy, related eclipse times in Armenia and Italy and was more accurate.[13] Although the data were flawed, the technique was solid; the errors may have been in timekeeping. But Ptolemy saw these methods of calculating longitude and latitude as the best proof of the viability of astronomical data, and that combining the two dimensions—a technique he may have invented—would provide a complete catalogue of localities.[14]

5   It is Necessary to Apply the Most Recent Researches Because of the Changes to the Earth Over Time

Having thoroughly made his case that proper geographical research is only viable by using astronomical calculations, Ptolemy stated what may seem obvious to modern readers: that it was necessary to use the latest available reports in order to create a treatise on world geography (*GG* 1.5.1–2). By emphasizing this he may have realized that the constant changes taking place on the surface of the earth, both geological and demographic, were not always apparent to many people. Learning about the extent of the inhabited world had been a steadily evolving process since prehistoric times, and the expanding limits of knowledge were apparent to anyone reading Homer, Herodotos, or about the expedition of Alexander the Great. Ptolemy was aware that the inhabited world was not fully known—due to its immense size—and, moreover, there were many errors in the presumed information about what was known, something that he attributed to ignorance.

In addition, the physical earth itself was changing: the fact that it was not static had been examined scientifically since the fifth century BC, when Xanthos of Lydia published a treatise on the topic.[15] Such changes were particularly visible with volcanic eruptions (significantly neither Pompeii nor Herculaneum appears on Ptolemy's map) and the siltation at the mouths of great rivers. Yet, in a Herodotean statement, Ptolemy emphasized the necessity of analyzing the received material in terms of what was plausible. These discussions allowed Ptolemy to move to the major topic of Book 1, his assessment of Marinos of Tyre. This analysis would occupy a large portion of the remainder of the book.

6   Concerning the *Geographical Guide* of Marinos

At this point Ptolemy introduced Marinos of Tyre and began his critique of that scholar's *Geographical Guide* (*GG* 1.6.1–4). Marinos, known from no other source, probably lived about a generation before Ptolemy produced his own *Geographical Guide*, whose adoption of the same name for his own treatise shows his intense debt to the earlier scholar. As with so many reviews of the work of an earlier scholar, both in ancient and modern times, Ptolemy began his analysis with praise and then turned to negative comments; the praise is brief and the criticisms lengthy and detailed.[16] Yet Ptolemy placed himself in the paradoxical situation of being heavily dismissive of a scholar that he was strongly dependent on, much like Hipparchos' attitude toward Eratosthenes three centuries earlier, as well as Strabo's feeling toward many of his predecessors.

Marinos, the reader is told, was a totally diligent scholar who was well acquainted with the bibliography on his topic and evaluated it carefully. Given Ptolemy's scant citation by name of literary sources (perhaps as few as four: Timosthenes, Hipparchos, Philemon, and Diodoros of Samos), this is solid evidence that an extensive amount of previous geographical literature was available to him but whose authorship was not mentioned. It is probable, however, that Ptolemy did not consult these sources directly (other than the four he named) but accessed them through Marinos.

Marinos' diligence was demonstrated by his publication of a revised edition of his work, after further—and evidently more critical—analysis of his sources. But Ptolemy found fault with his general presentation, and complained about his credulity as well as his lack of vigor in his diagrams, especially in terms of proportion, which was one of Ptolemy's particular concerns (*GG* 8.1.1–6). He assured his readers that his corrections would make Marinos' work easier to use. This is an interesting and significant point, because it suggests that Ptolemy's *Geographical Guide* was originally designed to critique, not replace Marinos' treatise (even using the same title), perhaps in the manner of Hipparchos' *Against the Geography of Eratosthenes*. But Ptolemy's efforts evolved beyond that, and the analysis of Marinos is no more than a portion of Book 1.

Ptolemy's first problem with Marinos was his dimensions of the inhabited world, which he believed were excessive. It was necessary to establish the terminology: that from east to west was the length (longitude) and that from north to south was the width (latitude). Moreover, it was assumed that the longitude was greater than the latitude: this was an idea that had originated by the fifth century BC through the efforts of Demokritos of Abdera, who believed that the length of the inhabited world was 1 1/2 times its width. It may have been suggested by the shape of the Mediterranean, which probably also led to the convention of placing north at the top of any map.[17] With such assumptions in place, Ptolemy was prepared to begin his critique of Marinos.

7   The Correction of Marinos' Width of the Known World Through Phenomena

Ptolemy believed that Marinos' north-south extent of the inhabited world was excessive, and set out to prove the fact. But interestingly all his corrections were to the south; he seems to have been content with Marinos' assumption that Thoule was the limit to the north, and that it was properly placed at 63 degrees north latitude.

Marinos' north-south limits of the inhabited world (Thoule to Agisymba and Cape Prason in Libya), a distance of 87 degrees or 43,500 stadia, had allegedly been determined by both astronomical data and traveler's reports. But Ptolemy rejected these figures on astronomical grounds. In the argument that follows Ptolemy cited Book 3 of Marinos' treatise, the highest number recorded, as well as providing the longest extant quotation from the work:

In the burned zone the entire zodiac passes overhead, so that the shadows alternate and all the stars set and rise, except the Little Bear appears to be completely visible only when one is 500 stadia north of Okelis. This is because the parallel through Okelis is raised 11 2/5 degrees. Hipparchos reports that the southernmost star of the Little Bear is 12 2/5 degrees from the pole. In addition, to someone going from the equator to the Summer Tropic, the northern pole is always raised above the horizon, but the southern is below the horizon. But for someone going from the equator toward the Winter Tropic, the southern pole is raised above the horizon but the northern is below the horizon.

Okelis was an important trading post at the southwest corner of the Arabian Peninsula; it is not obvious why Marinos chose it at his datum point.[18] The southernmost star of the Little Bear (α UMi) is known as Polaris today, but, as Marinos pointed

out, it was some distance from the pole in antiquity. Ptolemy, somewhat tenden-
tiously, complained that Marinos was creating a theoretical structure without any
data based on observation, especially south of the equator. Some of this material
was certainly derived from Hipparchos (his second and last mention in Ptolemy's
Geographical Guide).

In his continuing argument Ptolemy again quoted Marinos:

> Those who sail from Indike to Limyrike, as Diodoros of Samos says in his
> third book, see Taurus in the middle of the heavens and the Pleiades in the
> middle of its horns. Those who set forth from Arabia to Azania sail directly
> toward the south and the star Kanobos, which the locals call Hippos and which
> is in the extreme south. We do not have names for the stars visible to them,
> and the Dog rises before Prokyon and all of Orion before the summer solstice.

This is the only extant citation of Diodoros of Samos, who presumably was active
near the end of the first century BC, although his contrast between Limyrike (the
Malabar coast of India) and India (Indike) itself suggests that he may have been
relying on earlier material when the toponym Indike was still limited to the Indos
drainage. Obviously nothing is known about his career, or whether his material was
gathered through his own experience or an earlier literary source, but his treatise, at
least three books long, may have described the west coast of India. Marinos' quotation of him was probably limited to the single sentence about India.

Kanobos (Canopus, α Car) and Prokyon (Procyon, α CMi) retain their ancient
names; the Dog is Sirius (α CMa). Ptolemy quite rightly pointed out that some
of the data could have been gathered north of the equator, although there seems
to have been autopsy obtained from the sailing route along the east African coast
and in India. But Canopus is visible as far north as 37 degrees north latitude (that
of Sicily and southern Asia Minor), and again Ptolemy was being somewhat argu-
mentative, because Marinos made it clear that Canopus was merely being used as
the sailing direction from Arabia to Azania (the east African coast). Yet once again
Ptolemy's point is that Marinos could have made these calculations without resort-
ing to data received from people who had traveled south of the equator (of which
there were probably very few). Ptolemy has wandered rather far from his original thesis stated at the beginning of the section, which was to show that Marinos'
extent of the inhabited world was excessive.

8   The Same Correction of the Distance for Overland Journeys

Ptolemy then turned his attention to specific overland journeys catalogued by Marinos (*GG* 1.8.1–7). These involved the two Romans, Septimius Flaccus and, slightly
later, Julius Maternus, who were probably active around AD 100.[19] It was clear to
Ptolemy that Marinos' figures for these two expeditions were excessive—and this
is apparent to the modern reader—but Ptolemy's analysis was also hampered by
two factors he may not have realized: the distances were recorded by Marinos in
stadia whereas the two Romans would probably have used miles and Marinos'

conversion factor is unknown, and Marinos (and probably the travelers) followed the long-standing practice of using the ethnym "Aithiopians" for the peoples spread across the southern edge of the inhabited world, a concept established by the fourth century BC.[20] Aithiopians are mentioned seven times in this brief section, not always referring to the same peoples.

Flaccus set forth from a point on the North African coast, probably Leptis Magna, and went to the famous oasis of Gerama (modern Djerma in Libya). This locality had been known to the Romans since 20 BC when L. Cornelius Balbus reached the site.[21] A rich and fertile region, it served as the contact point between the Mediterranean world and interior Africa. Flaccus continued south for three months and reached the "Aithiopians"; the vagueness of the ethnym and a lack of detailed information makes his destination uncertain. Ptolemy reported that, as expected, his speed of travel was not constant.

A short time later Maternus set out from Leptis and also arrived at Garama. He then joined the local dynast on a campaign against unspecified Aithiopians, and in four months reached Agisymba, a mountainous territory where rhinoceroses gathered (*GG* 4.8.5). Agisymba is mentioned in no other independent source, and there have been many suggestions for its location, the most plausible being in the region of Lake Chad. The expedition seems to have had two purposes: to provide Roman military support for a southern expedition of the dynast of Garama, and to procure rhinoceroses for Roman spectacles, or at least to define their territory. The animal had been known to the Mediterranean world since the early third century BC when Ptolemy II collected a large amount of exotic fauna, and had become relatively common in Rome by the late first century AD, appearing on coins of Domitian.[22]

Ptolemy's interest was less in the ethnography and zoology of the expedition than the distances that they covered, and how this information could be used to understand the southern limits of the inhabited world. Marinos' original distance to Agisymba from Leptis placed it 24,680 stadia (perhaps 5000 km.) south of the equator, with a slightly longer distance by sea to the allegedly comparable latitude of Cape Prason (*GG* 1.9.4). Ptolemy accurately pointed out that this distance put Agisymba too far south (around 55 degrees south latitude), and, using the prevalent ancient theory that similar latitudes meant similar climate, further noted that this would place tropical Agisymba at a comparable latitude to the Skythian wastelands north of the Black Sea. Although the climate theory is flawed, the point is reasonable: the Roman expeditions could not have gone 55 degrees south of the equator.[23] But Marinos realized a problem, and in his second edition reduced the distance to approximately half because of the lack of a straight route and variable speeds. This was still problematic (placing Agisymba at the latitude of modern Zimbabwe). Yet at this point Ptolemy was less interested in the latitude reached than in emphasizing the inevitable problems of distance data recorded by overland expeditions. Somewhat fallaciously he added the argument about the Aithiopians extending across much of Africa as a way of arguing that Maternus' route could not have been in a straight line, but this made a valid point: that the expedition did not head straight to the south from Garama to Agisymba. He also commented that Marinos' distances did not account for pauses during the journey, thus implying that the original source of the figures was travel days, a further variable that would skew the numbers (*GG* 1.10.2). Moreover, the locals, who would have been the sources for some of the information, were either vague or creative in what they reported, a common problem: locals were often less than helpful to foreign expeditions.

In summation, Ptolemy realized the problems with Marinos' data as obtained from the two Romans. He could easily point out inconsistencies, yet he could not make any feasible corrections or apply more accurate astronomical data since there was no information beyond Marinos' critique of the Roman reports.

9   The Same Correction of the Distance for Sailing Voyages

Ptolemy then considered voyages along the East African coast as a way of further determining distances to the southern edge of the inhabited world. The region in question is from Aromata to Cape Prason: the former was the emporion at the Horn of Africa (*GG* 4.7.10, 40), and the latter somewhere on the coast of Mozambique, perhaps modern Cabo Delgado. A certain Diogenes, otherwise unknown, returned from his second voyage to India around AD 100 but was unable, presumably due to winds, to enter the Gulf of Aden (which leads to the Red Sea), and instead was driven south along the East African coast past the region called Trogodytike. After 25 days he reached "the lakes from which the Nile flows," slightly to the north of the trading post of Rhapta. Ptolemy was aware of the error regarding the lakes—the source of the Nile was not on the coast—but chose not to address it at this time. After Diogenes, a certain Theophilos made a similar but faster voyage in 20 days (although details for both seamen are vague). In fact, Theophilos used the formula of 1000 stadia (perhaps 200 km.) for a 24-hour sail, but a third sailor, Dioskoros, could only cover the 5000 stadia south of Rhapta to Cape Prason in "many days," allegedly because of wind issues near the equator. These sailing variables are no surprise, but Ptolemy, at some length, used these accounts to prove that the speed of sea voyages could be so different as to be of little use in determining distances.

As usual, Ptolemy was hardly interested in the ethnography and geography learned on these voyages, but his sparse account of these three travelers along the East African coast provides a certain amount of intriguing information. The name Rhapta is Greek rather than indigenous, meaning "sewn," describing the technique of the local boats,[24] so the emporion was established by Greek merchants; it is first documented in the middle of the first century AD but probably had existed well before that time, and was located at or near modern Dar es Salaam. One would expect that the traders used it for access to the interior, and in time learned about the lakes and mountains that were the source of the Nile, but Marinos, in stating that the lakes were on the coast, was obviously in error, or a transitional phrase that described an inland trade route had dropped out of his text. Ptolemy had more accurate information, perhaps from a source later than Marinos, and corrected the error later in the *Geographical Guide* (*GG* 1.17.6). An expedition commissioned by the emperor Nero went far up the Nile, in fact farther than any people before the nineteenth century, perhaps almost reaching the actual source, Lake Victoria Nyanza. This may have been designed to connect with the routes emanating from Rhapta, although this did not happen.[25]

Marinos also mentioned Cape Prason, which was the farthest point south reached by any of the East African explorers. Dioskoros may have attempted a reconnaissance south of Rhapta, investigating trade potential, but turned back after 5000 stadia, probably due to adverse winds and the lack of anything of interest. Nevertheless he was one of the few from classical antiquity to reach the southern hemisphere and the only one to provide a toponym from it.

Ptolemy reported that Marinos was concerned that his distances did not fit the theory of similar fauna and flora at the same latitudes, and that this was what led him to reduce his numbers. By doing so he made his data more accurate, but for a totally invalid reason. Yet his reliance on travel days and the assumption that they were of a constant length made his calculations inevitably incorrect, a problem that could not be rectified. To conclude the section Ptolemy also added a racial consideration: that Agisymba had to be between the equator and the winter tropic because of the characteristics of the people, who were similar to those at Syene, located at the summer tropic. This still placed Agisymba too far south, and is a rather feeble argument, but fit ethnographic theory of the era.[26]

10   The Aithiopians Should Not Be Placed South of the Parallel Opposite Meroë

Following on his previous argument, Ptolemy then concluded that Agisymba and Prason should be placed on the parallel opposite Meroë, or 16 5/12 degrees south latitude. Meroë, at the corresponding northern latitude, was the center of the Meroitic kingdom and had been well known to the Greco-Roman world since the fifth century BC.[27] Its location, on the upper Nile, was indisputable, and so connecting it geographically with Agisymba and Prason created a grounding for those more elusive places, and gave Ptolemy the figure of 80 degrees of latitude or 40,000 stadia (perhaps 8000 km.) for the north-south distance of the inhabited world. Despite the immense error in this calculation—both in assuming that Agisymba and Prason are on the same latitude and placing them too far south—it was what Ptolemy accepted, and did not have the same effect on the later history of geography as a similar error in the east-west extent of the inhabited world.

Yet Ptolemy did not dispute the reported distance between Leptis Magna and Garama, 5400 stadia (perhaps 1100 km.), which was a reasonably accurate figure. This was a well-known route, with numbers derived from Flaccus and Maternus (and perhaps earlier sources), but presumably converted from miles. But, to emphasize again the issues he had been discussing, Ptolemy pointed out that the travel time varied between 20 and 30 days, thereby proving his point of the impossibility of using such data to determine distances.

As a conclusion to his discussion of Marinos' handling of latitude, Ptolemy made an astute theoretical statement about the nature of the problem:

> One must be in doubt about large and little-known distances, or those that have not been travelled in an agreed manner. But those which are not large and are frequently travelled by many people in an agreed way can be trusted.

11   Concerning the Calculations That Marinos Incorrectly Made for the Length of the Inhabited World

Having made his arguments regarding latitude, Ptolemy now considered Marinos' handling of longitude (*GG* 1.11.1–8). Here the evidence was different, since there were established trade routes from the Mediterranean into central Asia that could theoretically provide distances. Ptolemy argued that Marinos had over-extended the east-west length of the inhabited world, defined as from the Blessed Islands to far eastern points such as Sera, Sinai, and Kattigara. The main parallel across the inhabited world, conceived by Dikaiarchos of Messana in the fourth century BC,[28] had been refined over the years: west to the Blessed Islands and east into Central Asia. The eastern expedition of Alexander the Great had helped Dikaiarchos in his determination, and in the Augustan period Isidoros of Charax published his *Parthian Stations*, still extant, which was a measured itinerary from the Euphrates to Arachosia. This route was measured in *schoinoi*, the unit used by Marinos and Ptolemy in their replication of it: Ptolemy mentioned *schoinoi* only three times in the *Geographical Guide*, all in this context (*GG* 1.11.4, 1.12.3, 1.12.8). The *schoinos* ("rope") was an Egyptian unit but could vary from 30 to 120 stadia, one of the many difficulties encountered by ancient geographers in attempting to calculate long distances.[29] Marinos and Ptolemy used a *schoinos* of 30 stadia.

The calculation of the east-west extent of the inhabited world was made, at least in its western portions, along the parallel of Rhodes (36 degrees north latitude), insofar as was possible. The more eastern portions were farther north. The parallel was in three distinct sections, with the westernmost from the Blessed Islands to the Euphrates, which could be based on the original parallel of Dikaiarchos and later modifications over the years. From the Euphrates, crossed near Hieropolis in Syria (an ancient cultic center at modern Membidi), the parallel continued to the Lithinos Pyrgos or Stone Tower, a distance of 876 *schoinoi* or 26,280 stadia (perhaps 5200 km.). This presumably was an established trade route, and its use of *schoinoi* probably means that the figures were based on Isidoros or a similar published itinerary. The third stage was from the Stone Tower to Sera, the metropolis of the Serians, a seven-month journey of 36,200 stadia (maybe 7200 km.) over exceedingly rough country. As usual, Ptolemy's concern was that these figures were excessive and indeed to a large extent unmeasurable: the journey east of the Stone Tower was subject to poor weather and could not easily be calculated.

The Stone Tower has not been located; the name (Tashkurgan) is common in central Asia, and probably marked a visible point where a trading rendezvous could occur. The route, especially east of the Stone Tower, had only recently been opened (insofar as the Mediterranean world was concerned) by agents of Maes Titianos, who traveled the seven months to the Serians and recorded the distance. But there were no details about the journey, and Ptolemy saw this as proof that the reported distance was a fantasy. In fact, it becomes impossible to determine how many stages composed the journey, or even how the information was collated. A close reading of Ptolemy's text implies that the data obtained from Marinos may only have been for the portion east of the Stone Tower, perhaps because the route west of the Stone Tower had already been recorded by others.

Ptolemy's statement about Maes is significant:

He [Marinos] says that a certain Maes, known as Titianos, a Macedonian man and a merchant like his father, recorded the distances but did not make the journey, sending others to the Serians.

So it is probable that Maes, the subject of so much modern study, never left home—perhaps somewhere in Anatolia or the Levant—but merely received reports from travelers he had commissioned, who themselves collated evidence from others and constructed a journey over several stages.[30] But the interesting point is that somehow the data Maes obtained was published in a way available to Marinos a few years later, most probably in the type of land itinerary that had existed since the time of Alexander the Great.

Because Maes was a merchant (*emporos*), Ptolemy digressed to make the casual but significant comment that Marinos did not trust merchants' reports, since they exaggerated distances. He believed this was a matter of boasting, but it is more likely an attempt to keep the sources of trade items secret from possible competitors, a phenomenon existing from ancient to modern times. As a further example of the unreliability of merchants, Marinos cited the case of a certain Philemon, who used merchants' reports to say that Ireland was 20 days across, which Marinos found implausible.[31] Philemon is little known, mentioned only here and by Pliny the Elder, and seems to have written about northern Europe in the early first century AD.[32]

12   The Correction of the Length of the Known World by Means of Land Journeys

Ptolemy took received data for the route to Sera and reduced it by half, primarily because of its twists and turns (*GG* 6.12.1–11).[33] He pointed out, rather ingenuously, that the climate along the route would not change, as opposed to the extensive variance along the Agisymba itinerary.

He first reduced the 876 *schoinoi* from the Euphrates to the Stone Tower, because of the failure of the route to follow a straight line, although he admitted that it was so well traveled the information was more accurate than farther east. He attempted to connect various portions of the route to parallels (that of Rhodes for the western part, Smyrna to the east, and then those of the Caspian Gates, Byzantion, and the Hellespont). Presumably he had the information to make this possible, but he thus fell into the same trap as Marinos in over-regularizing the itinerary. As far east as the Stone Tower topographic details were available: a climb up the Komedai Mountains (perhaps the Pamirs) and then a route through a gorge climbing to the Stone Tower itself, probably along the upper Oxos River system. At the Stone Tower the route left the mountains and entered the high plains of Xinjiang. Mention of Palimbothra (the Mauryan capital, modern Patna) at this point refers to a route from India and the Ganges to the Stone Tower and confirms its importance as a major gathering center: this route from the Ganges was alluded to by the author of the *Periplous of the Erythraian Sea*.[34] The rendezvous at the Stone Tower may have been described by Pliny, based on reports by a Taprobanian envoy to Rome and Roman merchants in the mid-first century AD:[35]

> When they arrived there, the Serians hurried to meet them. They exceed the normal height of men, with reddish hair and blue eyes. The sound of their voices is harsh and they have no language for commerce. The rest of the report agreed with our merchants: trade items were placed on the opposite side of a river next to the products they offered for sale, and if they were pleased by the exchange they took them away.

East of the Stone Tower nothing was reported until Sera was reached after a journey of seven months. Ptolemy, however, totaled the various distances (expressed in terms of degrees of longitude): from the Blessed Islands across the Mediterranean and to the Euphrates it was 72 degrees, and 105 1/4 for the route farther east, for a total of 177 1/4 degrees from the islands to Sera.[36] This was the famous elongation that caused such difficulty in early modern times: the actual distance is perhaps 110–120 degrees, depending on the location of Sera.

13   The Same Correction for the Length of the Sea Journey

Next Ptolemy summarized Marinos' account of the sailing route to the Chryse Chersonesos (Golden Peninsula, the modern Malay Peninsula) (*GG* 1.13.1–9). Again his interest was in reducing Marinos' distances. Since there are more topographical markers on a sea route that a land route, there is more information than what was available across the wastelands of Central Asia, but most places cannot be specifically identified. The sailing route began at a point in southeast India near the *emporion* of Paloura and referred to by Ptolemy merely as "the departure point for the sail to the Chryse Chersonesos" (*GG* 7.1.15): ships would have passed through the strait separating Taprobane from the Indian mainland at Cape Kory, and then headed toward the winter sunrise (east-southeast) across the Ganges Gulf (modern Bay of Bengal) to a place called Sada, located in the Silver Land and presumably on the coast of Burma. The route then continued east-southeast along the coast to the site of Tamala on the Golden Peninsula. Tamala was at the narrowest point of the peninsula, and here traders left their ships and crossed overland to the eastern side, where they could access the Gulf of Thailand and the South China Sea. The crossing was defined by the rare word *diaperama*, which is not documented before the first century BC and seems to have had only a geographical usage.[37] The distance across was reported as 1600 stadia (perhaps 300 km.), well above the actual 30–40 km., but there is no way of knowing exactly where the crossing was made, and it may not have been at the narrowest point. Moreover, this was perhaps a demonstration of Ptolemy's repeated assertions that Marinos' distances were excessive. Nevertheless there is little doubt that the Chryse Chersonesos is the Malay Peninsula.

Ptolemy reduced each of the distances on this route by one third, giving a total of 34 4/5 degrees of longitude from Cape Kory to the Golden Peninsula. He mentioned a constant reality of coastal sailing: that entering the numerous bays and inlets would add to the distance, but it was not known which of them the seamen used. The actual distance is perhaps 20 degrees of longitude, but the two locations are far from the same latitude.

14   The Sail from the Golden Peninsula Across to Kattigara

The final portion of the eastern sea trade route was to an enigmatic place called Kattigara (*GG* 1.14.1–10).[38] Marinos relied on a report from an otherwise-unknown Alexandros who recounted that it was 20 days' sail along a south-facing coast to a place called Zabai, and Kattigara was south-southeast of there. This was the final point on the eastern itinerary. But Ptolemy also placed Kattigara on the same latitude as Zabai, contradicting Alexandros and Marinos, and suggesting another source. Alexandros said that from Zabai to Kattigara was a sail of "some" (*tinas*) days but evidently Marinos turned it into "many" (*pollas*) days. Ptolemy, rather indignant at this, produced a rather inconclusive semantic argument about these two words and the actual sailing times involved. He seemed little interested in their actual location, other than their longitude, which in these remote places is less helpful than usual. It is impossible to determine their latitude. To be sure, Kattigara was near the metropolis of the Sinai, but this adds little.

The question is the direction of travel after crossing the Golden Chersonesos. The sources seem to imply that Kattigara was to the southeast, which suggests somewhere in Indonesia, a view supported by the nearby toponym Iabadiou, an island 10 degrees of longitude to its west (*GG* 7.2.29), which seems to be Java. Yet it may be that east of the Golden Chersonesos a route actually turned north toward the land of the Sinai, and information about it became entangled with Alexandros' material as well as other existing data about trade locations in the far east. It is even possible that the toponyms listed are simply a catalogue of known trade points, and are not to be connected in a single itinerary. Moreover, if Sinai and associated terms were just generic names for the peoples at the eastern edge of the inhabited world, it may have been assumed that all places in that region had some connection with Sinai. Clearly the material has become confused, but this has not kept generations of scholars from attempting to locate Zabai and Kattigara. Ptolemy did not help matters any by placing the two locales, as well as the Golden Peninsula, on the same latitude, perhaps indicative of his own confusion as to their position and his interest only in longitude. He was also somewhat misled by his belief that the Indian Ocean was an enclosed sea, writing: "The region to the west of Kattigara comprises unknown land as far as the Prasodic Sea and Cape Prason" (*GG* 7.3.6). *Prasodes* refers to seaweed or a greenish color; the term was applied to the western Indian Ocean and Cape Prason was the southernmost known point on the East African coast. But Ptolemy sought to prove that the Sinai was at 180 degrees east longitude, with all other places somewhat to the west, thereby reducing Marinos' dimension of the known world from 225 degrees of longitude to 180 degrees.

The location of Zabai is probably not recoverable, but Kattigara has been suggested to be in a wide range of places, from Indonesia in the south to the vicinity of

Hanoi in the north.[39] One intriguing suggestion is that it is the site of Oc-Éo in the Mekong Delta, where excavations have revealed Roman trade goods.[40]

15   Disagreements with the Details Set Forth by Marinos

Having established the perceived errors of Marinos regarding his routes to the extremities of the southern part of the inhabited world, and his issues regarding placement of localities along these routes, Ptolemy's next concern was to catalogue specific errors within the better-known parts of the world (*GG* 1.15.1–16.1), noting that Marinos was inconsistent and contradictory.[41] In particular, there was the matter of places that "lay opposite" (*antikeisthai*) one another. This was a term that had been used since the fifth century BC referring to localities on opposite coasts but on the same longitude, and was an early attempt to relate distant places to one another.[42] Ptolemy had several complaints about places that were said to be opposite to one another but in fact were not; the instances were geographically scattered through the Mediterranean world. Most seem minor to the modern reader. Interestingly, Ptolemy in this section twice mentioned Timosthenes of Rhodes (*GG* 1.15.2,4), the naval chief of staff of Ptolemy II, who produced a sailing manual in the first half of the third century BC. This is the only place in the *Geographical Guide* that Timosthenes was cited, seemingly to correct Marinos, a rare instance in which Ptolemy named a source in order to refute him.

Some of these errors may be simple mistakes in calculation, or vague use of the concept of "lying opposite." It is hard to believe that the relative position of Tergeste (modern Trieste) and Ravenna was not known. Placing Noviomagus (modern Chichester) both north and south of London is either a mistake, or a failure to reconcile two different sources. Since this is one of only two instances when Ptolemy (and presumably Marinos) used miles, there may have been a conversion problem. The only other use of miles is a few lines later (*GG* 1.15.6, 9), so this entire passage may be based on a unique source and subject to corruption.

The final entry in this section concerns the location of the source of the Nile. It was said to be the lakes reached by sailing down the East African coast from Aromata (the Horn of Africa). Ptolemy realized that there was a problem here, but for an incorrect reason: he believed that the Nile flowed directly from the south, and that Aromata was far to its east, so that it and the source of the river could not be on the same longitude. This is certainly true, and in fact Lake Victoria Nyanza, the source, is 500 km. from the coast. Ptolemy determined, essentially correctly, where it must lie, but this was based on fallacious interpretation of the data. Yet Victorian explorers searching for it saw the flaw and realized that it would be a journey of many days to reach the lakes.

16   Certain Matters that Escaped Notice Regarding the Boundaries of Provinces

This brief section is perhaps one of the least valuable in the *Geographical Guide* (*GG* 1.16). It describes several instances when provincial or ethnic boundaries seemed to have been erroneously stated by Marinos. Such boundaries were always fluid, and it is relevant that Ptolemy only offered general statements, no coordinates or distances. In the years between the publication of Marinos' *Geographical Guide* and that of Ptolemy, there may have been changes in the organization of the provinces in question (in a region generally extending northeast from the upper Adriatic to the Black Sea and north of the lower Danube). Hadrian was in this region in AD 118 and again four years later—in other words between the eras of Marinos and Ptolemy—and some adjustment to the boundaries was possible, since they often changed with current military or political needs.[43]

Even less reasonable are Ptolemy's complaints about the location of peoples north of India. These were poorly understood nomadic groups which cannot be precisely positioned. There also seems to be some confusion—hardly unexpected—about the relationship between the parallels north of the Imaon (Himalaya) Mountains within the territory of the Sogdians and Sakians and those through the Black Sea.

17   On the Disagreements Between Him and the Records of Our Time

It was noted that Marinos was not aware of the errors that Ptolemy had listed in the previous sections (*GG* 1.17.1–12). This was due to the size of his work and his failure to reconcile various portions. Marinos also failed to make a map for his second edition, which would have given him a visual basis for which to make corrections.

Yet Ptolemy moved beyond Marinos and made use of contemporary accounts to indicate that in some matters he was out of date. If the second edition of Marinos' *Geographical Guide* was published before AD 120, and Ptolemy's version by AD 160, this provides a window for Ptolemy's updating of Marinos' account. Ptolemy used more than one source in this process: at least two but perhaps more. With his usual diffidence he did not name them, but they seem to have been navigational accounts similar to that in the extant *Periplous of the Erythraian Sea*. There is no exact correspondence between Ptolemy's material at this point and any other extant source, and many of his toponyms are unique to this section.

The first itinerary was to India and points east as far as Kattigara. New information includes a definitive statement that although Kattigara is east of the Golden Chersonesos, the Seres, Sinai, and their metropolis are to the north of Kattigara. But there is a vague reference to territory even farther east where there are marshlands (*helodes*) with vegetation so dense that one can use it to cross them. One is at a total loss to suggest where these were, but they contradict the earlier assumption that there was nothing east of the Seres or Sinai, and indicate that by the middle of the second century AD travelers had penetrated into this unknown region. Ptolemy also mentioned an alternate route to the Stone Tower, from Palimbothra in India. This had been known since the mid-first century BC and probably earlier, but was not reported by Marinos.[44]

The greatest detail, however, is reserved for an account of the East African sailing route, from Arabia Eudaimon to Rhapta. There are a number of new toponyms, some of which are unique (Panon, Zingis, Mt. Phalangis, Essina, Sarapion, and Toniki).[45] There are also topographical and navigational details, including sailing times (expressed in days and nights), bays, headlands, and beaches. The tone of the account is similar to that in the *Periplous of the Erythraian Sea* but is unusual for the *Geographical Guide*. Ptolemy was using one or more otherwise unknown itineraries that had been composed near the middle of the second century AD, although it is always possible that this was unpublished material gathered from seamen in Alexandria.

18   Concerning the Difficulty of Making a Plan of the Inhabited World from Marinos' Composition

Ptolemy devoted the rest of Book 1 (1.18–1–24.2) to the methodology of drawing a plan of the inhabited world, on both a sphere and a flat surface. His complaint against Marinos at this point was that he did not provide the data to make this possible, because he separated his catalogues of parallels and meridians and there was no consistent recording of the data. Thus it became difficult to use his scattered information for map making, and in some way the work was broken up into pieces (*diesparmenon*). Ptolemy also revealed the interesting fact that others had attempted, evidently not very successfully, to make a plan based on Marinos' writings. Coastal locations were easier to plot, but those inland were difficult, and at times a meridian but not a parallel, or the reverse, would be all that was listed.

19   Concerning the Ease of Our Guide in Making a Plan

Ptolemy wanted to rely as much as possible on Marinos, but realized the need for corrections and a greater clarity, and thus used the research of those who had visited places relevant to his work, but did not specify them by name (*GG* 1.19.1–3). He also used plans that he found to be more accurate than those of Marinos; as usual, no details about their origin were provided. He recorded the boundaries of the provinces (as represented by parallels and meridians), their ethnic groups, and topographical features. His method of locating these features was to use parallels from the equator and meridians from the western edge of the inhabited world. With certain refinements, this was the type of grid system that had been created by Eratosthenes.[46] Yet he positioned his thousands of toponyms independently of any grid system, simply using their coordinates, although he did create a grid of meridians and parallels (*GG* 1.21.1).

20   On the Disproportionality of Marinos' Geographical Plan

One of Ptolemy's complaints was that Marinos had no sense of proportion on his map (*GG* 1.20.1–7). With some passing comments about the differences between a map on a globe and one on a flat surface, he credited Marinos with being careful about these issues, but at the same time being unable to create a map that was properly proportionate. Marinos was not alone: Ptolemy did not believe that any previous ones were satisfactory. He outlined Marinos' failures on this matter, with his meridians and parallels all straight lines and the meridians parallel to each other.

This was, to be sure, the accepted technique, but Ptolemy explained in detail that this was unsatisfactory. He used a comparison between the parallels of Rhodes and Thoule to make the error apparent.

21   What Should Be Observed on a Flat Map

Continuing his argument from the previous section, Ptolemy argued that it was best to keep the meridians straight but the parallels as circular segments, creating an illusion of the spherical nature of the earth with the meridians intersecting at the pole. Yet all parallels could not be properly proportionate, so it was best to preserve this for Thoule and Rhodes.

22   How to Draw the Inhabited World on a Sphere

Ptolemy next considered how to make a map on a globe. Globes had existed since that of Krates in the second century BC.[47] By Ptolemy's day, globes seem to have become a familiar tool of world cartography. He had already described a globe depicting the celestial sphere, and referenced this for the technique of making a globe of the earth.[48] It was noted that the larger the globe was, the more accurate it would be, but the concept was the same regardless of size. The parallels and meridians would be marked first, and then the actual toponyms. Interestingly Ptolemy's globe only showed the inhabited world; there seems to have been no attempt to draw the parallels and meridians on the remaining portion of the earth.

23   List of the Meridians and Parallels to Be Drawn On the Map[49]

Despite the title of the section there is actually no list of meridians preserved, just the statement that they will be drawn at 12-hour intervals (*GG* 1.23.1–23). In contrast there is a full list of 23 parallels, identifying them by length of the longest day and also the corresponding degrees of latitude. Using the longest day to assist in locating places was a technique from at least the late fourth century BC: Pytheas of Massalia used it on his voyage to Britain and the Arctic, and the process was refined by Hipparchos.[50] Ptolemy's parallels begin just north of the equator (oddly the equator itself does not seem to have been categorized as a parallel). There are 20 additional parallels to the north; the first 14 are at quarter-hour intervals (starting at about 4 degrees north latitude and eventually diminishing to 3), and then the intervals extend to half hours and then full hours. Only Meroë, Syene, Rhodes, and Thoule were mentioned by name, perhaps because these were all places of special importance in the history of geography. The remaining parallels have no toponyms attached.

There are two additional parallels, both south of the equator. The first was at half an hour (8 5/12 degrees) and was the parallel on which Cape Rhapton (just south of Rhapta) and Kattigara were located, probably more of an assumption than based on empirical evidence. This yielded the idea that the most southern places in the inhabited world (ignoring the vaguely known Cape Prason south of Rhapta and the problems involving Agisymba) were at the same latitude and created a sense of geographical regularity. It also suggested that their position was limited by the presumed enclosure of the Indian Ocean.

For completeness there was a second parallel farther south, defined as being as far south of the equator as Meroë was north of it. This would be at 16 5/12 degrees south latitude, but would be theoretical rather than real and well into the "unknown land" of Libya.

24   The Method of Drawing a Flat Representation of the Inhabited World Properly Proportioned to That of a Sphere

The final section of Book 1 is a lengthy and highly technical discussion of the actual process of making a map (*GG* 1.24.1–33). This section has little to do with topographic geography and has been discussed and analyzed in great detail else-where.[51] Ptolemy began with the method for creating a map on flat surface, includ-ing the construction of the meridians (which would have a common intersection) at 5-degree intervals. The parallels would then be plotted, and finally the toponyms. Ptolemy then proposed an alternate method of creating the map which would make it better proportioned and more similar to the globe.

The only toponyms mentioned in this section were the four important parallel points, Thoule, Rhodes, Syene, and Meroë, demonstrating their continued impor-tance in not only the history but also the technology of geographical representation. Upon conclusion of these technical details about map making, Ptolemy was ready for the topographical guide itself, his listing of the thousands of toponyms with their coordinates.

Notes

1 The earliest extant citation of the word is by Strabo (*Geography* 1.1.1), but this goes back to Eratosthenes (*Geography* F1); see also Roller, *Eratosthenes' Geography* 1–2.
2 Polybios 34.1.5; Francesco Prontera, "Geografia e corografia: note sul lessico della car-tografia antica," *Pallas* 72 (2006) 75–82.
3 Alexander Jones, "Ptolemy's Geography: Mapmaking and the Scientific Enterprise," in *Ancient Perspectives: Maps and Their Place in Mesopotamia, Egypt, Greece and Rome* (ed. Richard J. A. Talbert, Chicago 2012) 109–28.
4 This is repeated somewhat at *GG* 8.1.5.
5 Strabo, *Geography* 8.1.3.
6 Hipparchos would be cited by name at *GG* 1.4.2.
7 Baiton, *FGrHist* #119.
8 Ptolemy, *Mathematical Syntaxis* 5.1; Berggren and Jones, *Ptolemy's Geography* 59.
9 Eratosthenes, *Measurement of the Earth* M1–8; Roller, *Eratosthenes' Geography* 263–7.
10 Dicks, *Geographical Fragments* 193.
11 Hipparchos, *Against the Geography of Eratosthenes* F11; Strabo, *Geography* 1.1.12.
12 Dicks, *Geographical Fragments* 121–2.
13 Pliny, *Natural History* 2.180.
14 Klaus Geus, "Hellenistic Maps and Lists of Places," in *Hellenistic Astronomy: The Sci-ence in Its Contexts* (ed. Alan C. Bowen and Francesca Rothberg, Leiden 2020) 232–9.
15 Xanthos, *FGrHist* #765; Strabo, *Geography* 1.3.4.

16 Jones, "Ptolemy's Geography" 118.
17 Agathemeros 2; Bowersock, "East-West Orientation" 167–78.
18 *GG* 6.7.7; *Casson, Periplus* 157–8.
19 Berggren and Jones, *Ptolemy's Geography* 145–7; Serena Bianchetti, "Esplorazioni afri-
    cane di età imperiale (Tolomeo, *Geogr.*, I, 8, 4)," in *l'Africa Romana: Atti dell'XI con-
    vegno di studio Cartagine, 15–18 dicembre 1994* (ed. Mustapha Khanoussi et al., Ozieri
    1996) 351–9.
20 Ephoros F30a.
21 Strabo, *Geography* 17.3.19; Pliny, *Natural History* 5.36.
22 Athenaios 5.201c; Kitchell, *Animals* 161–3.
23 As a point of curiosity, this is well south of the southern tip of Africa.
24 *Periplous of the Erythraian Sea* 15–16.
25 Pliny, *Natural History* 6.181–7; Seneca, *Natural Questions* 6.8.3–5.
26 Strabo, *Geography* 2.1.14–15.
27 Herodotos 2.29.
28 Agathemeros 2, 5; Keyser, "Geographical Work" 353–72.
29 Strabo, *Geography* 17.1.24; Pliny, *Natural History* 12.53.
30 Nathanael Andrade, "The Voyage of Maes Titianos and the Dynamics of Social Connec-
    tivity Between the Roman Levant and Central Asia/West China," *MediterrAnt* 18 (2015)
    41–74. See also Irina Tupikova et al., *Travelling Along the Silk Road: A New Interpreta-
    tion of Ptolemy's Coordinates* (n.p. 2014), with many interesting maps.
31 The maximum distance across Ireland is 275 km.
32 Pliny, *Natural History* 4.95; 37.33, 36.
33 Berggren and Jones, *Ptolemy's Geography* 150–2.
34 *Periplous of the Erythraian Sea* 63–5.
35 Pliny, *Natural History* 6.88.
36 Berggren and Jones, *Ptolemy's Geography* 150–2.
37 Strabo, *Geography* 6.1.15; *Periplous of the Erythraian Sea* 32.
38 Berggren and Jones, *Ptolemy's Geography* 155–6.
39 Stückelberger and Graßhoff, *Klaudios Ptolemaios* 91.
40 Aujac, *Claude Ptolémée* 125.
41 Berggren and Jones, *Ptolemy's Geography* 157–62.
42 Herodotos 2.34.
43 Helmut Halfmann, *Itinera principum* (Stuttgart 1986) 190.
44 *Periplous of the Erythraian Sea* 64.
45 Casson, *Periplus* 134.
46 Roller, *Eratosthenes' Geography* 24–6.
47 Strabo, *Geography* 2.5.10; for a drawing of the globe, see Thomson, *History* 203.
48 Ptolemy, *Mathematical Syntaxis* 8.3.
49 For some adjustments in the ordering of the text at this point, see Berggren and Jones,
    Ptolemy's Geography 84–5.
50 Strabo, *Geography* 2.5.8; Dicks, *Geographical Fragments* 185–91.
51 Berggren and Jones, *Ptolemy's Geography* 84–93; Stückelberger and Graßhoff, *Klaudios
    Ptolemaios* 119–35, both with many diagrams.

3 Northern, Central, and Western Europe

Introduction

Northern, Central, and Western Europe are examined in Book 2 of the *Guide*. But by way of preface to the topographical catalogues, Ptolemy began with some issues of methodology (*GG* 2.1.1–10). He stressed that better-known places would be located more accurately than obscure ones, seemingly an obvious point but something that needed to be noted. The layout of toponyms would be from west to east and north to south. He also drew attention to the traditional division of the inhabited world into three continents: Europe and Libya were separated by the Straits of Herakles at the west end of the Mediterranean, and Europe and Asia by the Maiotic Lake (modern Sea of Azov) and the Tanais (modern Don) River. Libya (Africa) and Asia were divided by the Bay of Arabia (modern Red Sea) and the inlet at Heroonpolis (modern Gulf of Aqaba), with a line across to the Mediterranean, thus placing the Sinai Peninsula in Libya. This follows standard continental theory, which had developed in the Greek world as early as the fifth century BC. The Europe-Asia boundary that Ptolemy recorded was the common one,[1] but there had long been a dispute about Asia and Libya—whether they were to be separated at the Nile or the Red Sea—that was still current as late as the first century AD.[2] Ptolemy attempted to lay this to rest by being unusually explicit in his choice of the Red Sea boundary, arguing that seas rather than rivers were better used whenever possible. A connection between southern Africa and Asia, something Ptolemy probably learned from Hipparchos, was mentioned elsewhere (*GG* 7.3.6).[3]

Furthermore Ptolemy stated that he would use political boundaries as a way of distinguishing localities, both in the text and on the maps, but he had little interest in ethnography, a rule that he violated to some extent in the more remote and exotic parts of the inhabited world. Finally, he emphasized that this methodology would allow his map-making to proceed, but he would use parallel rather than converging meridians and straight latitude lines, thereby anticipating the navigators' projection created by Gerardus Mercator in the sixteenth century.[4]

After this introductory material, the topographical catalogues begin. Starting with Ivernia (Ireland), the material in Book 2 covers the first five of Ptolemy's maps of Europe, essentially the northern Atlantic, the Iberian Peninsula, Gallia (modern France and Belgium) and the territory west of the Rhenos, or Rhine, northern

DOI: 10.4324/9781003248590-4

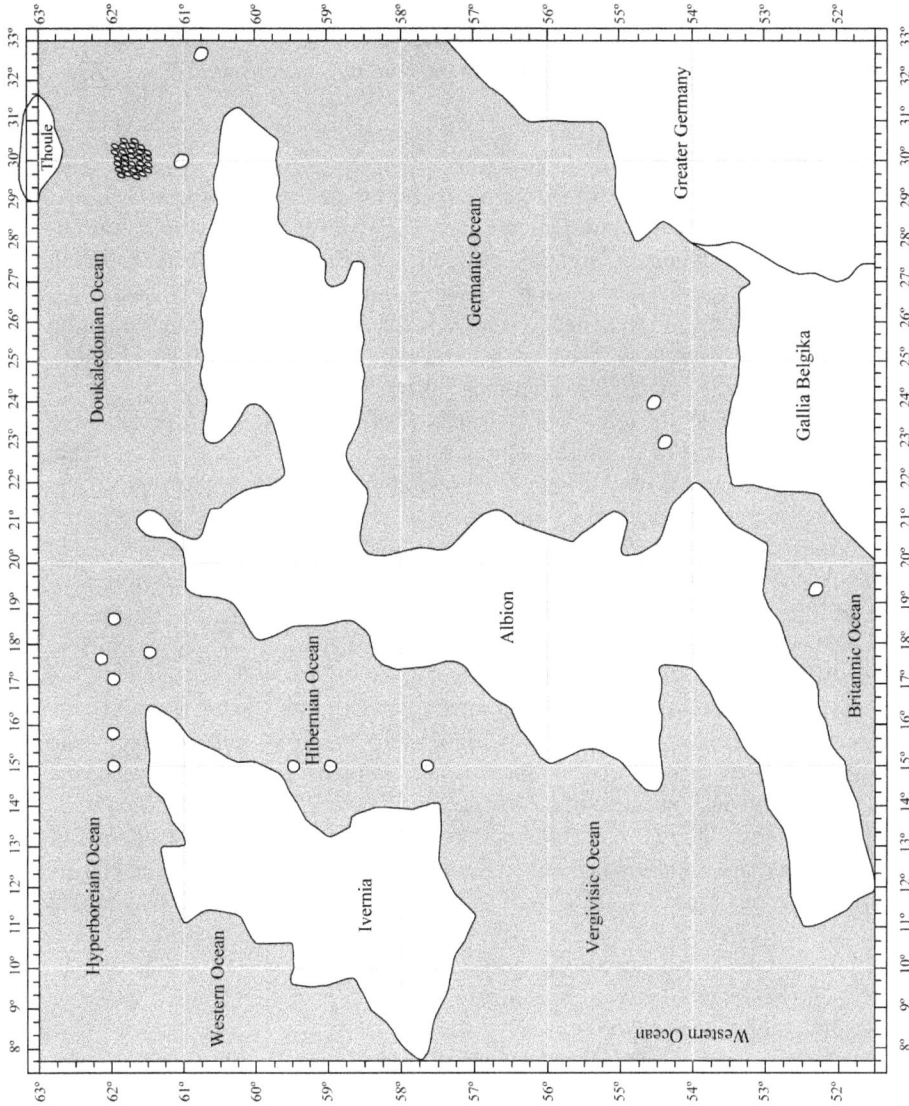

Map 3.1 Map 1 of Europe

Europe west of the Vistula, and south into the Alps and along the northern Adriatic coast as far south as Macedonia and east along the south shore of the upper Danube.[5]

Map 1 of Europe

The first map of Europe consists of Ivernia (Ireland) and Albion (Great Britain), with various surrounding islands, including the enigmatic Thoule (*GG* 2.2.1–3.33).

Ivernia

Ivernia (*GG* 2.2.1–12), the northwesternmost region of the inhabited world, was considered by Ptolemy the third largest island known.[6] His name, Ivernia, is probably based on the indigenous Iwernia—difficult to render in Greek—demonstrating that he had access to local sources, presumably those traders and merchants who had visited the island and were not connected to the Greco-Roman literary tradition except in the most basic way. There are several names for the island and its inhabitants, including Ierne and the more familiar Hibernia.[7]

Carthaginians may have reached Ireland as early as 500 BC but it remained at the fringes of the known world throughout much of classical antiquity: the island was believed to be a cold and disagreeable region that was barely inhabitable.[8] Yet traders and merchants regularly visited it, and Cn. Julius Agricola, active in the British Isles during the latter first century AD, considered invading the island and had contact with local rulers, perhaps even sending a brief expeditionary force.[9]

But Ptolemy relied almost totally on data from traders and merchants, perhaps collated by the obscure author Philemon. Ptolemy was thus able to catalogue a surprising number of places on Ireland.[10] There are a total of 31 toponyms and 16 ethnyms, the former classified as coastal features, river mouths, and towns.[11] For the latter Ptolemy consistently used the word *polis*, simply to mean a "settlement" without any of the ramifications of the term in Greek political theory. Half the toponyms in the catalogue are river mouths, demonstrative of the mercantile and trading origin of the data.

The description starts at the Northern (Boreios) Cape, at the northwest corner of the island, perhaps Bloody Foreland or Rossan Point. To its north was the Hyperborean ("Beyond the North") Ocean. The Hyperboreians had long been identified as an ethnic group living in the far north, beyond the Black Sea, but Ptolemy used the toponym merely as a descriptive generality.[12]

The first series of toponyms runs across the northern side of Ireland, as far as the Rhobogdion Cape at the northeast, perhaps Fair Head. Two ethnic groups in the north—the Venniknioi (to the west) and the Rhobogdioi (to the east)—are also named. The account then returns to the Northern Cape and proceeds south along the west coast, citing five river mouths and one settlement (Nagnata) before reaching the Southern Cape at the southwest corner, which cannot be identified because of the numerous promontories on the coast of County Kerry and County Cork. From here the *periplous* continues east to the Sacred Cape at the southeast corner, probably Carsore Point, and then back along the east coast to Cape Rhobogdion. River mouths dominate, with scattered ethnic groups in the interior, and two towns on the east coast, Manapia and Eblana. The list concludes with seven towns in the interior, listed roughly north to south. As one might expect, the heaviest density of toponyms of any sort is on the east side, facing the Roman territories in Britannia, and largely visible from them.

Practically all the toponyms have indigenous names except for three of the promontories, described in Greek. These would have been among the localities first identified by Greek mariners, before local names became known. In fact the southeastern corner, the Sacred (Ieron) Cape, is remindful of the name used by Avienus in his *Ora Maritima*, who made use of early material from western Greek and Carthaginian sources. He wrote of the Hierni, who lived in a land noted for its rich turf.[13] Whether there was confusion between the Hierni, Ieron, and various other names for Ireland is speculative but plausible. Two inland towns, both named Rhegia, seemingly a Latin term, may have been the seat of local chieftains. Yet few of Ptolemy's toponyms can be precisely identified, although most can be approximately located due to his coordinates.[14]

In addition to the places on Ireland itself, Ptolemy listed the oceans on each side, using indigenous names for the south and east (Vergivios and Ivernios, presumably variants of the name for Ireland), and Greek generic ones for the west and north (Dytikos and Hyperboreian, or Western and Beyond the North). This follows two common patterns for naming oceans: either after the territory on their coasts, or the direction in which they point.

To complete his description of Ireland, Ptolemy listed nine islands off its shores. The northernmost five were collectively called the Edoudai, with two of them actually having that name. These are the Hebudes recorded by Pliny,[15] and are without doubt the Hebrides, although both Ptolemy and Pliny had only a faint comprehension of their topographical complexity and number; in fact there are dozens of them, many of which are uninhabited. To the south, in the modern Irish Sea, were four more islands, two of which were deserted. Monaoida may be the Isle of Man, but all identifications are speculative.

Albion

The next region discussed is the island of Albion, or Great Britain (*GG* 2.3.1–33). Ptolemy made a distinction between Albion itself and the generalized Brettania Islands, which included Ireland and the surrounding islands. There was also the Brettanic Ocean (modern English Channel), south of Albion, but no specific toponym Brettania appears in the *Guide*. Unlike Ivernia, Albion was a well-known region in Ptolemy's day, and over 100 toponyms are listed, with, as usual, river mouths predominating.

Albion was believed to be the ancient name for the island, or a portion of it.[16] It was known to the Mediterranean world from at least the fourth century BC, or nearly a century earlier if the Carthaginian Himilko had gone this far and recorded the toponym.[17] It was more common as an ethnym, Albiones or Albioni, and may originally have referred to a specific ethnic group, probably encountered by Pytheas of Massalia in the late fourth century BC. He also used Prettanike as a more generalized term, a locality that had a circumference of 40,000 stadia; thus the toponym must refer to the entire island.[18] In time the name evolved into Brettanike, or, in Latin, Britannia.[19] These orthographic nuances, important for topographic history, astonishingly are often ignored by modern commentators. Ptolemy's choice of the older name Albion demonstrated that he was more attuned to popular than scholarly sources.

Even though the toponym Albion may have been highly localized at one time, Ptolemy used it for the entire island, from the Antivestaian Cape (probably Land's End) at the southwest to Cape Orkas (probably Dunnet Head) in the extreme north. The localities plotted are spread remarkably evenly across the island, and over 40 ethnic groups are included, as well as the usual assortment of towns and river mouths. There is also a reference to the Kaledonian Forest (*GG* 2.3.12) in the north, the first of many inland landscape features referenced in the *Guide*. Although there are issues with the positioning of Scotland, as noted in the following, its coordinates suggest that the forest is in the highlands east and south of Inverness.

Pytheas was the first from the Mediterranean to explore the Island of Albion.[20] Eventually, in 55–54 BC, Julius Caesar invaded the southern portion,[21] and in the first century AD part of the island was conquered and provincialized, with penetration into Scotland in AD 84.[22] In addition, there was a seaborne reconnaissance by an obscure Demetrios of Tarsos, who was probably the person who set up inscriptions at Eboracum (York), one of which was dedicated to the Ocean; these are rare examples of Greek inscriptions from the British Isles. His journey was perhaps in the AD 80s but is hardly known. Nevertheless his report may have been some of the most recent information about the coasts available to Marinos and Ptolemy.[23] Demetrios' seaborne data would have complemented the land information from Roman officials such as Agricola.[24] Yet Ptolemy seems to have had no knowledge of the Roman withdrawal from Scotland or the construction of Hadrian's Wall in the AD 120s, although even after this date traders would have continued to visit the far north.

The most peculiar issue regarding Ptolemy and Albion is the position of Scotland. England and Wales are well located, but north of the Vedra River (perhaps the Wear), which empties into the North Sea (Ptolemy's Germanic Ocean) (*GG* 2.3.6), the orientation takes a sharp turn to the east. Cape Orkas, the northernmost point, is over 11 degrees of longitude east of the Vedra, when it should be actually to the west (assuming Cape Orkas is Dunnet Head). Many reasons have been advanced for this anomaly. Erroneous data transmission is certainly a possibility, whether by Marinos or Ptolemy, but this would be unusually egregious. It is most likely that Marinos or Ptolemy wished to ensure space for places north of Scotland, such as Thoule, which—based on material from Pytheas—was located at 63 degrees north latitude. Nevertheless the positioning of Scotland remains one of the more peculiar elements of Ptolemy's topography.[25]

Many of the toponyms on Albion represent a Greek source, some as early as Pytheas, such as Kantion (modern Kent), specifically identified by the Greek explorer (*GG* 2.3.4).[26] But generally a more Roman perspective is represented, with Roman place names converted, sometimes awkwardly, into Greek versions. Examples include Noiomagos, or Roman Noviomagus (modern Chichester), or Darouernon, which is Roman Durovernum (modern Canterbury). On occasion his

Roman descriptive toponyms were directly translated into Greek: Aquae Calidae (modern Bath) became Hydata Therma (*GG* 2.3.27–8). In the north (Scotland), where there is a surprising density of ethnyms in this little-known region, Ptolemy may have had a list, perhaps obtained from the reports of either Agricola or Demetrios. Ptolemy was familiar with Tacitus and thus may have had information on Agricola's campaigns in the region in AD 81–83.[27] It is possible that the site of Victoria commemorates the Roman victory over a local coalition at Mt. Graupius.[28]

In addition, the Roman military presence was represented. Pteroton Stratopedon, at an unknown location in Scotland (*GG* 2.3.13), may be a Roman legionary fortress named Winged Camp (perhaps called Pinnata Castra), although it is possible that it was an indigenous hill fort. Orrea, to the west, is perhaps a Roman supply depot (Horrea, or "Granary"). But most demonstrative of Roman activity is the mention of three legions. The Legio VI Victrix was at Eboracum (York) (*GG* 2.3.17), but was also involved in the operation of Hadrian's Wall.[29] Farther south, Legio XX Valeria Victrix was at Deva (Chester) (*GG* 2.3.19), and had been in Britannia since Claudius' invasion of AD 43.[30] It was part of Agricola's command and then helped construct the wall. There was also Legio II Augusta, outside Iska (Exeter), also part of Claudius' forces (*GG* 2.3.30).[31] Mention of these three legions—and there are many others cited throughout the *Guide*—demonstrates that Ptolemy had a military source, probably Marinos, who provided up-to-date information about the legionary presence throughout the empire.

As was commonplace in the *Guide*, the catalogue of a major coastal region concludes with the surrounding islands. Seven are listed. The most enigmatic is Thoule (more commonly Thule), discovered by Pytheas at a point six days north of the British Isles.[32] Further details as to its position are lacking, but it became notable because of its far-north location and unusual morphology and climate. In fact, it evolved into a cultural paradigm, perhaps best defined by Vergil's Ultima Thule, a metaphor for the most remote place on the earth.[33] After Pytheas no one visited Thule again throughout classical antiquity although there were many attempts to locate it and many claims to have found it. It is most probable that Pytheas' Thule was Iceland, but this remains greatly disputed.[34] Eventually it was considered part of the British Isles. When Agricola sent a fleet around the northern coast of Scotland, it went to the Orcades (Orkneys), which are visible from the Scottish coast and had been known for at least half a century,[35] and then had a difficult passage north to what he called Thule.[36] This cannot be the place discovered by Pytheas—it is too close—and was probably one of the Shetlands, most likely Mainland, the southernmost of the group, or even Fair Isle, lying halfway between the Orkneys and the Shetlands. Agricola's report was probably Ptolemy's source, and he located Thule about 2 degrees of latitude north of the Scottish mainland, which would be at the Shetlands (*GG* 2.3.32). He placed it at 63 degrees north latitude, and provided four other coordinates that positioned it over approximately a degree of latitude and 2 2/3 degrees of longitude. This does not conform to the Shetlands in terms of detail, which are slightly farther north and oriented north-south rather than east-west, but there seems no doubt that they are Ptolemy's Thule.

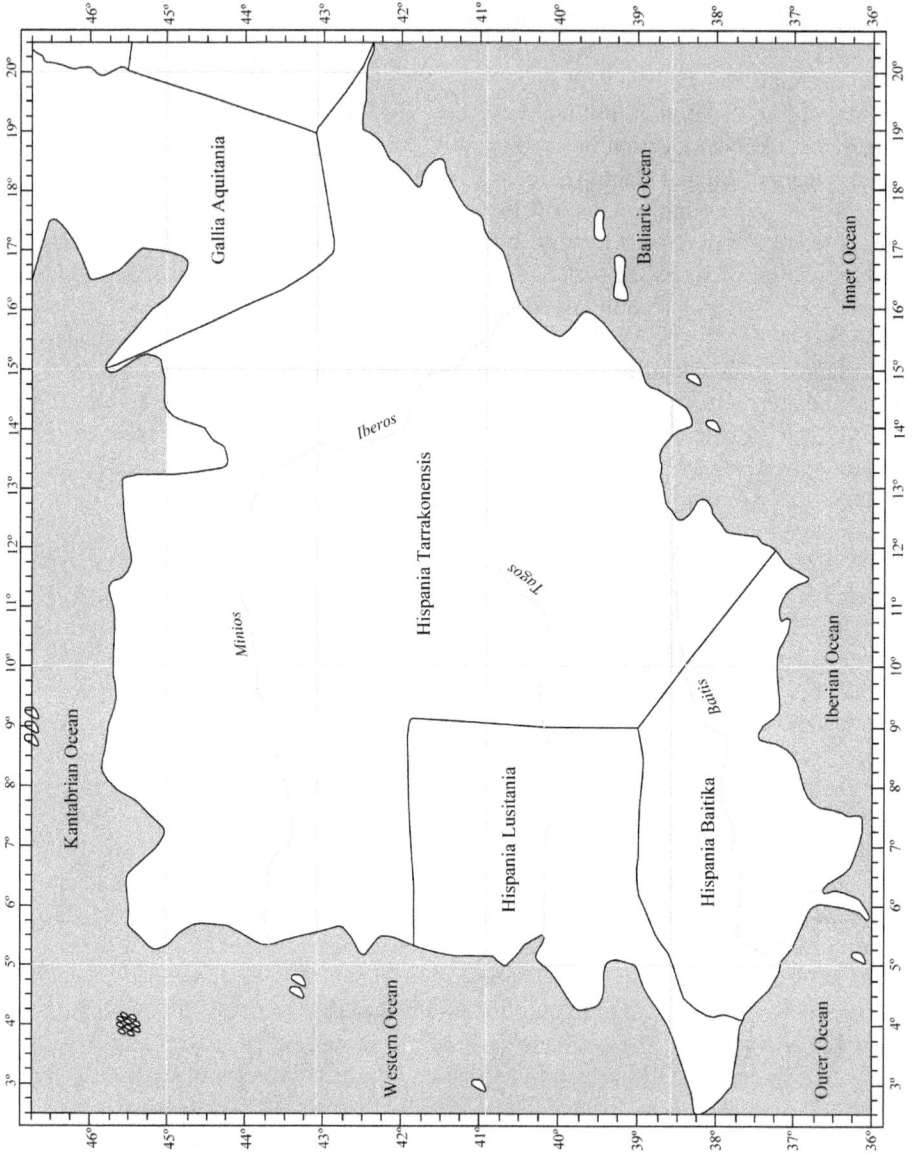

Map 3.2 Map 2 of Europe

The Orkneys also appear on Ptolemy's list (as Orkades), located, as expected, about halfway between the Scottish coast and Thule. South of them are two additional islands, Skitis (to the east) and Doumna (to the west) (*GG* 2.3.31). These may be the obscure islands of Pentland Skerries and Stoma, which lie between the Orkneys and the Scottish coast, and are to some extent where Ptolemy located them. Doumna was cited by Pliny, as Dumna, but in a Scandinavian context, which is of little help. Yet Skitis suggests Skye, west of Scotland proper, and would be more properly positioned if the turning of Scotland to the east is ignored. Doumna would then be an island to the west of Skye, probably one of the Outer Hebrides such as Lewis and Harris.[37]

Three other islands are located off Albion. Vectis, to the south, is the Isle of Wight. More problematic are Tanatis and Kounos (*GG* 2.3.33), which are placed east of the Tamesa (Thames) estuary, where there are no islands. If properly located, they may be promontories near the mouth of the Thames or tidal reefs no longer visible. These islands complete Ptolemy's survey of Albion as well as his first map of Europe.

Map 2 of Europe

Hispania

The second map of Europe represents the Iberian Peninsula (*GG* 2.4.1–6.78). Although not an island, the region is an enclosed unit, with the Pyrenees (Ptolemy's Pyrene) isolating it from the rest of Europe. The account is heavily romanized, with only limited vestiges of the earlier Phoenician, Carthaginian, and Greek toponymic history.

The ethnic history of Iberia is exceedingly diverse. Indigenous Iberians and Kelts had long occupied the region, with Phoenicians appearing as early as the eighth century BC if not before, and who established their own presence. Etruscans and Greeks arrived on the eastern coasts in the seventh century BC, and the Romans came in the third century BC.[38]

Although a few Greek toponyms, such as Tarrakon (Latin Tarraco, modern Tarragona) and Emporiai (modern Empúries) remain, especially on the east coast, the organization of the peninsula is purely Roman: Ptolemy used the name Hispania for its entirety, the standard Roman term, rather than the older Greek one, Iberia, still used by many Greek writers in his day and which for clarity he noted as an alternative. His account of the peninsula is based on the three Roman provinces created in the Augustan reorganization of 27 BC: Tarraconensis in the north and west, Lusitania in the southwest, and Baetica in the south (Ptolemy's orthography is Tarrakonesia, Lousitania, and Baitike).[39]

The density of Hispanic toponyms in the *Guide* reflects the long and rich history of the peninsula.[40] They are spread throughout the entire region, with the mouths of rivers again an important feature. A new element is the location of the source of the major streams, which will remain a feature of the *Guide* throughout the better-known regions. The six most important rivers of the peninsula (the Iberos, Minios, Dourios, Tagos, Anas, and Baitis, or the modern Ebro, Miño, Duero, Tajo, Guadiana, and Guadalquiver) all have both their mouths and sources provided. Some also have a midpoint designated. The sources of rivers are difficult to determine even today, but there would have been a need to provide some information about the course of a major river for the benefit of traders. The shorter streams have no coordinates except their mouths and were less important for access to the interior.

A new feature not previously encountered is the plotting of mountain ranges, another indication that Hispania was better known morphologically than Ireland and the British Isles, where even the mountains of Scotland do not appear in the *Guide*. Ptolemy located the ranges of Hispania by means of their end points—problematic at best—and on occasion their perceived midpoints. Obviously in such rugged territory as the Iberian Peninsula this assumed precision has an arbitrary quality and the data are variable: there is only a single point marking the Marianon Mountain (*GG* 2.4.15), the extensive uplands north of modern Seville which extend over several hundred kilometers. Some ranges are better plotted, especially the Pyrenees and the larger ranges of southern Tarraconensis. Yet even given the vague location of these features, it was probably received information that had helped traders and military forces orient themselves; mountain ranges would become a regular feature of the rest of the *Guide*.

Another new element is the plotting of shrines and sanctuaries, most notably the three points that mark the extremities of the peninsula. At its northeast, where the Pyrenees meet the Mediterranean, was the Aphrodite Sanctuary (*GG* 2.6.11, 2.10.2), originally a Phokaian shrine and located at modern Cape Béar on the French-Spanish border. This would be the first point reached in Hispania on the coastal sailing route from Italy.

The next coastal feature recorded is a temple of Hera, located at what Ptolemy believed was the southernmost part of the peninsula (*GG* 2.4.5), presumably the Promontory of Juno cited by Pliny and probably at modern Cape Trafalgar, which is not quite the southernmost point of the peninsula (Punta da Tarifa, to the southeast, is slightly farther south), but a prominent location nonetheless.[41]

Finally, at the northwest corner of Hispania were the Altars of Sestius (*GG* 2.6.3). These were the work of L. Sestius Quirinalis Albinianus, who, despite being one of the assassins of Julius Caesar, gained the confidence of Augustus and became consul in 23 BC. Afterward he was sent to Hispania Ulterior (as Baetica and Lusitania were called at that time), and set up the altars as a record of his conquests. Horace dedicated an ode to him.[42] The many rugged promontories of northwestern Spain make finding any precise location impossible, but the region of Touriñan is a possibility. These three sites—encircling Hispania from the Pyrenees to the northwest—were not only important navigational markers but a documentation of human control of the peninsula.

As was customary with coastal districts, the surrounding islands were catalogued at the end of the section, in this case after each province (*GG* 2.4.16, 2.5.10, 2.6.76–8). Most of these are obvious and fairly reasonably placed: the Baliarics to the east, Gades to the south, and Londobris (modern Berlonga) to the west. West of the mouth of the Limia River in northern Portugal are the Islands of the Gods (Theoi Nesoi), lying less than a degree of longitude off the coast. There are no islands at this location, but somewhat farther north, extending north from the Vigo estuary in Spain, are several lying slightly off the coast, from Cies in the south to

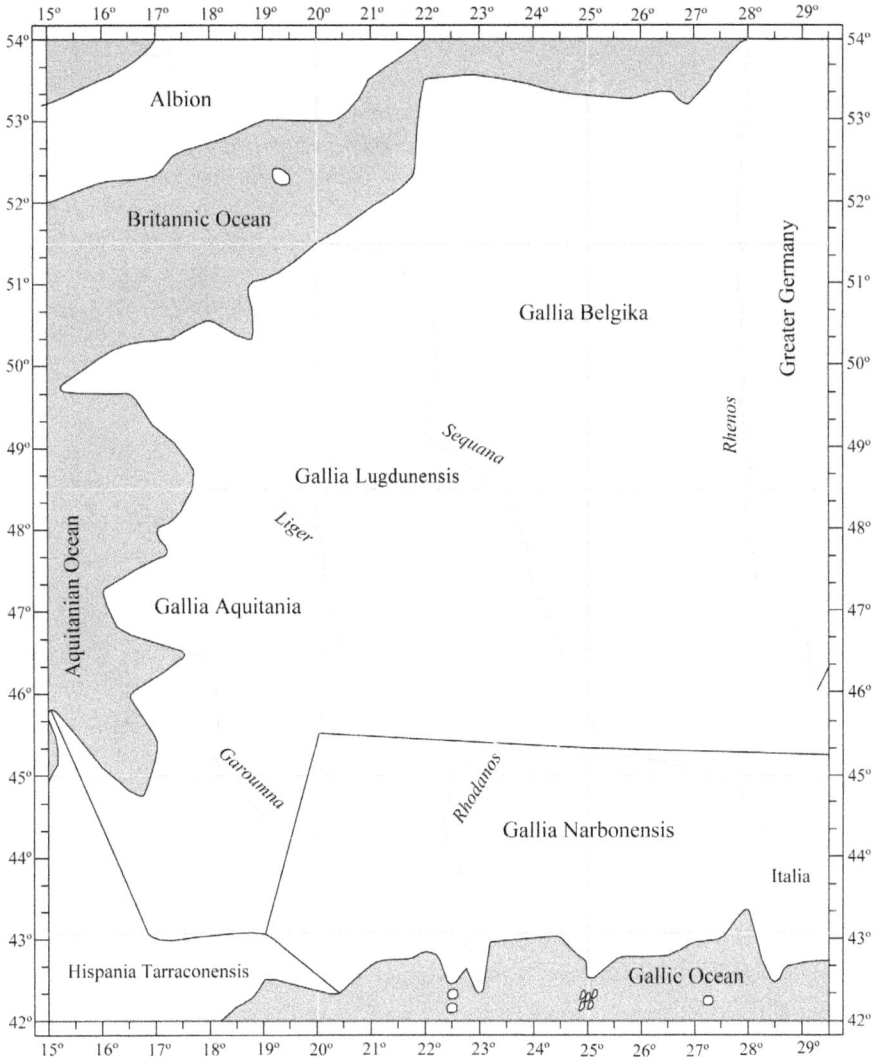

Map 3.3 Map 3 of Europe

Salvora in the north. This strip of islands is a national park today and a region of unusual geology and zoology, and could well be called the Islands of the Gods.

Farther north, according to Ptolemy, were the Kassiterides, somewhat over a degree of longitude west of the Altars of Sestius. Like Thule, they have become a major topographical mystery, a process that began in Greek antiquity. They had been known since the fifth century BC, but even Herodotos, the first to cite them, admitted that he knew little about them other than they were in the far west.[43] The name, meaning Tin Islands, is Greek (from Elamite) rather than indigenous, so presumably they were an early source of that metal, and they may have been discovered when traders from the Greek city of Phokaia penetrated the region after their foundation of Massalia (modern Marseille) near the mouth of the Rhone in the sixth century BC.[44] Traditionally the Tin Islands have been located off the southwestern British coast near Brittany, but all Ptolemy knew was that the sailing route to them began at the northwest corner of Hispania, something already reported by Pliny, and earlier by Diodoros.[45] Ptolemy's situation of the Kassiterides is a fine example of positioning a remote place by the beginning of the sailing route to it rather than its actual location.

Map 3 of Europe

Keltogalatia

This map shows the region that Ptolemy called Keltogalatia ("Keltic Gaul"), extending from the Pyrenees to the Maritime Alps and the Rhine, and north to the English Channel (*GG* 2.7.1–10.21). It included all of modern France, Belgium, and Luxembourg, the southern part of the Netherlands, and western portions of Germany and Switzerland. The rare term Keltogalatia for this territory may have been coined to distinguish the region from Galatia in Asia Minor, where Keltic peoples had settled in the early third century BC. The form follows the pattern of other hybrid ethnyms, especially in terms of the Kelts.[46] Yet it remained rare, and Galatia was the more usual form in Greek for the populations both in the west and east.

Greeks—first Phokaians and then Massalians—occupied the Mediterranean coast from around 600 BC, and penetrated up the Rhodanos (Rhone) and its tributaries into the interior. Using portages to the Liger (Loire) and Sekoana (Sequana or Seine), they reached the Atlantic and English Channel before the fourth century BC. The Romans arrived in the late second century BC, and the entire region came to be well known and documented, especially through the activities of Julius Caesar in the 50s BC. In 27 BC Augustus created four provinces: Narbonensis in the south, Aquitania in the southwest, Lugdunensis extending from the new city of Lugdunum (modern Lyon) north to the English Channel, and Belgica to its east. This was the pattern that Ptolemy followed.

The southern, western, and northern boundaries of Keltogalatia were easily plotted: Our Sea (the Mediterranean) in the south, the Atlantic (known locally as the Akouitanios, or Aquitanian Ocean), and the Brettanic Ocean (the modern English Channel) in the north. The eastern limits were more problematic. Ptolemy created a north-south line from the Mediterranean to the Brettanic Ocean at the mouth of the Rhenos (Rhine). Starting on the Mediterranean at the mouth of the

Varus (modern Var, the boundary between France and Italy until 1860), this line went north along the Maritime Alps and then the Adoula Alps (*GG* 2.9.5) as far as the Rhenos. It then followed that river to its eastern mouth (*GG* 2.9.4), which was placed less than a degree of longitude west of the mouth of the Varus. The mouths of the Rhenos had been explored by Tiberius in AD 5, and other expeditions went beyond, perhaps into the Baltic; Ptolemy's plotting of the multiple mouths of the river is unusually detailed.[47] Yet points in western modern Switzerland are much farther east than located, something Ptolemy may have realized since he placed the Iourassos Mountain (modern Jura) too far west. In addition, the vagaries of the course of the Rhenos are ignored. But he was following the technique of Eratosthenes in creating definitive boundaries for geographical regions, even at some expense to accuracy.[48]

These issues regarding the western Alps are the most notable topographical problems in this section, although obviously not every location throughout the map is placed accurately. Augusta Triberon (Augusta Treverorum, modern Trier) is located only one third of the north-south distance from Augusta Raurikon (Augusta Raurica, modern Augst near Basel) to Agrippinensis (modern Köln or Cologne), but Trier is much closer to Cologne than to Augst. Yet with the exception of the Alpine region such variances are relatively insignificant. But the Alps are only plotted as a line, running roughly north-south, along which the Rhodanos (Rhone) and its major tributaries have their sources, thus locating only the western edge of the mountains (essentially the Graian, Cottian, and Maritime Alps); consideration of their further extent is reserved for the Italian section (*GG* 3.1.1). Yet little was known about the interior of the mountains: for example the Isar (modern Isère) River, which joins the Rhone at Valence and has its source in the Graian Alps in the department of Haute-Savoie, is depicted as originating only half a degree of longitude south of the source of the Rhodanos, along the Alpine line plotted by Ptolemy. In actuality it is nearly 200 km. to the east and deep in the interior of the mountains. Thus even as late as the second century AD the Alps remained enigmatic, although Ptolemy was able to locate what he called Limene, modern Lake Geneva, and was aware that the Rhodanos flowed through it (*GG* 2.10.3).

Ptolemy divided Keltogalatia into four portions: Aquitania in the southwest, Narbonensis on the Mediterranean coast, Lugdunensis in the center between the Liger (Loire) and Sekoana (Seine) Rivers, and Belgica (Belgika), stretching from Narbonensis north to the Bretannic Sea and east to the Rhenos (Rhine). This concept of dividing Keltogalatia—Roman Gallia—into parts originated with Julius Caesar, but Ptolemy's units are from at least the Augustan period.[49]

There are no shrines or sanctuaries recorded in Keltogalatia, with the exception of the Aphrodite Sanctuary, previously noted, on the Hispanic border (*GG* 2.6.11, 2.10.2). The repertory is limited to towns, ethnic groups, and morphological features such as rivers, mountains, islands, and coastal promontories. The density of ethnic groups is large—nearly 70—reflecting the extensive amount of information provided primarily by Julius Caesar, and to a lesser extent by Strabo. There are over 100 towns and over 20 rivers; many of these not only have their mouths and sources plotted but points along their routes, especially at the junction of tributaries, demonstrating information gathered from traders who went up the streams.

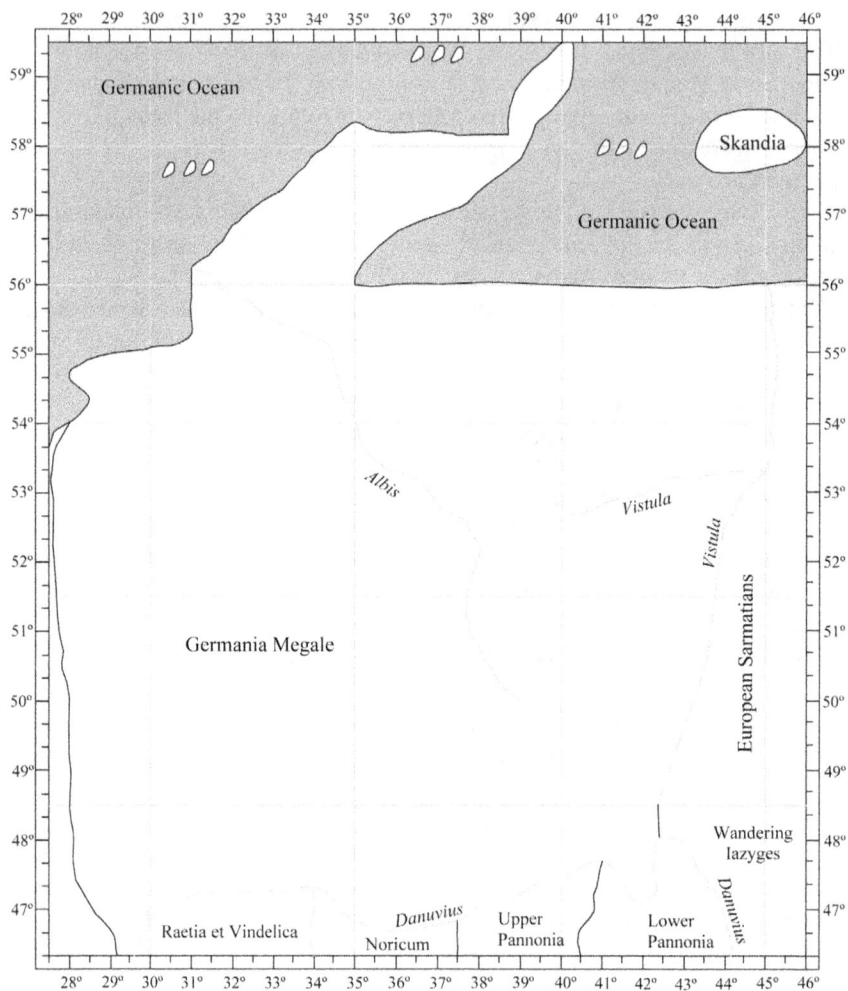

Map 3.4 Map 4 of Europe

There are also four legionary camps, all along the west side of the Rhenos (Rhine), reflecting the frontier defenses that had been established in the first century AD. The southernmost was the Legio VIII Augusta (Ptolemy here used the Greek translation of Augusta, Sebaste), stationed at Argentorate (modern Strasbourg) (*GG* 2.9.17). To its north at Mokontiakon (Mogontiacum, modern Mainz) was the Legio XXII Primigenia, and beyond the Legio I Minervia (Ptolemy called it the Atheniaka), at Bonna (modern Bonn) (*GG* 2.9.16), and finally the Legio XXX Ulpia at Vetera (modern Xanten), the last to be deployed, after AD 118 (*GG* 2.9.15).[50] Since Ptolemy's primary source, Marinos, does not seem to include material after around AD 114, mention of this final legion was probably an addendum by Ptolemy. These forces protected nearly 300 miles of the Rhine frontier.

In a region so heavily romanized as Keltogalatia most of the place names—other than the Greek settlements along the Mediterranean—were primarily known in Latin versions of the indigenous toponyms, which Ptolemy attempted to replicate insofar as Greek orthography allowed. But when there was an obvious Greek equivalent he would translate the name: hence Aquae Sextiae (in Narbonensis) became Hydata Sextia and Aquae Augustae in Aquitania was recorded as Hydata Augousta (*GG* 2.10.15, 2.7.9). But he could be inconsistent in his use of the imperial title, with forms of both Augustus and Sebastos appearing (e.g., *GG* 2.8.17, 2.9.17). He also used the rare word *phoros* as a transliteration of the Latin "forum," although making it masculine rather than neuter.[51] One of the most enigmatic localities with such a name is Phoros Tiberiou, or Forum Tiberii (*GG* 2.9.20), at an unknown location in the Helvetian territory of northwest Switzerland (within the Alpine region of problematic location of toponyms). Where it was, and how it related to the career of Tiberius, has long been a subject of inconclusive discussion. It may have been merely a trading post or a rendezvous point.[52]

Map 4 of Europe

Germania Megale

Map 4 depicts the region called Germania Megale or Greater Germany (*GG* 2.11.1–35), a typical toponym that reflects descriptive terminology which had been used since at least the fourth century BC—and probably far earlier—for the Greek portion of southern Italy.[53] Germania Megale was defined essentially as a square bounded by rivers (the Rhenos [Rhine], Danuvius, and Vistula) on the west, south, and north, and by what Ptolemy called the Germanic Ocean on the north, the adjacent parts of the North Sea and Baltic.[54] As usual, he was following the principle established by Eratosthenes in creating geometric forms for geographical units. The source of the Danuvius is close to the course of the Rhenos in the Black Forest of Germany, and the projected straight eastern and northern routes of the two rivers is a reasonable simplification. But Ptolemy seems to have been unaware of the actual course of the Vistula, although it may have been explored by Greek traders since at least the fourth century BC,[55] and an overland connection between the Vistula and Danuvius was assumed from the first century AD.[56] Yet Ptolemy actually had the Vistula flowing east from a source in eastern Germania Megale and then north into the Baltic (*GG* 2.11.4). His eastern boundary of the region ran north from the Danuvius along the Sarmatian Mountains, and then across a gap of over 2 degrees of longitude to the Vistula (*GG* 2.11.7). The actual source of the river is in extreme southern Poland, and it flows consistently northwest to the Baltic. The Sarmatian Mountains may be the Carpathians, but Ptolemy did not seem to realize that the source of the Vistula is far to the east (or he chose to ignore the matter in order to create a regular form for Germania Megale). The north side of the region, along the North Sea and Baltic, was better defined because of extensive explora-tion of those coasts since the time of Pytheas.[57] But Europe north of the Alps was only vaguely known to the early Greek world, although the amber route between the head of the Adriatic and the Baltic had been traversed since prehistoric times. There was some understanding of the vast forests of northern Europe as early as the fourth century BC, and Aristotle may have realized that there was a distant sea to the north.[58] But few Greeks went in this direction, and it was only with the Roman explorations of the first century BC that Germania became better known, starting with Julius Caesar's penetration east of the Rhine in the 50s BC. This was followed by the activities of the brothers Tiberius and Drusus later in the same century.[59] In the Augustan period, a Roman fleet may have entered the Baltic.[60]

Somewhat over a century later, when Ptolemy wrote, most of the region was not Roman territory. The only exception was the wedge between the upper Rhenos and upper Danuvius, established as part of the province of Germania Superior (with its center at Mogontiacum, or modern Mainz), and a small portion to the east and north of the Danuvius, as far as Castra Regina (modern Regensburg). This was generally administered from Augusta Vindelicum (modern Augsburg). An increas-ing number of fortifications, from the latter part of the first century AD, marked the frontier.[61] Yet Ptolemy showed no interest in the Roman organization of even this portion of Germania Megale and treated the entire region, from the sources of the Rhenos and Danuvius to the northern ocean, as a unit.

Although towns and ethnyms dominate, there are a number of other features. A region designated as the Alps extends from southwest to northeast between the Rhenos and Danuvius (*GG* 2.11.7, 10); this presumably is not the Alps proper but the Black Forest. Other mountains are less easily identified, but are some of the ranges and summits of central Germany. These include the Abnoba Mountains run-ning north-south to the east of the Rhenos, and the Melibokon Mountains, probably the Harz. The name of the Soudeta Mountains is preserved in the Sudetes region of the German-Czech Republic border zone. The Askibourgion Mountains are in the northeast of the region and are probably the Tatras on the Polish-Slovakian border, but somewhat misplaced due to the topographical problems in this area.

An unusual feature of this map is the plotting of four forests, designated by the traditional Greek word *hyle*. All are in the southeast of the region. The Gabreta is the most southwestern, with the Luna to its east, the Orkynian (Herkynian in Latin) to the north, east of the Soudeta Mountains, and the Semanos north of the Soudeta (*GG* 2.11.5–7). None of these can be specifically identified, but would be portions of the thick woodlands of Bohemia and Thuringia. Greeks, coming from the arid Mediterranean, wondered at the thick forests of northern Europe, which were documented by Eratosthenes and known to Julius Caesar as the most fertile part of Germania.[62] Particularly noted were the Orkynian or Herkynian (the modern Bohemian) forests, where a certain Marobodos, after a visit to Rome in his youth (probably in the first century BC), returned to his royal seat of Bouiaimon—a early version of "Bohemia"—and became the most powerful king in the region.[63]

Two human constructions other than towns were documented by Ptolemy. In the northeast, near the source of the Suebos River (perhaps the Oder, but this is far from certain), was the Limis Grove (*GG* 2.11.4). It was located on the north side of the Askibourgion Mountains; if the river in question is the Oder, the grove would be in the vicinity of Ostrava in the Czech Republic, or perhaps at Olomouc. Presumably it was an ancient sacred grove of the type described in detail by Tacitus but neither its exact location nor the reason Ptolemy singled it out can be determined.[64] To the west, near Louppia (probably modern Leipzig) was the Monument of Drusus (*GG* 2.11.28). This is probably the tumulus erected by the brother of Tiberius during his German campaign of 12–9 BC, presumably a prominent enough construction to be noted a century or more later.[65]

Topographically the most interesting features recorded in Germania Megale were at its northern limits. A treatise by the little-known geographical author Philemon, seemingly about the northern oceans, was probably Ptolemy's source for these regions, since he was one of the few predecessors that he named (*GG* 1.11.8). The Kimbrian Chersonesos extended north into the Germanic Ocean (*GG* 2.11.2, 11). Its shape and position indicates that it is Jutland, which had been rounded by a fleet sent by Augustus that saw the Baltic opening up before it.[66] There were also a number of enigmatic islands to the north of Germania Megale. To the west of the Kimbrian peninsula and near the mouth of the Albis (modern Elbe) were the three Saxonoi, probably some of the Frisian islands, which had been explored by several Roman expeditions from 12 BC to AD 5 (*GG* 2.11.31–2).[67] Farther north were the three Alokiai. There are many islands extending along the northwestern European coast from the Netherlands to Denmark, and it is impossible specifically to determine which six Ptolemy had in mind.

Perhaps more interesting are the islands in the Baltic, east of the Kimbrian Chersonesos. Most notable is the large island of Skandia, which was nearly 3 degrees of latitude across and home to seven ethnic groups (*GG* 2.11.35). The toponym, needless to say, is what evolved into Scandinavia. Pytheas in the late fourth century BC may have been the first from the Mediterranean to reach this region and record the toponym, which came to be known (in varying forms) to Pomponius Mela and Pliny.[68] Yet the exact location of Ptolemy's Skandia is more uncertain than one might expect. Its location and size suggest Bornholm, yet the existence of seven ethnic groups on an island whose maximum dimension is scarcely more than 30 km. seems improbable. But it is not uncommon for mainland promontories to be seen as islands by early travelers, and it is more likely that Skandia is the southern tip of Sweden (a region still known as Skanör), with the ethnic groups actually living to its north in interior Sweden and beyond, whose traders came to the southern end of Scandinavia in order to connect with peoples farther south. Particularly worthy of note among these groups are the Finnoi (actually Phinnoi to Ptolemy, Fenni in

Latin), who would seem to be none other than the Finns, recorded by Tacitus as the northernmost ethnic group. They used sharp bone, possibly from reindeer antlers, for arrow points and could run exceedingly quickly, perhaps a reference to ski travel.[69] It would seem that some of them came far to the south to trade and described some of their customs.[70] Other ethnyms on Skandia may also represent far northern peoples. Four small islands west of Skandia and called the Skandiai are part of the same environment and are perhaps some of the Danish islands (*GG* 2.11.33).

Map 5 of Europe

Map 5 of Europe represents a sprawling region extending from the Alps to the northern edge of the Greek peninsula (*GG* 2.12.1–16.4). Ptolemy was unable to reduce this to any geometric form, or to provide a collective name for the area, identifying it instead by the relevant Roman provinces. At its western edge it adjoins Map 3 of Europe in the Adoula Mountains (*GG* 2.12.1) near the presumed source of the Rhenos, and then extends east between the southern Alps and the Danuvius to its junction with the Savos (modern Sava) at Taurounon (modern Zemum in Croatia, near Belgrade) (*GG* 2.15.5, 2.16.7). Then the boundary of the map follows an arbitrary line toward the southeast to Mt. Skardon (*GG* 2.16.1, 6), the modern Šar Planina in Macedonia. From that point another arbitrary line heads westerly to the Adriatic south of Lissos (modern Lezha in Albania) (*GG* 2.16.5). On its western side the region is bounded by the Adriatic and its offshore islands until it reaches a point south of Emona (modern Ljubjana) (*GG* 2.14.7). Running north to the city itself, the limit of the map returns to the Adoula Mountains along the southern side of the Alps; this line is also the northern edge of Map 6 of Europe, representing the Italian peninsula. Greeks had been in the coastal regions of Map 5 since early times, and exploited the trade routes north from the Adriatic. Greek settlement at the mouth of the Istros (Danube) from at least the seventh century BC allowed access at least some distance upriver,[71] but the interior was little explored until the Romans arrived in the second century BC.

Raetia and Vindelica

In Ptolemy's day the region of Map 5 consisted of four Roman provinces. The westernmost was Raetia and Vindelica, which in theory extended east from near the source of the Danuvius (in the Black Forest of Germany) and Brigantium (modern Bregenz in Austria), but as usual the topography of the Alpine regions is uncertain (*GG* 2.11.7). To the east the province went as far as to the vicinity of the Ainos River (the modern Inn), with its northern boundary at the Danuvius (except for an artificial line from its source to the Adoula Mountains). The southern edge was the main range of the Alps, but needless to say this was poorly defined, as was much of the province. It was established in the Augustan period with its administrative center originally at Cambodunum (modern Kempten), eventually moving northeast to Augusta Vindelicorum (modern Augsburg), perhaps because it was closer to the Danube frontier.[72]

The interior Alps were still hardly known in Ptolemy's day, and the mountains were seen as an east-west line running through several ranges north of the Italian peninsula, connecting at the west with the north-south range that marked part of

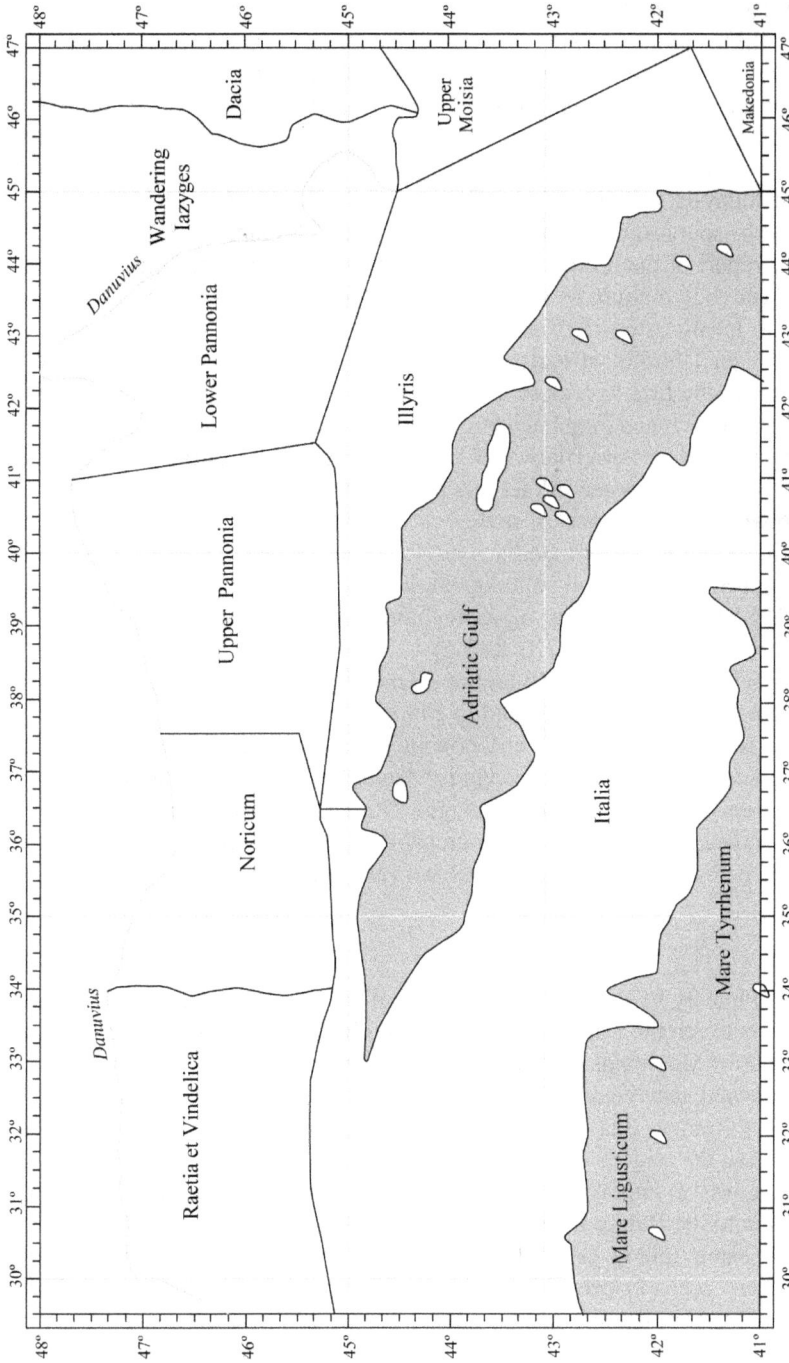

Map 3.5 Map 5 of Europe

eastern Keltogalatia. Their true depth and complexity were not comprehended, and this led to topographical issues. The Alps had been known as a toponym since the fifth century BC, but merely as a tributary of the Danube system,[73] yet it was not until the brothers Tiberius and Drusus crossed them in 16 BC that there was anything but the vaguest information.[74] Their route through the mountains, presumably somewhere in the canton of Graubunden, is not known, but they reached the Rhenos somewhere near or below Chur and followed it down to the lake that today is known as Lake Constance or the Bodensee. Oddly it is not named in the sources, nor does it appear on Ptolemy's map, but Strabo (probably using the report of the brothers) provided dimensions and noted that the Rhenos flowed through it, making its identification certain. Tiberius and Drusus probably visited the local center of Brigantium (modern Bregenz) at the east end of the lake, and then Tiberius, at least, headed north to the source of the Danube. He was probably the first to realize that this was the same as the great and famous Istros River, which had long been known to the Greek world as the largest stream emptying into the western Black Sea.[75] Ptolemy placed the source of the river due north of Brigantium, when in fact it is nearly 2 degrees of longitude to the west. This probably reflects Tiberius' route, who would have gone north from the lake at the village of Tasgaetium, perhaps modern Stein-am-Rhein (*GG* 2.12.5), passing through Bragodurum (at an unknown location), reaching the Danube near its source. Yet he reported this as a journey north from Brigantium to the river, and thus it appeared to Ptolemy (*GG* 2.11.5).

The province of Raetia and Vindelica extended to the east of the line of Tiberius' expedition. A total of 18 villages and 11 ethnic groups were documented, as well as the Likias River (the modern Lech), flowing north from the Alps into the Danube. All the towns have local names except for Augusta added to the seat of the Vindelici (at modern Augsburg), and Drusomagus (*GG* 2.12.5), at an unknown location to the west of the Likias River. This was probably a local settlement renamed in honor of Drusus, perhaps when he died a few years after the expedition.

Noricum

Noricum (Ptolemy's Norikon) was to the east of Raetia and Vindelica, and along the same axis between the Danuvius and the Alps (*GG* 2.13.1–4). Its eastern boundary was the Ketios Mountains, a term for various summits and ranges extending from the Wienerwald near Vienna in a southerly direction toward Emona (modern Ljubljana). The Romans had had some contact with the region since the second century BC,[76] and the trading center known today as the Magdalensberg, northeast of Klagenfurt, was an important early emporium, since it lay on the amber route from the Adriatic to the Baltic. Its ancient name is unknown. Noricum became a province at the same time as its neighbor to the west, with its administrative center at Virunum (*GG* 2.11.27), probably modern Zollfeld near the Magdalensberg. Towns such as Iulium Carnicum (at Zoglio in Italy, on the amber route) and Claudivium (at an unknown location on the Danuvius), as well as the lack of a significant military presence, indicate a relatively high degree of romanization.

Upper Pannonia

To the east of Noricum and also on the south side of the Danube were the two provinces of Upper and Lower Pannonia, so named because of their relative position
along the river, with the upper province farther upstream and adjoining Noricum on
its west (*GG* 2.14.1–15.8). Due to local instability, the region had been provincialized in the early first century AD,[77] but was not divided into two until the end of the
century, with the establishment of a fortified frontier on the Danube by early in the
following century.[78] The division was an indistinct line south from Brigetio (modern
Szöny in Hungary, on the Danube) to a point in the uplands south of the Savus (modern Sava) River, running west of the north-south stretch of the Danube that extends
from Budapest to Belgrade. Ptolemy called the southern terminus the Bebioi or Biblioi Mountains, which cannot be located beyond being somewhere in northeastern
Bosnia and Herzegovina. The southern edge of Upper Pannonia then returned west
through mountainous territory to the edge of Noricum in the vicinity of Emona.

There were several important cities in the province which served at various
times as administrative centers: Poetovio (at modern Ptuj in Slovenia), Carnuntum
(at modern Bad Deutsch-Altenburg just downstream from Vienna), and Savaria
(modern Szombathely in Hungary). All the towns in the province had local names
except for Praetorium, presumably a military encampment, at an unknown location
in the eastern portion (*GG* 2.14.6). The northern edge was heavily fortified and
Ptolemy located the three legions assigned to it (*GG* 2.14.3). The Legio X Gemina
was at Vindobona (Vienna) from early in the second century AD, and the Legio XIV
Gemina was downriver at Flexum (probably modern Mosonmagyaróvár), posted
at the end of the first century AD. Brigetio was the location of Legion I Adiutrix,[79]
in position from the early second century AD, situated where the Arabon (modern
Raab or Rába) River joined the Danube and the effective dividing point between
the upper and lower province.

Lower Pannonia

From Brigetio, Lower Pannonia extended south of the Danube as far as Taurounon,
modern Zemum near Belgrade (*GG* 2.15.1–8). Then the boundary of the province
followed an arbitrary line west-northwest to the Bebion Mountains and the edge
of Upper Pannonia. Since the Danube turns sharply south just above Budapest, the
river was effectively both the northern and eastern boundary of the province. It was
separated from Upper Pannonia at the beginning of the second century AD and had
its administrative center at Aquincum, modern Budapest. This is where the Legio II
Adiutrix was stationed, beginning in AD 106, yet oddly its presence was not noted
by Ptolemy. Six ethnic groups and 24 towns were plotted, most with local names,
although Salinon (or Vetus Salina) (*GG* 2.15.4), at modern Adony on the Danube
downstream from Budapest, may indicate the location of salt deposits.

A curiosity is that Ptolemy positioned the assumed northernmost point of the
Danube (*GG* 2.15.2) just downstream from Kourta, an unidentified location but
easily placed between Solva (modern Esztergom) and Aquincum. This point, at 48
degrees north latitude, is in fact not the northernmost point on the river (which is at

Regensburg in Germany), but where it makes the sharp turn to the south, running in that direction as far as Belgrade. Thus it was an important navigational marker.

Illyris

The final region examined for Map 5 of Europe, as well as for Book 2 of the *Guide*, is Illyris, a vast sprawling district extending from Emona at the eastern edge of Italy to Lissos at the north end of Macedonia (*GG* 2.16.1–14). Its northern boundary is along the southern edge of both Pannonias, from Noricum to Taurounon near modern Belgrade. Then the boundary extends south to Mt. Skardon (modern Šar Planina in Macedonia), and thereafter westerly to the Adriatic at Lissos (modern Lezha in Albania), and back along the Adriatic coast to a point south of Emona and the boundary of the Italian peninsula. A short arbitrary line returns the boundary to Emona itself. This creates a long and narrow region that is oriented along the eastern coast of the Adriatic, subdivided into Liburnia (in the north) and Dalmatia (in the south). Greeks had been in the area that they called Illyris since the fifth century BC, and the Romans arrived in the second century BC as their interests spread north from Macedonia.[80] By the time of Ptolemy, Illyris (Illyria in Latin) was more a geographical than political term, and Dalmatia was more widely used for the entire region.[81] The administrative center was at Salonae, modern Solin near Split in Croatia.

There are serious issues about the orientation of Illyris in the *Guide*, since it is tilted too much toward an east-west axis. Lissos and Emona, the extremities of the region, are separated by 8 1/2 degrees of longitude, but are in fact only 5 degrees apart. To some extent this is the problem caused by parallel meridians, but it also reflects a similar error in the orientation of the Adriatic and Italy itself, replicated in Illyris, even though its north-south dimension is a reasonable 4 degrees of latitude.

Illyris is largely a karst environment, so there are few rivers to be depicted on the map. The major one is the Dreinos, flowing north from an unnamed mountain into the Danube at Taurounon. In fact there is no such river, unless Ptolemy has confused it with the Margos (modern Morava), which joins the Danube farther downstream, or the Drilon (modern Drin), which flows from the same region as the unnamed mountain of the Dreinos into the Adriatic at Lissos. Ptolemy also noted a third river, the Naron (modern Neretva), also flowing from the mountain into the Adriatic at Narona (modern Vid), the largest stream entering the sea in Illyris. The fact that all three of these streams originate from a mountain whose name was not known demonstrates how interior Illyris was little understood, and is suspiciously suggestive of an attempt to create geographical regularity based on scant evidence.

The only mountain named in Illyris is Skardon (modern Šar Planina), which marks the junction of Macedonia and Moesia (Mysia to Ptolemy). Otherwise the topography, in addition to the few rivers, is limited to settlements, ethnic groups, and islands. Most of the towns are near the coast, although this is not apparent in their locations on the map, which has them spread rather evenly throughout the province. For example, Nedinon (modern Nadin in Croatia) is located at the eastern edge of the district, close to Taurounon on the Danube but in fact is less than 15 km. inland. Likewise Asseria (modern Podgradje in Croatia) is placed near the border of Lower Pannonia but is also only a few kilometers inland and not far from Nedinon. This suggests that when exact data were lacking Ptolemy may have indulged in a tendency to spread uncertain locations evenly within a region, a technique hinted at the beginning of Book 2 (*GG* 2.1.2).

To complete his outline of Illyris, Ptolemy listed some of the offshore islands. There are dozens of them along the eastern coast of the Adriatic, and it seems that by listing only eight he was making a conscious selection of the most important, perhaps those with prominent towns, since nine settlements are scattered through the northernmost six islands. Yet there are a number of issues regarding the placement of these islands.

The catalogue runs from northerly to southerly, allowing for the peculiar orientation of Illyris. The northernmost is Apsoros (more commonly Apsyrta, modern Cres in Croatia), the largest of a group (the Apsyrtides) located in the bay between Istria and Croatia. Two more are located along the Liburnian coast: the more northern is Kourikta (modern Krk), which is placed over a degree of longitude southeast of Apsoros but is actually between it and the mainland.

Farther southeast is Skardona, with two towns, Arba and Kolenton (or Colentum). Ptolemy's information was especially confused, since Skardona (modern Skradin) is actually a town on the Croatian mainland, an important early center that in his day had become a major Roman settlement.[82] But Arba (modern Rab) is an island just south of Kourikta, and Kolenton (modern Murter) is another island located approximately where Ptolemy placed his island of Skardona, and which has a similar shape. Clearly Ptolemy was a victim of particularly egregious misinformation—perhaps surprising in this well-known area—or there was an attempt on his part to space the islands regularly along the coast.

Problems continue to the southeast. The island of Issa (modern Vis) was placed a short distance off the coast at the mouth of the Titos (modern Krka) River. Yet this is actually the location of the island of Kolenton, and Issa in fact lies well offshore and is about 90 km. to the south of the mouth of the Titos: it is the farthest of the Adriatic islands from the coast. The opposite problem exists with the next island: Tragourion (modern Trogir) is placed far out to sea but actually it is only separated from the mainland by a narrow channel. It is possible that some of the data on Issa and Tragourion have been reversed.

The last three islands, Pharia (modern Hvar), Korkyra Melaina (Korčula), and Melite (Mljet) are essentially properly placed. It may be that Ptolemy's information for them came from another source than that for the previous five. Listing of these islands closes Book 2.

Notes

1 Herodotos 4.45; *Airs, Waters, and Places* 13.
2 Strabo, *Geography* 1.2.25; Pliny, *Natural History* 6.177.
3 Hipparchos, *Against the Geography of Eratosthenes* F4; Polybios 3.38; Strabo, *Geography* 1.1.9.
4 George Kish, "Mercator, Gerardus," *DSB* 9 (1974) 309–10.

5 Florian Mittenhuber, "Die Länderkarten Europas," in *Klaudios Ptolemaios: Handbuch der Geographie, Ergänzungsband* (ed. Alfred Stückelberger and Florian Mittenhuber, Basel 2009) 268–81.
6 The two larger were Taprobane and Albion: *GG* 7.5.11.
7 Avienus 110–12; Caesar, *Gallic War* 5.13; Strabo, *Geography* 1.4.4.
8 Avienus 117, 383, 412; Strabo, *Geography* 1.4.4; 2.1.13; Duane W. Roller, *Three Ancient Geographical Treatises in Translation* (London 2022) 142–3.
9 Tacitus, *Agricola* 24.
10 J. J. Tierney, "The Greek Geographic Tradition and Ptolemy's Evidence for Irish Geography," *PRIA-C* 76 (1976) 257–65.
11 Grigory Bondarenko, "Goidelic Hydronyms in Ptolemy's *Geography*: Myth Behind the Name," in *Periphery of the Classical World in Ancient Geography and Cartography* (ed. Alexander Podossinov, Leuven 2014) 147–54.
12 Herodotos 4.13; Timothy P. Bridgman, *Hyperboreans: Myth and History in Celtic-Hellenic Contacts* (New York 2005).
13 Avienus 111; Freeman, *Ireland* 29–30.
14 For suggestions as to identification, see Freeman, *Ireland* 71–84.
15 Pliny, *Natural History* 4.103.
16 Pliny, *Natural History* 4.102.
17 Avienus 112–19.
18 Strabo, *Geography* 2.4.1.
19 Duane W. Roller, *Through the Pillars of Herakles: Greco-Roman Exploration of the Atlantic* (New York 2006) 70.
20 Pytheas T1–2, 5, 8, F2–3 Roseman.
21 Caesar, *Gallic War* 4.20–36; 5.7–23.
22 Tacitus, *Agricola* 29–38.
23 Plutarch, *On the Obsolence of Oracles* 18.419e; Roller, *Through the Pillars* 124.
24 It has been suggested that Agricola and Demetrios worked together; Demetrios had an imperial commission for his expedition (David J. Breeze, "The Ancient Geography of Scotland," in *In the Shadow of the Brochs: The Iron Age in Scotland* [ed. Beverly Ballin Smith and Iain Banks, Stroud 2002] 10–14).
25 Barri Jones and Ian Keillar, "Marinus, Ptolemy and the Turning of Scotland," *Britannia* 27 (1996) 43–9.
26 Pytheas F3 Roseman.
27 *GG* 2.11.27, from Tacitus, *Annals* 4.73.
28 Tacitus, *Agricola* 29–38; John C. Mann and David J. Breeze, "Ptolemy, Tacitus and the Tribes of North Britain," *PSAS* 117 (1987) 85–91.
29 David J. Breeze, "Auxiliaries, Legionaries, and the Operation of Hadrian's Wall," *BICS Supplement* 81 (2003) 147–51.
30 Nigel Pollard and Joanne Berry, *The Complete Roman Legions* (London 2015) 101–3.
31 Pollard and Berry, *Complete Roman Legions* 89.
32 Pytheas F2 Roseman.
33 Vergil, *Georgics* 1.30.
34 Roller, *Through the Pillars* 78–87.
35 Pomponius Mela 3.54.
36 Tacitus, *Agricola* 10.
37 Pliny, *Natural History* 4.104; George Broderick, "Some Island Names in the Former 'Kingdom of the Isles': A Reappraisal," *JSNS* 7 (2013) 1–28.
38 Michael Dietler, "Colonial Encounters in Iberia and the Western Mediterranean: An Exploratory Framework," in *Colonial Encounters in Ancient Iberia* (ed. Michael Dietler and Carolina Lopez-Ruiz, Chicago 2009) 5–13.
39 Strabo, *Geography* 3.4.20.
40 Olivier Defaux, *The Iberian Peninsula in Ptolemy's Geography* (Berlin 2017); Arthur Haushalter, "L'Ibérie de Ptolémée, entre géographie mathématique et procédés empiriques," *GA* 26 (2017) 61–73.

41 Pliny, *Natural History* 3.7.
42 *PIR* S436; Horace, *Ode* 1.4; Pomponius Mela 3.13; Pliny, *Natural History* 4.111.
43 Herodotos 3.115.
44 Pliny, *Natural History* 7.197; Roller, *Through the Pillars* 11–14.
45 Diodoros 5.38.
46 Diodoros 5.32.5; Strabo, *Geography* 1.2.27, 4.6.3.
47 Augustus, *Res gestae* 26; Thomson, *History* 239–40; Klaus Geus and Irina Tupikova, "Von der Rheinmündung in der Finnischen Golf . . . Neue Ergebnisse zur Weltkarte des Ptolemaios, zur Kenntnis der Ostsee im Altertum and zur Flottenexpedition des Tiberius im Jahre 5 n. Chr.," *GA* 22 (2013) 125–43.
48 Eratosthenes, *Geography* F66.
49 Caesar, *Gallic War* 1.1.
50 Pollard and Berry, *Complete Roman Legions* 66–77.
51 See also Acts 28:15.
52 Hans Lieb, "Forum Tiberii," *BAPR* 31 (1989) 107–8.
53 Timaios F13; Valerius Maximus 8.7.ext. 2; Strabo, *Geography* 6.1.2; Friedrich E. Grünzweig, "Gross-Germanian," in *Klaudios Ptolemaios: Handbuch der Geographie, Ergänzungsband* (ed. Alfred Stückelberger and Florian Mittenhuber, Basel 2009) 305–11.
54 On the possible origins of Ptolemy's representation of Germania Megale, see Gudmund Shütte, "A Ptolemaic Riddle Solved," *C&M* 13 (1952) 236–84.
55 Roller, *Through the Pillars* 87–90.
56 Pomponius Mela 3.33.
57 Pytheas T21, 25 Roseman.
58 Aristotle, *Meteorologika* 1.13.350b.
59 Roller, *Ancient Geography* 162–4.
60 Augustus, *Res gestae* 26; Pliny, *Natural History* 2.167.
61 *BNP Historical Atlas of the Ancient World* 210–11.
62 Eratosthenes, *Geography* F150; Caesar, *Gallic War* 6.24.
63 Strabo, *Geography* 7.1.3, 5.
64 Tacitus, *Germania* 39; H. Reichert, "Limios alsos," *Reallexicon der germanischen Altertumskunde* 18 (2001) 448–50.
65 Florus 2.30; see also Dio 55.1.
66 Pliny, *Natural History* 2.167.
67 Roller, *Through the Pillars* 118–20.
68 Pytheas T23 Roseman; Pomponius Mela 3.54; Pliny, *Natural History* 4.96, 103.
69 Tacitus, *Germania* 46.
70 J. Svennung, *Skandinavien bei Plinius und Ptolemaios* (Uppsala 1974) 219–45.
71 John Boardman, *The Greeks Overseas: Their Early Colonies and Trade* (fourth edition, London 1980) 247–50.
72 Velleius 2.39; Suetonius, *Augustus* 21.
73 Herodotos 4.49.
74 Strabo, *Geography* 7.1.5, 7.3.13; Pomponius Mela 3.24; Pliny, *Natural History* 9.63.
75 Hesiod, *Theogony* 339; Strabo, *Geography* 7.3.13.
76 Livy 43.5.
77 Velleius 2.96, 116.
78 Franz Schön, "Limes V: Danube," *BNP* 7 (2005) 578–80.
79 Pollard and Berry, *Complete Roman Legions* 186–96.
80 Herodotos 1.196; Livy 43.1.
81 Velleius 2.116.2; Pliny, *Natural History* 3.141.
82 M. Zaninović, "Scardona," *PECS* 812–13.

4   Italia and Eastern Europe

Introduction

Book 3 of the *Guide* covers Italia (Italy) and eastern Europe. There are no preliminaries; it begins with Italia and then includes the islands west of the peninsula, followed by the text to a series of maps east of those in Book 2, which are the final European maps. These begin at what Ptolemy called the Sarmatic Ocean (the southeastern Baltic) and continue south to the Greek peninsula, including the western Aegean islands and Crete. The eastern boundary of these maps runs from the unknown lands east of the Baltic to the Maiotic Sea (modern Sea of Azov), the western Euxeinos Pontos (the modern Black Sea) and the western Aegean. Ptolemy, following long-standing Greek geographical tradition, considered this line to be the eastern boundary of Europe. Maps 6 through 10 of Europe are included in this book.

Map 6 of Europe

This map is devoted to the Italian peninsula, Kyrnos (Corsica), and the intervening islands (*GG* 3.1.1–80). All of its boundaries are the seas surrounding these territories, with the exception of the northern limit (of Italy), which is the Alps and Map 5 of Europe.

Italia

As expected, the Italian peninsula has a high density of toponyms, mostly towns and villages, over 200 in all. In addition there are 24 rivers and the major ranges of the Alps, extending across the northern end of the region from Ligystike (Liguria) to the Adriatic and Illyris. The main chain of the Appennines (Apennines) is plotted from its junction with the Alps north of Nikaia (modern Nice) to its end at Cape Leukopetra, essentially the southern end of the peninsula (modern Punta di Pellaro). The only other mountain shown is the Garganos (modern Gargano), the promontory of the southern Appennines that protrudes into the Adriatic.

Other features include three of the Alpine lakes. Bainakos (modern Lago di Garda) and Larios (modern Lago di Como) are well known. There is also an unidentified one, Lake Poinina, at the source of the Pados, the modern Po, perhaps a vanished glacial lake (*GG* 3.1.24). In addition, there are two human constructions. The

DOI: 10.4324/9781003248590-5

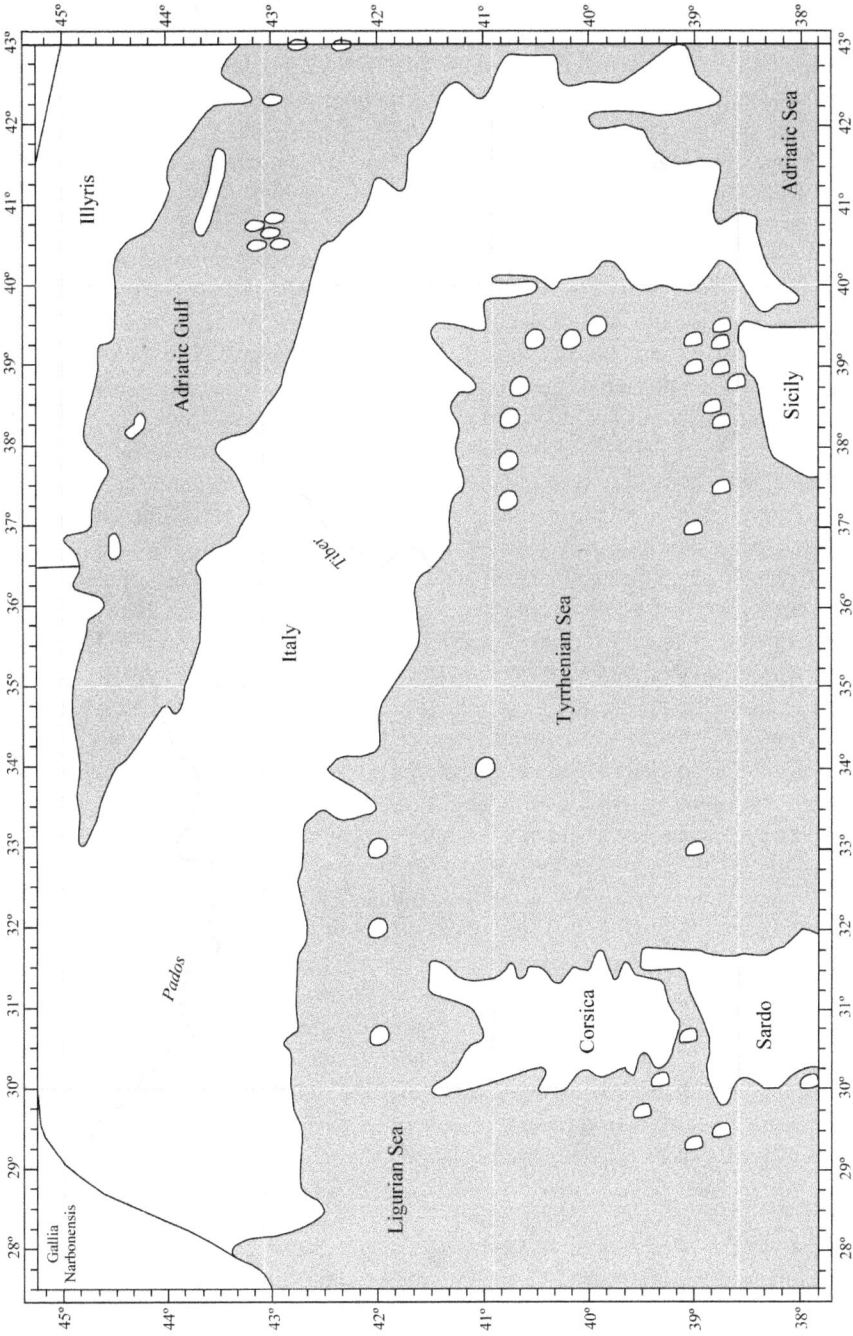

Map 4.1 Map 6 of Europe

Tropaea Augusta (Ptolemy's Tropaia Sebastou) lies on the French-Italian border; its impressive remains, known as La Turbie, are still prominent today above Monaco (*GG* 3.1.2).[1] A Sanctuary of Herakles was on the coast just north of the mouth of the Arnus (modern Arno) River and south of the quarries at Luna (modern Carrara) (*GG* 3.1.4). This is not mentioned elsewhere and cannot be specifically located.

Although the historic Greek cities in the south of Italy—the region long called Megale Hellas or Magna Graecia (*GG* 3.1.10, 75)—were presented by Ptolemy in their original Greek forms (e.g., Taras, Kroton, Metapontion, and many others), his map of Italy generally reflects a romanized aspect of the early second century AD. This is most apparent in the appearance of two Trajanic toponyms, Traiana and the Traianos Limen (Trajanic Harbor) (*GG* 3.1.4, 52). The former was in the bay southeast of Populonium, and the latter was inland to the south of Ancona. Neither is documented elsewhere, nor can they be located, but both provide a date of after AD 98 for the Italian portion of the *Guide*.

As was normal, a number of offshore islands were plotted. The only ones in the Adriatic are the five Diomedeiai (modern Isole di Tremiti) (*GG* 3.1.80), famous as the final home of the hero Diomedes.[2] Ptolemy placed them north of the mouth of the Aternos (modern Pescara) River, but they are actually farther south, just north of the Garganos Mountain. On the other side of Italy, where there are numerous islands, Ptolemy, as he had done with those on the eastern coast of the Adriatic, made a selection of the most important (*GG* 3.1.78–9), from Kapraria (or Caprasia, modern Capraia, between the north end of Corsica and the mainland) and Ilva (modern Elba) in the north to Kaprea (modern Capri) and the Sirenousai (modern Li Galli) in the south.

Most of these islands are fairly accurately placed, but an island named Aithale, located north of Corsica, where there are no islands, is a doublet for Elba. Here Ptolemy used the Greek name, not realizing that Aithale (usually Aithalia) and Ilva were the same island in two different languages.[3] There may also have been some confusion with the obscure island of Urgo (modern Gorgona), northeast of Corsica.[4] Farther south is another non-existent island, Parthenope (located between Pandateria and Prochyte); the toponym is actually an ancient name for Neapolis, or Naples.

The major issue regarding Ptolemy's placement of the Italian peninsula, and one that affects the *Guide* as whole, is that Italy is turned too far toward the east. The actual orientation of the peninsula is toward the south-southeast, but it is positioned more to the southeast on Ptolemy's map. This means that Italy is spread through 14 1/2 degrees of longitude (from Nikaia to Brundisium), when the span is only 11 degrees. Ptolemy's tendency to elongate the east-west dimensions of the inhabited world became a problem in the Renaissance with the first attempts to seek land west of the Pillars of Herakles, since the distance to eastern Asia appeared shorter than it was.

Kyrnos

On the same map is the island of Kyrnos (modern Corsica), the fourth largest island in the Mediterranean (*GG* 3.2.1–7). It had a diverse ethnic history—indigenous, Greek, Etruscan, and Carthaginian—before the Romans arrived in the third century BC. Ptolemy's representation of the island is indebted to the ancient *periplous*

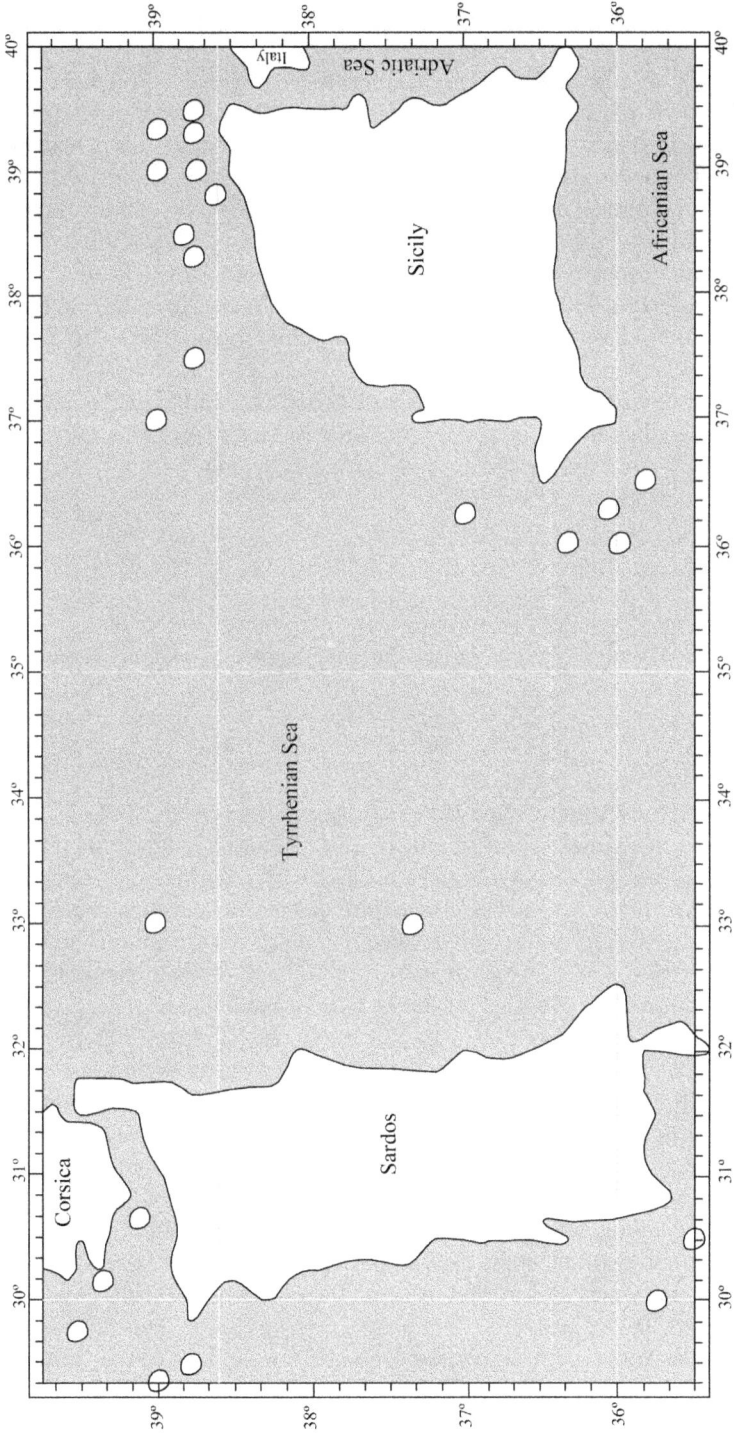

Map 4.2 Map 7 of Europe

format, and may go back to the first Greek navigators to explore it, probably in the seventh century BC.[5]

Since it is an island, Kyrnos is positioned by the seas surrounding it. The Ligystikian (Ligurian) is to the west and north, the Tyrrhenikian to the east, and an unnamed sea on the south between it and Sardo (Sardinia); this is actually the narrow strait known today as the Bocche di Bonifacio. On the island are 23 coastal features (promontories, river mouths, one mountain, 13 towns, and one sanctuary) stretched along a mere 1046 km. of coastline. The highest peak on the island, Chrysoun Mountain (modern Monte Cinto), seemingly "Gold Mountain" but probably an indigenous name, is also recorded. At 2706 m. high it is exceedingly prominent, and visible from coastal France to central Italy; thus it would have been of great importance to navigators.

By contrast there are only 14 towns in the interior. There are also 11 ethnic groups scattered throughout the island and one constructed feature, the Altar of Tutela, the Roman tutelary goddess. It was placed on the east side of the island, where it would be visible on the approach from Italy, and was presumably a guardian of sailors.

Map 7 of Europe

Map 7 of Europe includes Sardo (Sardinia), lying immediately south of Kyrnos, Sikelia (Sicily), and a number of surrounding islands. The boundaries are the appropriate seas.

Sardo

Sardo is the second largest island in the Mediterranean (*GG* 3.3.1–8). Remaining largely outside the Greek sphere of influence, it had a rich indigenous culture (the builders of the famous *nuraghi*), and Phoenician and Carthaginian settlers. The Romans arrived in the second half of the third century BC and soon provincialized the island. As with Kyrnos, Ptolemy located it in terms of the surrounding seas: the Tyrrhenikan to the east, the African to the south, the Sardoan to the west, and the unnamed strait (modern Bocche di Bonifacio) between it and Kyrnos.

Although the topography of Sardo has few errors and its relation to Kyrnos is correctly rendered, the improper orientation of Italy—tilted too far to the east—affects its position. The north end of Kyrnos is accurately located on a parallel with Populonium in Italy, but the north end of Sardo is opposite Campania when it should be opposite Latium. Its south end, actually opposite Bruttium, is positioned south of the southern latitude of Sikelia (Sicily). Thus most of Sardo and Sikelia are on the same latitude, when all of the former is north of the latter. Moreover, this also affects the positioning of the northernmost portion of the continent of Africa (in modern Tunisia), which lies west of Sicily but which Ptolemy has placed 2 1/2 degrees to the south.

The *Guide* is the major source for the topography of Sardo. Ptolemy's catalogue emphasizes the coasts at the expense of the interior, with 39 coastal features but only 12 interior ones. Nevertheless the diverse ethnic history of the island is shown through the placement of 17 ethnic groups in this relatively small region. Also noted is the most notorious point on the island, the Mainoumena (Raging) Mountains, visible as one approaches from the east. It was here in 202 BC that the consul Tiberius Claudius Nero suffered a disaster to his fleet;[6] whether this gave the mountains their name, or it was a previously applied toponym, is not known. In Latin they were called the Insani. Although the mountains have not been exactly located today, Sardinia is especially rugged, with Punto la Marmora reaching 1834 m.

Ptolemy located three thermal springs on the island, again unusual in a limited location. The best known is the Hydata Lesitana, in the eastern portion, still active as a spa today called Sorgenti di Benetutti. There was also a sanctuary of Hera (or Juno) in the northeast, but not specifically located. In the southwest was the sanctuary of Sardopater, at Antas (near modern Iglesias), where remains of a temple of the third century AD have been discovered.[7] Although this is a century after Ptolemy, it presumably was at the location of an ancient indigenous shrine, perhaps from as early as the sixth century BC, and honoring Sardos, the eponymous hero of the island.[8]

The island was rich in mineral resources, especially lead, concentrated in the southwestern district where the Temple of Sardos was located.[9] Ptolemy's Molibodes (Lead) Island, just to the southwest, reflects this. This is modern San Antioco, now joined to Sardo proper. A town in the north, Ploubion, may also be named after local lead mines, using a hellenized form of Latin *plumbum*. Other islands are scattered around Sardo and southern Kyrnos, roughly in their proper region, but following Ptolemy's tendency to locate them more regularly than they actually are.

Sikelia

Sikelia (Sicily) is the largest island in the Mediterranean (*GG* 3.4.1–17). Situated at its narrowest part, less than 150 km. from North Africa, it has been a cultural crossroads from ancient to modern times. The indigenous population was followed by Iberian, Greek, Carthaginian, Italian, and Roman invaders and settlers. It became the first Roman province in the mid-third century BC, and became part of unified Italy in 1860, but today still retains a strong sense of an independent ethnic identity.

Ptolemy's biggest problem in depicting Sicily was its shape. It had long been known that the island was roughly triangular, and its ancient name, Thrinakia or Trinakia, was said, perhaps erroneously, to describe its outline.[10] But Ptolemy attempted to regularize this to excess. The eastern coast, from Cape Peloros (modern Capo Peloro) south to Cape Pachynos (modern Capo Passero) was plotted directly north to south, which is accurate. But the southern coast, west to Lilybaion (modern Marsala) is projected along an east-west line, with a slight southern dip at the west end. In fact the coast runs to the west-northwest, and by ignoring this Ptolemy was left with a wobbly line heading to the north and then east in order to connect back to Cape Peloros, even though much of this side of the island runs west to east.

Given its complex history, it is not astonishing that there is a density of towns in Sicily, over 50 in all, yet only five ethnic groups, indicative of a homogenization and urbanization of the population in the Roman period. But some of Ptolemy's material was ancient and reflects early exploration of the coasts, probably by Greeks. There are the mouths of 20 rivers and a dozen promontories; many of the rivers are hardly

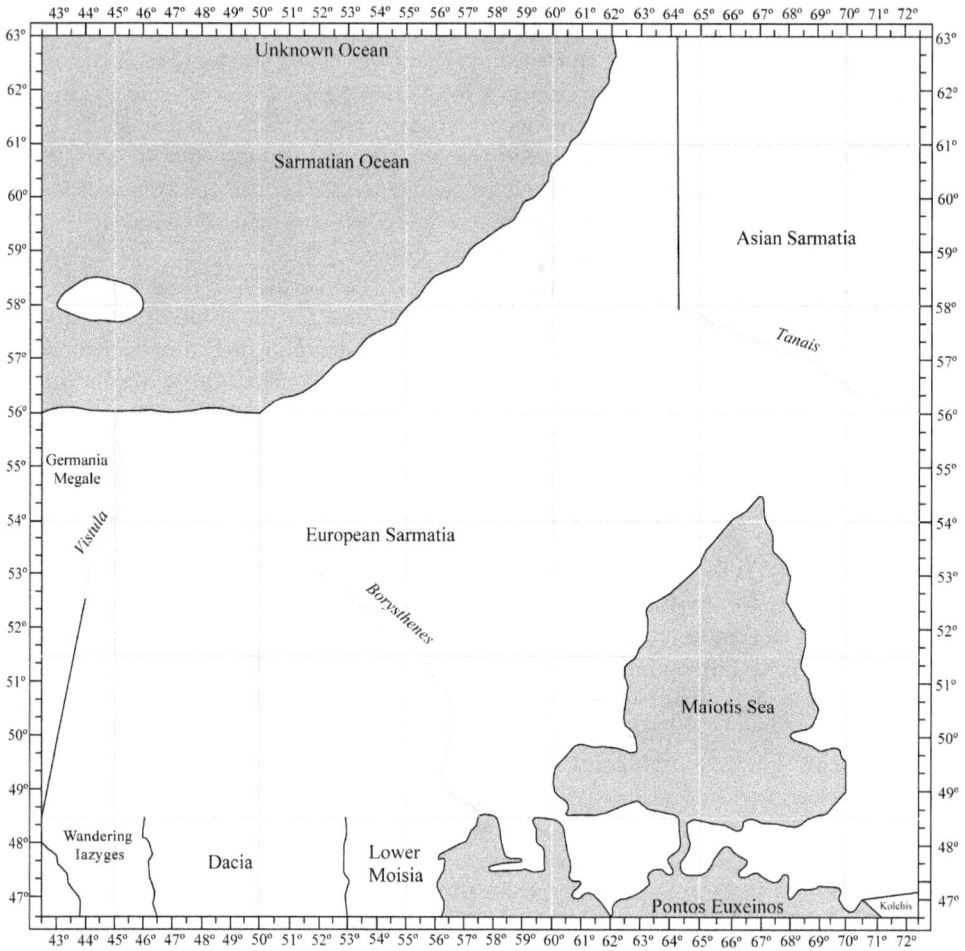

Map 4.3 Map 8 of Europe

noticeable but would have been recorded by the early Greek navigators, who had come to Sicily by the eighth century BC. One feature of particular interest is Cape Odysseus (probably modern Punta delle Formiche) (*GG* 3.4.7), unlikely to be an early name but reflecting the long-standing opinion that Sicily was part of the world of the hero, especially its southeastern portion—where the cape is located—and where the trajectory of the *Odyssey* suggests that he might first have approached the island. At the very least, early Greek settlers of a later date made the same landfalls, and, well acquainted with the epic tales of the hero, located and named places where they believed he had been, toponyms that survived hundreds of years later.[11]

The only other features on the map of Sicily are two mountains. Aitne (modern Etna) is perhaps the most conspicuous mountain on the Mediterranean littoral, and has been volcanically active since prehistoric times (*GG* 3.4.10). In the southern part of the island Ptolemy located Mt. Kratas, not easily identified but perhaps the uplands of the Gemelli Colli north of Agrigento and the second highest point on Sicily.

A total of 15 islands were plotted around Sicily. To the north were the eight islands of the group known as the Aiolians or Lipari (*GG* 3.4.16). Although there are some questions about relative placement, all can easily be recognized. Two isolated islands lie to the west. Oustika (modern Ustica) is about 60 km. north of Palermo, and Aiolos is to its west-northwest, probably a phantom doublet for the Aiolian group. West of Sicily are five additional islands, three of which are among the modern Egadi, directly west of Trapani. A fourth island within this group, Pakonia, cannot be identified. The remaining island on the map is Osteodes, west of Panormos (modern Palermo), whose name suggests a doublet for Oustika, with a possibly Greek name ("Bony"). This was where 6000 recalcitrant Carthaginian mercenaries were deposited and left to starve, probably in the third century BC; their bones were visible for many years.[12] Other islands south of Sicily, even though culturally part of it, were plotted on the second map of Libya (*GG* 4.3.47).

Map 8 of Europe

Most of this map was devoted to the region called Sarmatia in Europe, but it also included the small peninsula of the Tauric Chersonesos, on the north shore of the Black Sea.

Sarmatia in Europe

Sarmatia in Europe was practically all of northern Europe east of the Vistula and north of an arbitrary line heading west from the northern Black Sea to the eastern boundary of Greater Germany southeast of the Herkynian Forest. Sarmatia was actually divided into two parts, with the western called European (*GG* 3.5.1–31) and the eastern Asian (*GG* 5.9.1–32). The dividing line was the Tanais (modern Don) River, from its mouth at the innermost point of the Maiotic Sea (modern Sea of Azov) to its presumed source at 58 degrees north latitude, and then an assumed line to the Baltic (whose southeastern portion Ptolemy called the Sarmatian Ocean). The map ends at 63 degrees north latitude with unknown sea and land beyond. Except for information gathered by traders along the Baltic coast, and those moving inland along the rivers emptying into the northern Black Sea, this region was little known to the Greco-Roman world. Pytheas of Massalia, however, may have crossed its western portions in the late fourth century BC, following established trade routes.[13]

The Tanais River had long been accepted as the division between Europe and Asia,[14] and this line continued southwest through the Maiotic Sea and the Kimmerian Bosporos (the modern Strait of Kerch) into the Black Sea (*GG* 3.6.1).[15] The alignment from the Baltic to the Black Sea formed the eastern boundary of Map 6 of Europe.

The Sarmatians were a nomadic group that originally lived north of the Black Sea. They had been known (as the Sauromatai) to the Mediterranean world since at least the late sixth century BC, in an environment closely associated with the Amazons.[16] To reach them, one journeyed three days up the Tanais from its mouth and then three days to the east. But from the third century BC they began to move west and south, perhaps seeking contact with the increasing Greek presence on the north shore of the Black Sea, in time ending up in the regions north of the Istros (Danube) River. One of their groups, the Rhoxolanoi, were at an early date believed to be the farthest north of peoples, although this detail was quite reasonably contradicted by Ptolemy (*GG* 3.5.19).[17] But to him the territories of the Sarmatians extended from the lower Danube far to the east and north.[18]

Ptolemy's line at 63 degrees north latitude is farther north than he thought—it is north of Oslo or Stockholm—and demonstrates how little known these regions were. It is in fact an arbitrary line corresponding to the latitude of Thoule, the assumed northernmost point of the inhabited world (*GG* 3.5.3). Even so, Ptolemy was well aware that if one reached this latitude both the sea (the upper Baltic) and the land to its east (essentially Finland and Karelia) were unknown (*GG* 3.5.1). Nevertheless traders had followed the coast of the Baltic far beyond the mouth of the Vistula and reached the people known as the Karbones, who, along with other peoples, lived north of the Chesinos River (*GG* 3.5.2, 22). The ethnym Karbones suggests charcoal makers (hellenized from Latin Carbones), but, as always, it may be an indigenous name that sounds Greek.

The coast of the Baltic is depicted as a smooth curving line heading more and more northerly from the Vistula, with a single toponym on the coast, the Ouenedikos (Venedicus) Gulf, probably Gdańsk Bay (*GG* 3.5.1). This relatively unindented Baltic coast indicates that Ptolemy's information ended before the great bays of the Gulf of Riga and the Gulf of Finland. There are four rivers documented between the Vistula and the northern end of the map (Chronos, Roubon, Tourountos, and Chesinos), and one expects these can be equated with the rivers of the Baltic states (particularly the Prego and Neman, and perhaps the Western Drina, which empties into the Gulf of Riga). Thus the Karbones would probably be in Latvia or Estonia, the limit of Mediterranean knowledge of trade routes in this direction; in fact, it was known that if one ventured too far north on the Baltic it became frozen.[19] Ptolemy's source for much of the material was probably the obscure Philemon, one of his few predecessors that he cited by name, and who wrote in the first century AD.[20]

North of the Black Sea the eastern border of the region generally followed the Tanais River (*GG* 3.5.10, 31). As the accepted division between Europe and Asia it had special significance. The river is 1870 km. long with its source 375 km. south-east of Moscow near Tula. It provided trading access to the remote regions north of the Black Sea, a "rough wilderness with forested valleys" according to Pliny.[21] By the seventh century BC Greek traders had gone at least 400 km. upstream, but this was less than a quarter of its course.[22] But Ptolemy was aware of its sharp turn to the northwest (as one goes upstream) near modern Volvograd, and plotted the river from there through 8 1/2 degrees of latitude and 2 of longitude to a presumed source southeast of the Baltic. From there an arbitrary line running due north marked the eastern edge of the map, to the assumed limit of human habitation at 63 degrees north latitude. This was just east of the Karbones and 2 degrees of longitude from the Baltic, perhaps at the southern edge of the Gulf of Finland around Tallinn, although such an assumption is speculative.

The southern part of the eastern border of the region was more easily determined, since it followed the historic division between Europe and Asia through the Maiotic Sea and the Kimmerian Bosporos into the Black Sea. But European Sarmatia did not include the peninsula of the Tauric Chersonesos (the modern Crimea), probably because of its heavy Greek settlement, and so Ptolemy placed the southeastern border of Sarmatia in the marshy area of the channels that separated the Chersonesos from the more mainland regions to the northwest. This was an isthmus about 100 km. across that made a distinct geographical boundary, extending west from what was known as the Byke Lake (an indentation in the Maiotic Sea) to the Karkinitis Gulf of the Black Sea (*GG* 3.5.10).

Yet the southern border of European Sarmatia, west of the Black Sea, was problematic, not because this was little-known territory like that to the east and north, but because of serious questions regarding where to draw the line. The obvious boundary would have been the Istros/Danube River, west from its mouth on the Black Sea to a point in Pannonia, perhaps around Aquincum (modern Budapest), where a connection could be made with the eastern boundary of Greater Germania and Map 4 of Europe. But this had the difficulty of placing the portions of Roman territory that lay north of the Danube within Sarmatia, perhaps not politically correct, since it would link two Roman provinces more with the barbarian north than the Mediterranean world. The province of Dacia had been established by Trajan in AD 106 and extended an uncertain distance north of the river, perhaps approximately to the northern limit of modern Romania. Moesia Inferior, established in AD 86 along the lower river, by Ptolemy's day included some lands north of the river near the Black Sea coast.[23]

Thus in this instance Ptolemy abandoned his traditional procedure of having his maps conform to recognizable geographical units, and simply drew a line for the southern border along 48 1/2 degrees north latitude, keeping the Roman provinces out of Sarmatia. It extended from the mouth of the Borysthenes River (modern Dnieper in Ukraine) west to a point just north of the great turn of the Danube above Budapest in the so-called Sarmatian Mountains, probably part of the Carpathians and at the edge of Greater Germania (*GG* 2.11.7, 3.5.50). Thus he was able to separate Sarmatia from the Roman provinces of the lower river, placing them in Map 9 of Europe. From the Sarmatian Mountains the western boundary of European Sarmatia was the same as the eastern boundary of Greater Germania, in other words, north to the Vistula and then along it to its mouth, which erroneously made the river run due north-south.

Within these boundaries were a variety of geographical features, ethnic groups, towns, cities, and monuments. As expected, their density is greatest along the Black Sea coast (from the mouth of the Borysthenes east to the beginning of the Tauric Chersonesos) and some distance inland, especially along the rivers that empty into the sea. But there were no Greek cities far into the interior, and a decreasing number of ethnic groups and topographical features toward the north. The rivers—whether emptying into the Baltic or the Black Sea system—tend to have their presumed origins in mountains, which, except for the Karpatos (Carpathians) in the south (*GG* 3.5.6), are ephemeral and based on the geographical theory that most rivers must have their source in mountains, hardly applicable to the vast plains of north-eastern Europe. In fact, one of the ranges cited, the Rhipaia (*GG* 3.5.15), had a long history in Greek thought, perhaps because people from the rugged Mediterranean could not conceive of a region without mountains and recorded their existence even where there were none. Located by Ptolemy near the source of the Tanais, the Rhipaia had been assumed as early as the sixth century BC, when Anaximenes of Miletos believed that there were high mountains at the northern edge of the earth that hid the sun.[24] Although entirely mythical, they became a standard feature of Greek geography, and to Ptolemy they were also the probable source of the Chesinos river, which flowed northwest into the Baltic (*GG* 3.5.2).

In the same region Ptolemy placed another feature with a peculiar history, the Altars of Alexander (*GG* 3.5.26). Assuming that they were dedicated to Alexander the Great, they were in a region hundreds of kilometers from any point that the king had reached. But those in his entourage who chronicled his eastern expedition were notorious for their manipulation of topography in order to enhance his reputation, in this case applying the name "Tanais" to the Iaxartes River (modern Syr Darya in south-central Asia east of the Caspian Sea), which Alexander reached in 329 BC. The rationale for this was since the Tanais had long been considered the boundary between Europe and Asia, if Alexander were to cross it, he could say that he had gone beyond the limits of Europe. When he crossed the Iaxartes—now called by him the Tanais—he set up altars.[25] Ptolemy or his source, unaware of this confused history, assumed that there were Altars of Alexander somewhere near the true Tanais, although it is doubtful that anyone saw such a monument in this region.[26] Also enigmatic are the Altars of Caesar, 5 degrees of longitude and slightly to the south of those of Alexander. It is not impossible that a Roman trader erected them in the first or second century AD while on the upper river.

Other than the general uncertainty of the northern portions of European Sarmatia, the major topographical problem on Map 5 of Europe is the size and positioning of the Maiotic Sea (the modern Sea of Azov). According to Ptolemy, its main axis (from the Kimmerian Bosporos to the mouth of the Tanais) runs almost due south to north; in fact the sea is oriented toward the east so that the longitude of the Tanais mouth is approximately 200 km. east of that of the Bosporos. In addition, the dimensions of the sea are greatly exaggerated, so that Ptolemy made it the same size as the Aegean, which is actually five times larger. This problem is mostly due to a cumulative error of data along the coasts of the Maiotis, since there was probably no detailed survey of them but simply sailing reports from point to point on the journey from the Bosporos region to the mouth of the Tanais. Between the Byke Lake and the mouth of the Tanais there are two towns, three river mouths, and a shrine recorded on the western (European) side (*GG* 3.5.13), along what was probably the primary sailing route, as well as other towns and river mouths on the east (Asian) side (*GG* 5.9.2–5). A failure to coordinate many individual reports would have resulted in an overestimation of the size of the sea.

The Tauric Chersonesos

The one portion of Map 8 of Europe that does not depict European Sarmatia is the region at its southeastern corner known as the Tauric Chersonesos (the modern Crimea) (*GG* 3.6.1–6). This is a diamond-shaped district that is separated from European Sarmatia by the marshes and channels that run across the 100 km. between the Byke Lake and the Black Sea. Greeks had moved into here by the early sixth century BC, establishing towns and trading posts, often at the site of indigenous villages, and sending traders far up the local rivers, such as the Borysthenes (modern Dnieper), Hypanis (modern Bug), and Tanais.[27] By the following century a Bosporanian kingdom had been established with its primary city at Pantikapaion (Pantikapaia to Ptolemy, modern Kerch), and which became a major exporter of grain to the Aegean world.[28] This kingdom, although going through many convulsions, was still in existence when Ptolemy wrote and would last into the fourth century AD as one of the longest-surviving states in the ancient world.[29] The long history of Greek settlement is demonstrated by the density of population, with Ptolemy recording over 20 towns in the limited region of the Chersonesos.

His emphasis was on the local demography. Except for a number of promontories around the perimeter of the Chersonesos, which would have been important for sailing directions, and the mouth of one river, the Istrianos, a short stream in the southeastern part of the region that is not mentioned elsewhere and cannot be identified, the toponyms are totally the local towns and villages. Ignored completely are the rugged Tauric Mountains, which extended across the southern part of the peninsula and reach 1500 meters in elevation only a few kilometers from the coast. Given the ephemeral ranges that are scattered across the northern portions of Map 8, this seems an unusual omission.

Map 9 of Europe

This map lies directly south of Map 8 of Europe and east of Map 5 of Europe. It completes the mapping of Europe except for the Greek peninsula, covered in Map 10 of Europe. The northern boundary of Map 9 is the arbitrary line at 48 1/2 degrees latitude that Ptolemy established as the southern boundary of European Sarmatia, running from the Black Sea at the mouth of the Borysthenes to the Sarmatian Mountains near the great curve of the Danube above Aquincum (Budapest). As noted, this was created to separate Sarmatia from the Roman provinces to its

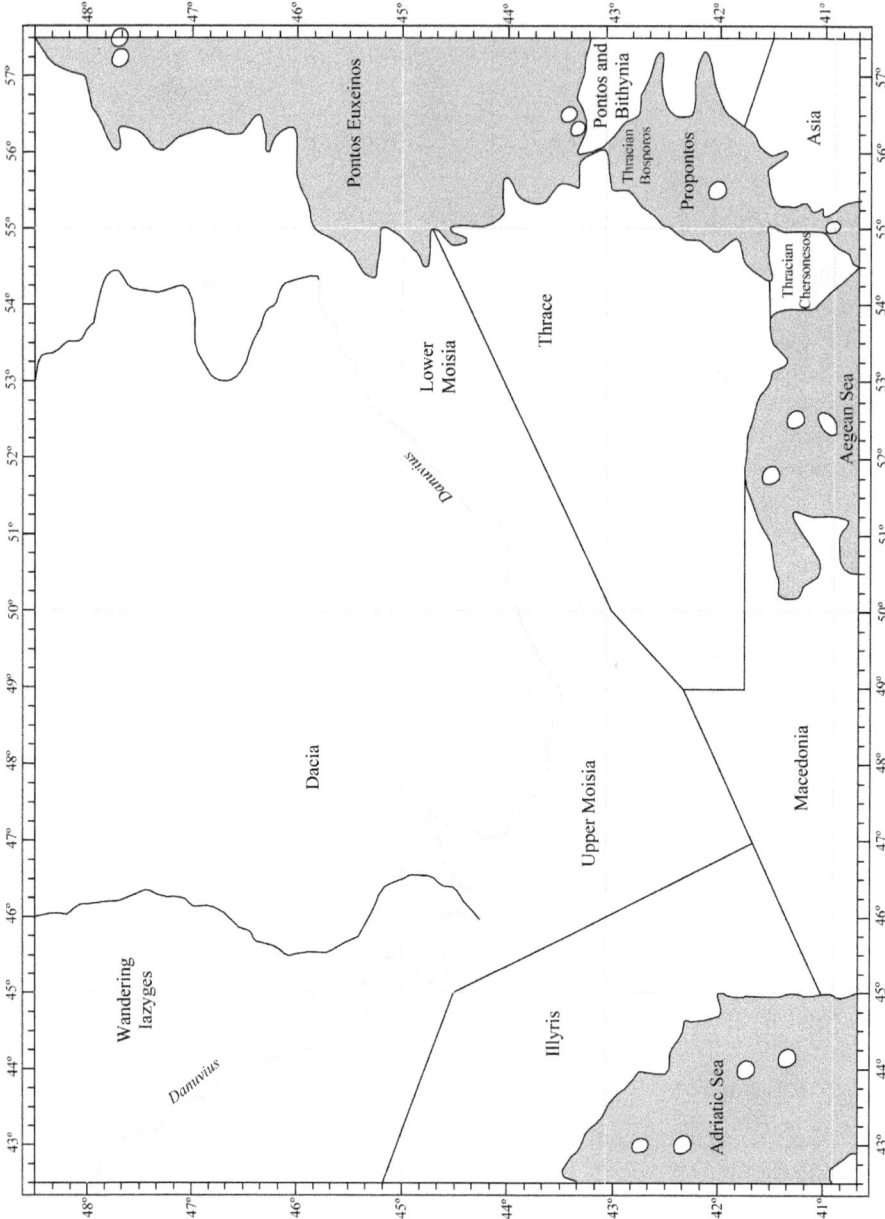

Map 4.4 Map 9 of Europe

south, and thus the two provinces north of the river—Dacia and Lower Moisia—were included in this region.

The western boundary is the same as the eastern boundary of Map 5 of Europe; in other words the Danube from its curve downstream to Taurounon (modern Zemun) and then southeast to the Skardon Mountains, along the eastern boundary of Illyris. From those mountains a rough line runs northeast and then south and east, along the northern boundary of Macedonia, using as its major reference point the Orbelos Mountains in southwestern modern Bulgaria (*GG* 3.11.1). The boundary then keeps north of the Strymon River and reaches the Aegean at the mouth of the Nessos (usually Nestos) River (*GG* 3.11.2). From there the border of Map 9 of Europe can be easily followed along the upper coast of the Aegean, through the Hellespont, Propontis, and Thracian Bosporos, and along the west coast of the Black Sea to the mouth of the Borysthenes and the boundary of European Sarmatia.

Depicted on this map are four Roman provinces, Dacia, Lower and Upper Moisia, and Thrace (Ptolemy's Thrake), and what Ptolemy saw as an unorganized region between the Danube and Tibiskos (probably the modern Tisza in Hungary and Croatia) Rivers, occupied by the Wandering Iazyges. Since the map covered an extensive region—a territory from roughly the upper Tyras River (modern Dniester in Ukraine) to the Aegean, a distance of roughly 1000 km.—its demographic realities varied immensely. The Aegean coast had been a region of Greek settlement since earliest times, and outposts stretched along the Black Sea coast and inland along the major rivers, especially along the Istros (Danube). Many of these originated in the seventh century BC. In the interior were a number of indigenous states, often becoming hellenized, especially after Alexander the Great traveled north of the Istros in 335 BC.[30] The Romans penetrated much of the region beginning in the second century BC, but their provincial organization extended north of the Istros only from AD 106, when Dacia was established in the western part of modern Romania.[31] The northern side of the lower Istros remained outside Roman territory.

The Wandering Iazyges

The Wandering Iazyges were a Sarmatian population that settled in the wedge between Lower Pannonia and Dacia, a district that in Ptolemy's day had been left out of the Roman organization of the region (*GG* 3.7.1–2).[32] As their name implies, they were nomadic, and known in several places under varying names. They may have been the Irykai encountered by Greeks as early as the fifth century BC, although at that time they were more to the north.[33] They were certainly known to Ovid from his viewpoint at Tomis,[34] and Tacitus provided some details of their social structure.[35] Ptolemy had scant topographical information about their territory, only listing eight cities scattered in the portion of modern Hungary east of the Danube (which runs north-south in this region). Other than the Karpaton (Carpathian) Mountains and the two rivers, the Danube and Tibiskos, that marked their limits, no other features were noted. Even the Tibiskos—the western boundary of Dacia—cannot be identified with certainty: its junction with the Danube suggests the modern Timiş, but that river originated well to the east of its mouth and within Dacia, which does not conform to Ptolemy's orientation. The modern

Tisza, a much longer stream (906 km., over twice that of the Timiş), flowing from the Carparthians southwest of the Timiş, is better located for Ptolemy's Tibiskos.

Dacia

Dacia (Ptolemy's Dakia) was the region north of the lower Danube (*GG* 3.8.1– 10).[36] The indigenous Dacians had had contact with the Greek cities on the Black Sea coast from the third century BC, and by the middle of the first century BC a vigorous monarch, Byrebistas, created an extensive kingdom that was centered at his city of Sarmizegetusa and which threatened the Greek settlements.[37] Although his kingdom began to collapse with his assassination in 44 BC, his successor Dikomes offered to assist Kleopatra VII and Antonius at Actium.[38] Thus the region was brought to the notice of the Greco-Roman world, and when a new consolidated Dacia emerged again in the late first century AD under Dekebalos, direct Roman intervention became inevitable and Roman forces were sent, as depicted on the Column of Trajan in Rome. This resulted in the establishment of the province of Dacia in AD 106, the environment that Ptolemy reflected.[39]

Ptolemy's Dacia, like the Roman province, was completely north of the Danube. The western boundary was the Tibiskos River. Beyond the river, farther west, was the territory of the Wandering Iazyges, living in the unorganized rectangle between Dacia and Upper Pannonia. The northern boundary of Ptolemy's Dacia was his arbitrary line at 48 1/2 degrees north latitude that separated the province from Sarmatia. He believed that part of this line was along the Tyras (modern Dniester) River, but it is unlikely that there was a sharply defined frontier on this side of the province, especially in its early days. The northwest corner was at the Karpaton (modern Carpathian) Mountains, but since they extend over several hundred kilometers it is not easy to determine what point Ptolemy had in mind other than somewhere in the vicinity of the source of the Tibiskos, which, if the Tisza, would be in the mountains of southwestern Ukraine.

It is probable that the original northern boundary of the province extended east to the Black Sea, in part along the lower Tyras, but Ptolemy, perhaps more realistically, placed it along the Hierasos River (probably the modern Siret in eastern Romania), which joins the Danube where it curves to the east after having run north for some distance. Thus the Danube (ancient Istros) completed the boundary of Ptolemy's Dacia in the east and south. This meant that the region between the Istros, Hierasos, Tyras, and Black Sea was considered outside the region, and Ptolemy actually assigned it to Lower Moisia. He did not seem to be aware of the efforts of Q. Marcius Turbo, who was sent as praetorian prefect to Dacia in AD 117 or 118, and who, over the next decade or more, divided the province into upper and lower regions.[40] This suggests that Ptolemy's data (probably from Marinos) were acquired before the late AD 120s.

In fact, Ptolemy's information for Dacia is remarkably scant for a rather large area. The only topographical features (in addition to the Karpaton Mountains at the northwest corner) are two rivers, the Rhabon (probably the modern Jiu) and the Aloutas (the modern Olt). The latter, at 615 km. in length, was the major river of the province (and of modern Romania). Both flow southward into the Danube and would have been major access routes to the north. No other features are shown, despite the existence of the Transylvanian Alps, which run east-west across the middle of the province.

There are a large number of ethnic groups (15 in all) and over 50 towns, most of which are indigenous. Eleven of these are located along the Aloutas River, suggesting that Ptolemy's information was based on reports from traders going upriver rather than any Roman organization. These included a place called Hydata ("Waters"), probably a spa, and Salinae ("Salt Works"), presumably the location of that mineral. Knowledge of both of these would have been important to traders in a remote area.

Yet a locality called Praetoria Augusta in the north central region suggests a Roman military base. Sarmizegetusa, near modern Gradiştea Muncelului, was a royal town ("Basileios"), and was the seat of Byrebistas and his successors, an important archaeological site today. A short distance away was Ulpia Traiana Sarmizegetusa, the provincial capital, established by Trajan (also an important site today), but this was not mentioned by Ptolemy, demonstrating again the early nature of his information.[41] Yet the Roman presence is demonstrated by Ulpianon (Oulpianon), perhaps a military base, at an unknown location in the northwest.

A place with the Greek name Zeugma ("Bridge"), in the southwest at an unknown location, may be a faint memory of Greek traders in the region, but it may also be Ptolemy's rendering of Pontes, the bridge or bridges across the Danube downstream from modern Belgrade.

Upper Moisia

The territory south of the Danube and east of Illyris as far as the Black Sea was known as Moisia (Latin Moesia). In Ptolemy's day it was divided into an upper (western) and lower (eastern) region, with the dividing line at the Kiabros, or Cebrus River (the modern Tsibrica), a short stream in northern modern Bulgaria that flows into the Danube. Upper (Ptolemy's Ano) Moisia was more isolated than Lower Moisia, and was a small region bounded by Illyris, Macedonia, and Thrake (Thrace) in the south and west, and the Danube in the north (*GG* 3.9.1–6). Its southwest corner was the Skardon Mountains (modern Šar Planina) (*GG* 2.16.1, 6), at the junction with Illyris and Macedonia. The south corner was at the source of the Kiabros River and the western end of the Haimos range of central Bulgaria. Where Upper Moisia, Macedonia, and Thrace came together there was another mountain, Orbelos; it is clear that Ptolemy was using specific mountain points—not easy to locate in this rugged region—to define boundaries.

There was essentially no Greek knowledge of Upper Moisia, and little Roman penetration until the first century BC, especially with the expedition of M. Licinius Crassus in 29 BC.[42] Originally an outpost of the province of Macedonia, Moisia does not seem to have been provincialized until the time of Claudius and then was divided into two provinces later in the first century AD.[43]

Ptolemy's information on Upper Moisia reflects the period after the division, but when the Roman presence was still limited. Two legions were cited, both in the northwest part of the province. The Legion IV Flavia Felix was organized by Vespasian and eventually sent to Singidunum (modern Belgrade), where remnants of its facilities are still visible. The Seventh Legion, presumably Legion VII Claudia (although Ptolemy did not provide the surname), which dated back to the time of Julius Caesar, was downstream at Viminacium (at modern Kostolac, also a site with impressive remains).[44] To the south, in the vicinity of modern Ćuprija, was a locality merely identified as Horrea ("Granary"), perhaps a legionary supply depot. Raitiaria, at modern Arçar, was an indigenous settlement on the Danube that became a major Roman base, already established as a *colonia* at the time Ptolemy wrote. The Roman presence was further documented by the site of Ulpianum, probably at modern Grančanica north of Skopje. Otherwise, Ptolemy's localities are limited to a handful of indigenous settlements, of which the most important are Skoupoi (modern Skopje) and Singidunum (modern Belgrade). A few indigenous ethnic groups were also worthy of note, from the eponymous Mysoi in the east to the Trikornioi (who perhaps had acquired a Latin name) in the west.

Lower Moisia

Lower Moisia was the region east of Upper Moisia and the Kiabros River, between the Istros and Mt. Haimos (the modern Balkan Range), and east to the Black Sea (*GG* 3.10.1–17). Because part of it was coastal, it was less isolated than Upper Moisia, and Greek settlements were established as early as the seventh century BC, including Tomoi (at modern Constanza), famous as the place of exile of Ovid.[45] Greeks penetrated into the interior, especially up the Istros. In the late sixth century BC, accompanied by a substantial Greek contingent, Dareios I of Persia led an expedition that passed through coastal Moisia (not so identified as yet) and crossed the Istros into the little-known lands beyond.[46] Nearly two centuries later Alexander the Great replicated this expedition.[47] Thus areas near the coast, at least, were well known to the Greek world by the time the Romans arrived in the first century BC, eventually to establish the province of Moisia and divide it into Upper and Lower portions.

In describing Lower Moisia, Ptolemy spent a remarkable amount of time considering the mouths of the Istros (*GG* 3.10.2–6). Multiple river mouths seemed to hold a particular fascination to Greek and Roman geographical scholars, and the literature is replete with discussions of the outlets of the major rivers, including the Rhodanus, Pados, Nile, and Rha as well as the Istros/Danube. In the latter case, Ptolemy's six mouths compare with the two of Eratosthenes and the seven of Strabo.[48] The variances are due to changes in flow and different ways of counting a mouth, yet such information was a valuable navigational tool, especially for those entering the river from the sea. One mouth was called the Pseudostomos—False Mouth—presumably a warning to seamen. Ptolemy also located a unique feature, a marsh called the Northern Thiagola or Thiagos, on the northernmost mouth, perhaps another hazard. The names of the mouths are Greek, as well as a cape to the immediate south, Pteron ("Winged"). Its location is between the Greek settlement of Histria and the southernmost outlet of the river, the Hieron (Sacred) Mouth, but it is impossible to locate any promontory along this coast, due to changes in the landforms. Yet these names demonstrate that even as late as the second century AD Greek influence on the toponymic map was still profound.

The Greek cities on the coast had existed for centuries when Ptolemy wrote. There was also a Greek presence some distance into the interior: the settlement of Axiopolis (at modern Hinog), located where the river makes its turn to the north and due west of Tomoi, may have been the upper limit of Greek settlement. Yet beyond it was still largely a region of indigenous population. Some of these had been assimilated but their topographical vestiges remained: Rhegianon (at modern Kozlodui), in the southwest, was probably the home of a local dynast, perhaps of the Triballoi, which had been one of the most powerful ethnic groups since early times but had become virtually extinct by Ptolemy's time.[49] But their memory remained, and Oiskos (at modern Gigen), to the northeast of Rhegianon and on the Istros, was still known as a Triballian town.

By the time Ptolemy wrote, several legions were stationed along the Danube in Lower Moisia. The Legio I Italica was at Novae (near Svištov), a town presumably founded when the province was created in AD 45: it was posted in AD 70 and remained as long as there was a local Roman presence. Farther downstream was the XI Claudia, at Durostorum (modern Silistra), posted at the beginning of the second century AD. At Troesmis (modern Igliţsa), at the upper end of the Danube delta, was the V Macedonia. These three legions, extending along approximately 300 km. of the river, from its delta well into the interior, demonstrate the Roman frontier defenses of the second century AD.

Other settlements in Lower Moisia have indigenous names, and there were remnants of various local populations scattered across the region. A persistent onomastic problem is the matter of the northernmost of these peoples, whose name varies in the manuscripts between Troglodytai or Trogodytai. The former would mean "Cave Dwellers"; the latter is an ethnym that occurs several places in the ancient world. Neither name is totally satisfactory.[50]

Although the province and region of Moisia probably did not extend north of the Istros/Danube, Ptolemy was left with a topographical problem. He had failed to account for the territory north of the river and east of the Hierasos (the eastern boundary of Dacia). This was because of his arbitrary line of 48 1/2 north latitude, which ran west from the mouth of the Borysthenes (modern Dnieper) River and was designed to separate Sarmatia from the Roman territories to the south. It created a rectangle of unassigned land between the Istros and Borysthenes, which Ptolemy considered part of Lower Moisia. Yet this is unrealistic, since the Istros and Hierasos were clearly the limits of Roman territory. There is no obvious name for this region: "Skythia" is perhaps the best, given the expeditions of Dareios and Alexander, which had Skythia as their goal, yet since the fourth century BC the toponym had been a generic term for the entire northeastern portion of the inhabited world, so it was hardly useful.[51] Thus this anomalous region was, in some sense, a victim of the persistent need to regularize the topography of the inhabited world, an issue that had existed ever since the time of Eratosthenes. But Ptolemy's confusion is apparent since he placed the Tyrangeitian Sarmatians in the northern part of the district— along the Tyras River, as their name implies—a contradiction to his attempt to separate the Sarmatians from the perceived more civilized territories to their south.

Regardless of how this region was to be categorized, it was a hostile marshy area of little habitation except for a few Greek cities along the Black Sea coast. In fact there are only five inland localities and four ethnic groups. There are also two islands: one, Borysthenes Island, is badly misplaced, located a degree of latitude east of the mouth of the Tyras. This is actually the small island known today as Berezan, at the entrance to the Borysthenes estuary, perhaps a peninsula in antiquity and settled by Greeks in the seventh century BC.[52] Ptolemy placed it 200 km. southwest of its actual location, which may be because of a confusion with the Island of Achilles, or Leukos (White) Island, modern Zmeinij, famous as the burial place of the hero and noted for its precipitous white cliffs.[53] It also is improperly placed, since it is about 80 km. south-southeast of the mouth of the Tyras. Yet Ptolemy located both islands side by side, almost adjacent (with the Island of Achilles to the east), directly east of the mouth of the Tyras. The Island of Achilles is isolated, although its location is not as erroneous as that of Borysthenes Island; it is probable that the information for both islands became entangled in some way. Moreover, isolated islands are always difficult to position, and Ptolemy may have found it convenient to put them adjacent to one another.

Thrake

Thrake, more commonly Thrace, was the territory at the head of the Aegean, extending from Macedonia to the Black Sea, and north to the Haimos Mountains that marked the southern edge of Lower Moisia (*GG* 3.11.1–14). Since earliest times it had been part of Aegean culture, known to Homer.[54] Phoenicians also probed the Thracian coast. After the collapse of Persian ambitions in Europe, a number of hellenized local dynasties emerged, although there was also a Macedonian presence in the region from the fourth century BC. As the Romans came into contact with the Macedonian kingdom in the following century, they became more involved in Thracian internal politics, first supporting the aspirations of the local kingdoms. But increased instability meant that eventually, during the reign of Claudius, the territory was provincialized.

Although some of the coastal regions were at times attached to Macedonia, Ptolemy's Thrace basically follows the boundaries of the Roman province, east from the mouth of the Nessos (more commonly Nestos) River and north to Mt. Orbelos at the border of Upper Moisia. The northern boundary runs east along the Haimos Mountains (modern Balkan Range) to Mesembria (modern Nesebur), a Greek settlement from around 600 BC on the Black Sea.[55] Following the Black Sea coast, Ptolemy's boundary continues south to the Thracian Bosporos and then along the European coast through the Propontis, crossing over to the Aegean at the interior end of the long peninsula called the Thracian Chersonesos, which he chose to treat separately (*GG* 3.12.1–4).

Within Ptolemy's Thrace there was a density of topographical features, an expected element of this historic and well-known region. A unique feature, not catalogued elsewhere in Europe, is the listing of 14 *strategiai*, spread evenly through the province. These were military commands, presumably Ptolemy's translation of the Roman concept of prefectures.[56] It is not certain why Ptolemy chose to list these—presumably they were data from his source—but their existence would have reflected the recent turbulent history of the region: assassination and rival dynasts had become a standard element of the Thracian kingdoms, and when Rhoimetalkes III was eliminated in AD 46, the emperor Claudius saw no alternative but provincialization.[57]

Other features in Thrace include rivers, coastal beaches and promontories, and mountains. There is a large number of towns, many of which reflect the early Greek history of the region. The Hellenistic and Roman periods are less well defined, with Traianopolis (at an unknown location on the upper Hebros River) the only representative of settlement recent to the era of Ptolemy.

Several islands were included. Thasos and Samothrake (Samothrace), the two northernmost in the Aegean, are a short distance from the Thracian mainland and have similar histories to Thrace proper, although both retained a measure of independence in the Roman world. The Kyaneai were two groups of islands on either side of the Black Sea mouth of the Bosporos; Ptolemy considered the western group part of Thrace, since they lie just a short distance off its easternmost point. They were an important navigational marker, because it was necessary to round them in order to head north along the Black Sea coast. The islands were known hazards: the name Kyaneai means "Dark Ones," suggesting that they were hard to see. A more significant alternate name was the Symplegades ("Clashers"). The Argonauts were said to have cautiously passed through them.[58] Ptolemy also included the island of Proikonesos (usually Prokonesos, modern Marmara), in the middle of the Propontis, as part of his Thracian section. Historically and culturally, however, it was more connected to the Asian mainland and thus the Roman province of Asia.

The Chersonesos

Following the well-established practice of seeing the inhabited world in terms of distinct geographical units, Ptolemy devoted a separate chapter to the Chersonesos (generally called the "Thracian" to distinguish it from other uses of this common Greek toponym meaning "peninsula"). It is a strip of land, 100 km. long and between 10 and 30 km. wide, extending from northeast to southwest at the northeastern corner of the Aegean. Its southern coast is the north side of the Hellespont. Culturally it was Thracian, with a similar history, although affected by its location along the Hellespont and involvement from earliest time in events concerning movements between the Aegean and the Black Sea.

Noting that there were no interior towns in the Chersonesos, Ptolemy listed several navigational points and six coastal towns, all early Greek settlements except for Lysimacheia, founded by Lysimachos in 309 BC.[59] Ptolemy also took note of the Long Wall (Makron Teichos) that had been built by the fifth century BC to keep the locals out of the peninsula, although it was misplaced well to the north of its course (which was between Lysimacheia and Paktye).[60] The Thracian Chersonesos is the last region on Ptolemy's ninth map of Europe.

Map 10 of Europe

The final European map covers the Greek peninsula, as well as Crete, the islands to the west of the mainland, and most of those in the Aegean. The Aegean islands near the coast of Asia Minor were reserved for the first map of Asia (*GG* 5.2.28). Map 10 includes everywhere south of Illyris, Upper Moisia, and Thrace, and as far east as the mouth of the Nessos (Nestos) River. One feature of it is that, although the problems of orientation (too much to the east) of Italy and the Dalmatian coast have largely been corrected, and the Greek peninsula is generally presented in its

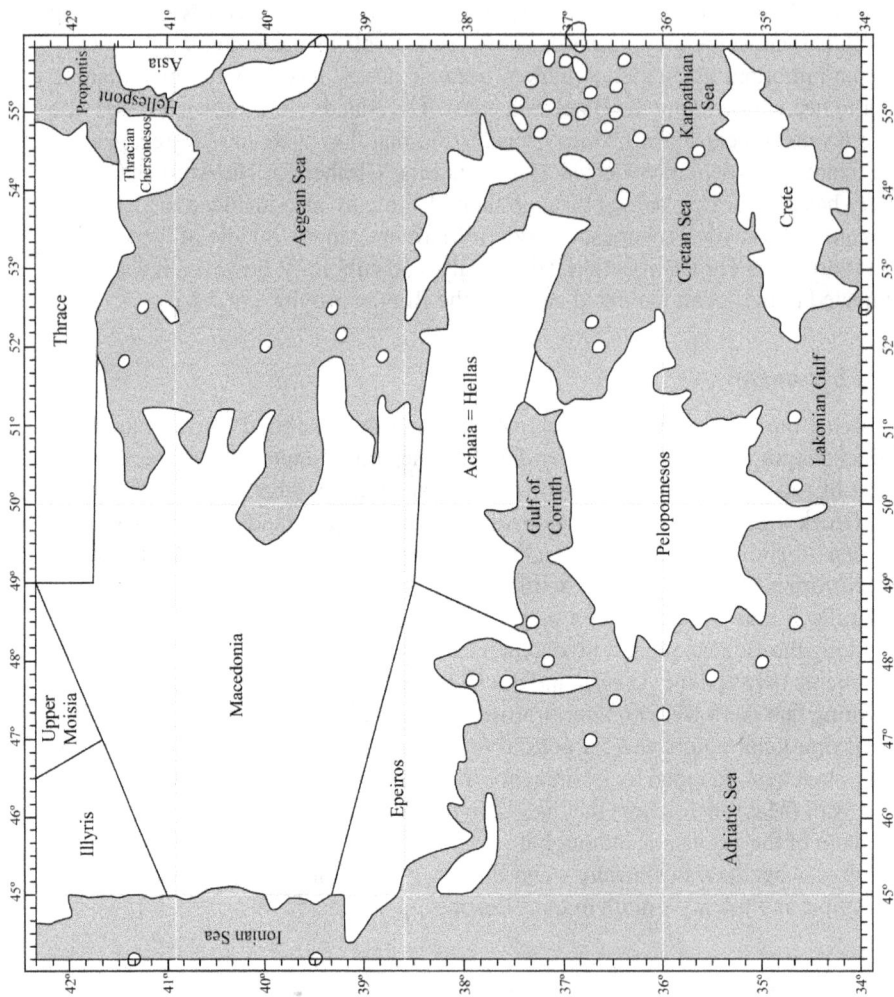

Map 4.5 Map 10 of Europe

proper north-south direction with Sparta and Amphipolis on the same longitude, as is proper, there are still anomalies of detail, and the damage has been done, with Athens lying at 52 3/4 east longitude when in fact it should be at no more than 40 degrees. The east-west error will continue to compound toward the east, but not so much because of problems of orientation as a lack of proper data.

The Greek Peninsula

As would be expected, the density of toponyms for the Greek peninsula is vast (*GG* 3.13.1–17.11). There are nearly 30 rivers, over 30 mountains, dozens of towns and cities, and a large number of coastal features, including prominent coastal sanctuaries. There are a handful of ethnic groups, limited to northwest Greece, the most remote area of the peninsula. In northeast and central Greece, and the Peloponnesos, the traditional regional divisions were noted (e.g., Boiotia, Arkadia, or Thessaly). As in Italy, the toponyms are no less numerous in the interior than on the coasts, but the vestiges of *periploi* are still apparent, with points such as capes and river mouths plotted every few kilometers along the entire coast from the boundary of Illyris in the northwest around to the Nessos River in the northeast.

The political organization reflected on the map generally follows that created in January 27 BC, with the provinces of Macedonia in the north, Epeiros in the west, and Achaia in Central Greece, although Ptolemy also noted the ancient name Hellas that was applied to this region (*GG* 3.15.1, 14).[61] An anomaly is designating the Peloponnesos as a separate region, a term that was not used administratively by the Romans (it was part of the province of Achaia), but perhaps cited in recognition of its long history in Greek affairs.

The towns are the historic settlements of the Greek world, established from prehistoric times through the Hellenistic period. The only indication of a Roman presence, beyond the administrative divisions, is Nikopolis near the site of Actium, the victory city founded by Augustus shortly after the battle. Ptolemy did not cite the Roman renaming of many ancient Greek cities, such as Corinth becoming Colonia Laus Julia Corinthus, and the demography of Greece generally reflects the late Hellenistic period. There are minor errors of placement, but the most notable is that the Pindos Mountains, the main north-south chain through Greece north of the Peloponnesos and which separate Macedonia from Epeiros, are positioned extending from west-north-west to east-southeast, a vestige of the erroneous orientation of Italy and the Adriatic to its west. The result is that part of the southern coast of Epeiros is turned more to the east than it is in actuality, and thus the islands off its coast (primarily Kerkyra, Leukas, and Kephallenia) are to the south, not the west. Yet there are a remarkable number of islands plotted off the west coast of the peninsula, including such obscurities as Lotoa,[62] and Theganousa (modern Venetiko, at the south end of Messenia), indicative of the navigational precision of the sailing routes between Italy and Greece.

The Aegean Islands

A number of islands were plotted in the Aegean, from Lemnos in the north to Crete in the south. Their exact positioning can be confused, especially in the Northern Sporades, where Syros is placed to the north of its actual position, and

Peparethos too far to the south. One of the islands is named Skopelos, which is a name used for Peparethos, not a separate island; Ptolemy's Skopelos is actually at the location of Peparethos, east of Skiathos. Again, there was a probable failure to blend sources.

To the south of the Northern Sporades are the Kyklades. Ptolemy identified 22 of them, from Andros and Tenos in the north to Melos in the south. Most are reasonably well placed, but the group extends south almost to the coast of Crete, in part due to the erroneous placement of that island, as noted in the following. In addition, two features are noted within the Kyklades, Cape Phorbia on the east side of Mykonos and Cape Sounion at the north edge of Paros. These two capes are in direct line with one another along a channel that takes one through the eastern Kyklades and marked the route from Crete to Chios and other points near the coast of Asia Minor.

Moreover, Thera and Therasia, which are adjacent and part of the same volcanic caldera, are separated by over a degree of longitude, with the latter to the east of the former, a reversal of their actual relationship. Therasia may actually have been confused with Anaphe, which was not named by Ptolemy.

Crete

The placement of Crete, the fifth largest island in the Mediterranean, is in error and is another vestige of the turning of Italy and the Adriatic, since Ptolemy located it to the east of the Peloponnesos, not in its true position to its south (*GG* 3.17.1–11). It also is oriented slightly from southwest to northeast rather than due west-east. The result is that much of the Cretan Sea—the body of water between Crete and the Kyklades— has vanished and the name applied to the sea off the eastern Peloponnesos, traditionally called the Myrtoan Sea. But that name is not used in this region by Ptolemy, and the toponym is located between the eastern Kyklades and Karia, in the southern part of what traditionally was called the Ikarian Sea (*GG* 5.2.9). It is perhaps simplistic to assume these oddities are merely errors, because any sea in the Mediterranean can have a variety of names. Yet the failure to identify the Cretan Sea as between Crete and the Kyklades has meant that the southwestern Kykladic island, Melos, lies only a quarter of a degree of latitude north of the Cretan coast (when in fact it is over 100 km. away) and was considered by Ptolemy to be a Cretan island. And Dia, a small island merely 10 km. north of Crete, is north of the latitude of Melos.

Nevertheless Crete has a high density of settlements (31 in all). The length of Ptolemy's list is similar to that of Pliny's, but the names are not identical.[63] In addition, Ptolemy listed several coastal features and rivers, and three of the four major mountains of the island (oddly the missing one is Aigaion, the highest). Immediately south of Crete are two small islands, Kaudos (to the west) and Letoa. With the description of Crete the tenth and final map of Europe comes to an end, as does Book 3 of the *Guide*.

Notes

1 C. Goudineau, "Tropaeum Alpium," *PECS* 936–7.
2 Strabo, *Geography* 6.3.9; Duane W. Roller, *A Historical and Topographical Guide to the Geography of Strabo* (Cambridge 2018) 329–30.

3 Strabo, *Geography* 5.2.6; Pliny, *Natural History* 3.81.
4 Pomponius Mela 2.122.
5 G. Moracchini-Mazel and R. Boinard, *La Corse Selon Ptolémée* (Bastia 1989).
6 Livy 30.39.
7 D. Manconi, "Antas," *PECS* 58–9.
8 Pausanias 10.17.2.
9 Alfred Michael Hirt, *Imperial Mines and Quarries in the Roman World* (Oxford 2010) 79.
10 Homer, *Odyssey* 11.107; Thucydides 6.2; Strabo, *Geography* 6.2.1.
11 E. D. Phillips, "Odysseus in Italy," *JHS* 73 (1953) 53–67.
12 Diodoros 5.11; Pomponius Mela 2.120; Pliny, *Natural History* 3.92.
13 Strabo, *Geography* 2.4.1.
14 Herodotos 4.45; *Airs, Waters, and Places* 13.
15 Strabo, *Geography* 7.4.5.
16 Herodotos 4.110–17.
17 Strabo, *Geography* 2.5.7, 7.3.17.
18 T. Sulimirski, *The Sarmatians* (New York 1970).
19 Pliny, *Natural History* 4.95.
20 *GG* 1.11.8; see also Pliny, *Natural History* 4.95, 37.33, 36.
21 Pliny, *Natural History* 6.19.
22 Herodotos 4.20–1; Strabo, *Geography* 11.2.2; Boardman, *Greeks Overseas* 244.
23 *BNP Historical Atlas of the Ancient World* 177.
24 Aristotle, *Meteorologika* 1.13.350b; Strabo, *Geography* 7.3.1.
25 Pliny, *Natural* History 6.49; Arrian, *Anabasis* 3.30.6–9.
26 Stückelberger and Graßhoff, *Klaudios Ptolemaios* 305.
27 Boardman, *Greeks Overseas* 250–4.
28 Duane W. Roller, *Empire of the Black Sea* (Oxford 2020) 66–7.
29 *BNP Names, Dates, and Dynasties* 112–14.
30 Arrian, *Anabasis* 1.1–13.
31 *BNP Historical Atlas of the Ancient World* 176–7.
32 Strabo, *Geography* 7.2.4; Stückelberger and Graßhoff, *Klaudios Ptolemaios* 812–13.
33 Herodotos 4.22.
34 Ovid, *Epistulae ex Ponto* 1.2.77.
35 Tacitus, *Annals* 12.30; *Histories* 3.5.
36 Dan Dana and Sorin Nemeti, "Ptolémée et la toponymie de la Dacie (I), *C&C* 7.2 (2012) 431–7; "Ptolémée et la toponymie de la Dacie (II–V)," *C&C* 9.1 (2014) 97–114; "Ptolémée et la toponymie de la Dacie (VI–IX)," *C&C* 11 (2016) 67–93.
37 *SIG* 762; Richard D. Sullivan, *Near Eastern Royalty and Rome, 100–30 BC* (Toronto 1990) 147–8.
38 Plutarch, *Antonius* 63.
39 Dio 68.6–14.
40 *PIR* M179.
41 L. Marinescu, "Ulpia Traiana," *PECS* 946–7.
42 Dio 51.23.2.
43 Jan Burian et al., "Moesi, Moesia," *BNP* 9 (2006) 115–19.
44 Pollard and Berry, *Complete Roman Legions* 175–6, 184–6.
45 Boardman, *Greeks Overseas* 247–50.
46 Herodotos 4.87–140.
47 Arrian, *Anabasis* 1.1–8.
48 Eratosthenes, *Geography* F148; Strabo, *Geography* 7.3.15.
49 Herodotos 4.49; Thucydides 2.96; Appian, *Illyrike* 3–5.
50 Roller, *Historical and Topographical Guide* 378–9.
51 Ephoros F30. It was originally an ethnym but came to be used topographically.
52 Boardman, *Greeks Overseas* 250.
53 Strabo, *Geography* 7.3.16.

54 Homer, *Iliad* 2.844–5, 10.434–5.
55 Boardman, *Greeks Overseas* 247.
56 Stückelberger and Graßhoff, *Klaudios Ptolemaios* 331.
57 Synkellos 631.
58 Euripides, *Andromache* 795; Eratosthenes, *Geography* F117; Roller, *Eratosthenes* 207–8.
59 Getzel M. Cohen, *The Hellenistic Settlements in Europe, the Islands, and Asia Minor* (Berkeley 1995) 82–7.
60 Herodotos 6.36; Strabo, *Geography* 7F21.
61 Strabo, *Geography* 17.3.25; Dio 53.12.
62 This is perhaps the Letoia of Pliny, *Natural History* 4.55.
63 Pliny, *Natural History* 4.59, with 39 towns.

5   Libya

Introduction

Book 4 of the *Guide* is the only portion devoted to the continent of Libya, also known since Roman times as Africa. Libya was its ancient name, first cited in Greek literature by Homer, and originally referring to the territory west of Egypt.[1] As geographical knowledge expanded to the west, Libya came to define the territory as far as the Atlantic Ocean and north of the Sahara. In time Greeks—and eventually some Romans—applied the term to the entire continent known today as Africa.[2]

Yet by contrast Africa was originally merely a regional toponym referring to the central part of the Mediterranean littoral of the continent (essentially modern Tunisia). It was probably an ethnym that became well known when applied to the new Roman province established in the region in 146 BC after the final defeat of Carthage. By the following century the toponym had expanded to mean (to the Romans) the entire continent, but Greek authors regularly continued to use Libya except when referring to the Roman province.[3] This was Ptolemy's terminology, with Libya for the continent—used throughout the *Guide*—and Africa merely for the province and essentially limited to Book 4 (*GG* 4.3.1–47). Occasional citation of the name "Africa" in some of the headings to other portions of Book 4 may not be Ptolemy's diction.

Three of Ptolemy's maps of Libya follow the Roman provincial organization as it had existed since the middle of the first century AD. There were five provinces, from Mauretania Tingitana and Mauretania Caesariensis in the west, and Africa, Cyrenaica (Ptolemy's Kyrenaika), and Egypt to their east.

The fourth map depicted territory to the south of the Roman provinces, from the Atlantic to the Red Sea and the Indian Ocean. This region Ptolemy titled Inner Libya in the west and Aithiopia Below Egypt in the east. South of this, and again extending across the continent, was a vast area that he called Inner Aithiopia, as far as approximately 16 degrees south latitude, and which was considered "unknown land."

Map 1 of Libya

Map 1 covers the sprawling area called Mauretania, which extended along the northwestern Libyan coast for over 1500 km., from the Atlantic to the Ampsaga River (probably the modern El-Kebir in eastern modern Algeria): the eastern boundary of Mauretania approximates the modern Algerian-Tunisian border. The region

DOI: 10.4324/9781003248590-6

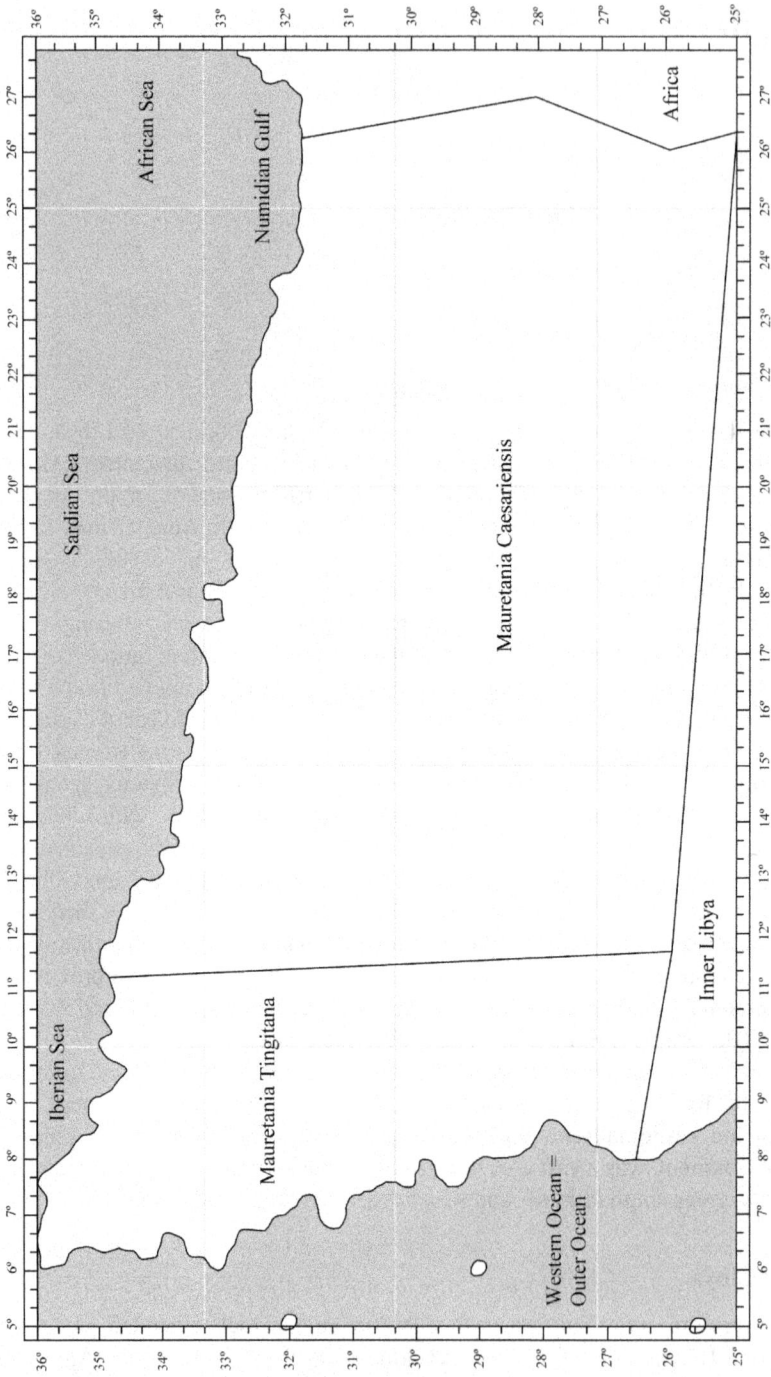

Map 5.1 Map 1 of Libya

had long been ruled by indigenous dynasts, who were influenced by their more powerful neighbors, beginning with the Carthaginians and followed by the Greeks and Romans.[4] In time the region was split between two cousins, Bocchus II and Bogudes II, with the former in the east and the latter in the west. The boundary between their kingdoms seems to have been at the Muluccha River, probably the modern Molouya, near the modern frontier of Algeria and Morocco, presumably Ptolemy's Maloua (*GG* 4.1.7). Both kings became increasingly involved in Roman events of the first century BC: Julius Caesar visited Bogudes and had a relationship with his wife.[5] But by the time of the Battle of Actium in 31 BC both kings were dead, and after an unorganized period, a new figure came to rule both portions of Mauretania. This was the dispossessed heir to the Numidian throne (the territory just to the east), Juba II, who was the husband of Kleopatra Selene, the daughter of Kleopatra VII and Marcus Antonius.

The careers of the joint monarchs, and Juba's career as explorers, scholar, and political figure, have been discussed in detail elsewhere. When Juba, who survived Kleopatra Selene, died in AD 23 or 24, the kingdom passed to their son Ptolemy, who ruled for nearly 20 years until a dispute with his cousin, the emperor Gaius Caligula, resulted in his death. After a period of internal convulsions the territory was provincialized, probably in AD 41, but divided once again into the two portions of the old kingdoms of Bocchus and Bogudes, now called Mauretania Tingitana and Mauretania Caesariensis. This was the political organization recorded by Ptolemy.

Mauretania Tingitana

Mauretania Tingitana (Ptolemy's Mauritana Tingitania), extended from the Kotes Promontory (modern Cape Spartel, the northwest corner of Africa) east to the Maloua River, and south to the Atlas Mountains (*GG* 4.3.1–16). As is customary, Ptolemy emphasized the coastal features, beginning with those on the Atlantic, where there were few settlements but a large number of river mouths, as might be expected, and then westward along the Mediterranean coast from the Maloua to the Atlantic and the Kotes Promontory.

The Atlantic coast had been explored at least as far south as the Atlas, the highest mountains in North Africa. There were 12 river mouths and several settlements and topographical features. Phoenician and Carthaginian explorers had been along this coast since early times, and Ptolemy (or Marinos) probably had access to at least two reports on this region. There was that of the Carthaginian Hanno, from about 500 BC, which survives today in a summary form but existed in a fuller version as late as the early second century AD.[6] In addition, Polybios traveled this coast after the fall of Carthage in 146 BC and published an account, no longer extant.[7]

The Carthaginians, in particular, were heavily involved in trade along the Atlantic coast—as many as 300 of their settlements were reported—and one locality of interest recorded by Ptolemy was Emporikos Kolpos, the "Trading Bay," near the Sala River (perhaps the modern Oued Bou Regreg), a rare survivor of these outposts.[8] In the northern interior of Mauretania Tingitana over 20 settlements were listed, generally indigenous or Carthaginian foundations. The four most important were Tingis, Zilia, Lix, and Volubilis. Tingis (at modern Tangier) was on the coast only a few kilometers east of Kotes. It was an indigenous city that due to its location as the closest point to Iberia became a contact point between the two continents, a role that it still plays, as well as a royal residence for the indigenous monarchy, and in 38 BC its inhabitants were given Roman citizenship because of their support for Octavian.[9] Later it became the administrative center of the Roman province. Ptolemy called it Tingis Kaisareia, a rarely used name that may have been bestowed on it by Juba II. In Ptolemy's scheme, Tingis was the northernmost point of the continent (although it is actually farther south than Carthage); this was perhaps to emphasize its cultural closeness to Iberia.

South of Tingis was Zilia (Zulil or Zelis), at modern Asilah.[10] This was another indigenous town; it became a Roman veterans' settlement and in fact was legally attached to the province of Baetica. About 35 km. farther south was Lix (more commonly Lixos), near modern Larache and at the mouth of the Lixos River (modern Loukos). The settlement was in a fertile river estuary and was developed by Hanno around 500 BC as the major outpost on this coast.[11] It remained an important city into late antiquity, and has extensive visible remains.

The only inland city of note in the province was Volubilis, at the northern edge of the Atlas uplift and about 120 km. from the Atlantic coast. It had been a royal residence and became an important Roman city and a secondary capital of the province. Its remains are the most extensive in Morocco. Otherwise toponyms in the interior of the province are limited to a few settlements, ethnic groups, and mountains, as well as the speculative upper courses of rivers whose mouths were well known. The one exception is the Pyrrhon Pedion (*GG* 4.1.10). If not an indigenous name, it could mean Red Plain, but its location on Ptolemy's map, at the northern edge of the Atlas, makes it difficult to reconcile with the actual topography. The ultimate source for the toponym was probably the report of C. Suetonius Paulinus, who was propraetor of Mauretania in AD 41 and led an expedition that crossed the Atlas and went far to the south, perhaps reaching the Niger River. He was the first Roman to reconnoiter this territory, and was impressed by the flora and fauna both at high altitudes and in the tropics.[12] He was Pliny's major source for interior southern Mauretania and his report, now lost, was probably also available to Marinos and Ptolemy.

There are a number of mountain ranges appearing on the map. The Diour, just north of Volubilis, may be an ephemeral construct for geographical regularity, since four rivers, leading either to the Atlantic or the Mediterranean, were said to originate on its slopes. Otherwise, given the general ruggedness of interior Mauretania, the several mountains cannot be identified among the many summits and prominent peaks of the region. There are only two exceptions. The complex of the Heptadelphoi (Seven Brothers) and adjacent Abyle is at the northern point of Mauretania where the Mediterranean narrows to the straits that lead to the Atlantic. The peaks were often identified as the southern Pillars of Herakles, and are known today as Jebel Mousa.[13]

At the southern end of Mauretania was Mt. Atlas, not only the theoretical boundary of the Roman province but the northern edge of Ptolemy's Inner Libya. It was the highest mountain known to Mediterranean antiquity before the Alps and

Caucasus were explored, although in fact it is less a mountain than a range extending over 2000 km. and separating the coastal regions of northwest Africa from the desert. Its highest peak, modern Toubka, rises to 4167 m. and was probably the summit that in antiquity was considered to be Mt. Atlas. As noted, Ptolemy used the report of Suetonius Paulinus for positioning the mountain, but applied the name only to the western portion (perhaps suggestive of the Roman route through them), which he called the Larger (Meizon) Atlas. As recorded, the mountains extended over 2 degrees of longitude east of the Atlantic. But there were also various isolated ranges toward the east, well into southern Mauretania Caesariensis, as well as the Lesser (Elatton) Atlas on the coast southwest of Volubilis. These and other scattered mountains identified on Ptolemy's map are representative of summits spread through the region.

To complete his outline of Mauretania Tingitana, Ptolemy named two islands in the Atlantic, which remain enigmatic: Paina and Erytheia. The Atlantic coast of Mauretania is generally devoid of islands except for the islets at modern Essaouria (west of modern Marrakesh), where a small group lies up to 2 km. offshore, but others may have become attached to the mainland since antiquity. These islands may be where Juba II established an industry for processing Gaetulian purple dye.[14] One of these may be Ptolemy's Erytheia, although he placed it nearly 2 degrees of longitude offshore, which need not be an issue. The name Erytheia ("Red") may suggest the production of the purple dye, but Erytheia (in varying forms) is a common toponym in the ancient world, and there was a more famous Erytheia Island at Gades in southern Iberia.[15] Yet Ptolemy did not mention this locale, and it is quite possible that his Erytheia is a misplacement of the Gadeiran island.

More uncertain is Paina Island, a unique toponym located 3 degrees of latitude north of Erytheia and farther offshore. There is no evidence of islands in this region; the Canaries are much farther south and appear on Map 4 of Libya as the Makarioi Nesoi (*GG* 4.6.34). There were ephemeral islands in the estuaries of the great rivers of western Mauretania, and one of these, no longer in existence, is certainly a candidate for Paina. Moreover, Ptolemy had a tendency to position islands that were immediately offshore far out to sea: this was the case with Kerne and Gades (*GG* 2.4.16, 4.6.33). But the most likely possibility for Paina is one of the Madeira group, which had been visited by the Carthaginians and entered vague Greek knowledge after 300 BC. Yet they remained more a fantasy locale in the far west than an accessible place.[16] The name Madeira is not documented before the fifteenth century, and so Ptolemy's Paina may be the only survival of an ancient toponym.

Mauretania Caesariensis

To the east of Mauretania Tingitana and the Maloua River was the other Mauretanian province, Caesariensis (Ptolemy's Mauritania Kaisarensia), continuing east to the province of Africa at the Ampsaga River, or essentially the extent of modern Algeria (*GG* 3.2.1–35). Its north side was the Mediterranean, Ptolemy's Sardoan Sea, so named because it was south of Sardo (Sardinia).

Ptolemy's south side of the province was the boundary of the region called Inner Libya. Unlike Tingitana, however, the southern boundary of Caesariensis was not marked by any topographical features but merely an arbitrary line running from the east end of the Greater Atlas (in Tingitana) to a point south of the alleged source of the Ampsaga River at the southwestern corner of the province of Africa at 25 degrees north latitude.

Like its companion province to the west, Mauretania Caesariensis had been organized in 41 BC after the elimination of King Ptolemy. It was a region well documented due to Juba's ethnography, *Libyka*, which provided extensive detail since his capital city, Kaisareia, was on the coast in the center of the district. Thus the density of toponyms is unusually great, with nearly 100 towns and 14 rivers recorded, as well as a number of mountains, coastal promontories, and ethnic groups. Nearly a third of the towns are on the coast.

The cultural heart of the province was Kaisareia (at modern Cherchel). Originally named Iol, it was a Carthaginian outpost that had become the dynastic seat of Bocchus II, who died in 33 BC. Eight years later Juba II and Kleopatra Selene arrived to establish it as their dynastic capital, renaming it Kaisareia (the form Iol-Kaisareia was also in use), following the contemporary practice of naming towns after the emperor Augustus. The monarchs transformed it into an innovative architectural and cultural city, of which there are many physical remains. After the end of the monarchy it continued as the administrative center of the new province, and was given status as a *colonia*. A small island offshore, noted but not named by Ptolemy, today contains a lighthouse and is the only island in the province cited in the *Guide*.

Although Juba's *Libyka* was probably the primary source for Ptolemy's map of Mauretania Caesariensis, certain elements were updated to reflect a later era in the Roman period, such as the status of Iol-Kaisareia as a *colonia*. There are a few Roman toponyms (although some of these could predate the province), such as Portus Magnus, the large bay at modern Bettioua, east of Oran, a major Roman port city, with perhaps earlier but less significant occupation. It lay in a sheltered location near the western end of the province and was perhaps developed in the early years of provincialization.[17] In the vicinity of Iol-Kaisareia were several Roman establishments, including Victoria, Oppidium Novum (Ptolemy's Oppidion Neon), and Oppidium. The exact location of these is unknown except that Oppidium Novum is at modern Ksar el-Kabir. These were probably veterans' settlements and were near the legionary fortress of Castra Germanorum, west of Iol-Kaisareia and slightly inland on the east side of the modern Oued Damous, and a name that suggests a settlement or encampment of German auxiliaries.[18]

About 30 km. southeast of Iol-Kaisareia was the *colonia* that Ptolemy called Hydata Therma, but which had the Latin name of Aquae Calidae (*GG* 4.2.26), located at modern Hammam Righa, still a spa.[19] This cluster of Roman toponyms within 50 km. of Iol-Kaisareia shows the effect of the establishment of a Roman presence after AD 41, with romanization concentrating on the historic heartland of the district.

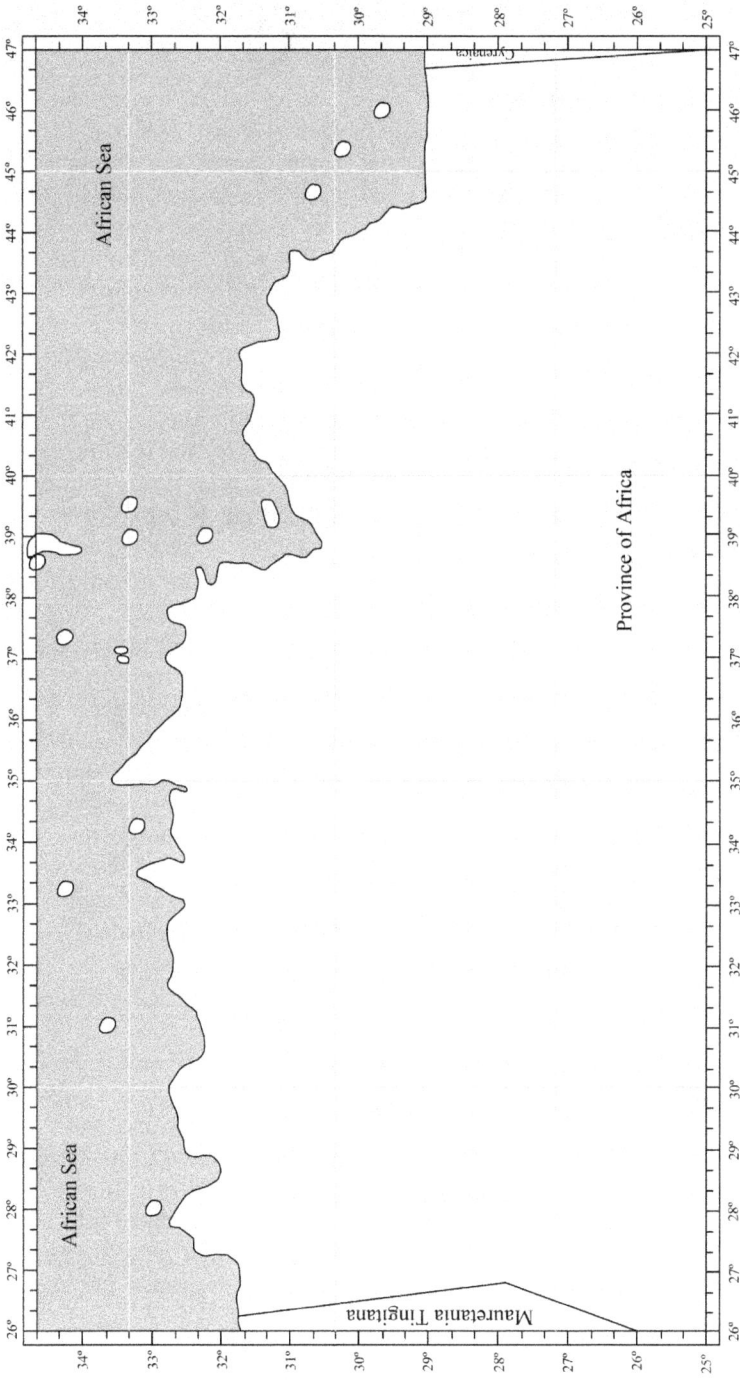

Map 5.2 Map 2 of Libya

Map 2 of Libya

Africa

This map depicts the Roman province of Africa (Afrika to Ptolemy), from the Mauretanian boundary at the west to that of the Cyrenaica (Ptolemy's Kyrenaika) at the east (*GG* 4.3.1–47). The western edge of the province was at the Ampsaga (modern el-Kebir) River. The map depicts territory in eastern modern Algeria, Tunisia, and western modern Libya as far east as the Altars of Philainos (or Philainoi, near modern Ras el-Aali), named after the Carthaginians who determined the eastern boundary of that state's territory, and located at the southernmost point of the Mediterranean. From this point the eastern edge of Map 2 follows a line south to the east-west line that extends east from the southern boundary of Mauretania (along approximately 25 degrees north latitude), south of which is Inner Libya.

The northern limit of Map 2—other than some offshore islands—is the Mediterranean coast; Ptolemy called the adjoining sea the African. The alignment of the coast is badly represented, since Ptolemy was unaware of the magnitude of its north-south direction between the vicinity of Carthage and the Altars of Philainos. This passes through nearly 7 degrees of latitude, but Ptolemy recorded only slightly over 4. The problem is largely due to the issues regarding the placement of the opposite European coast and Sicily, and the eastward tilt of Italy.

An additional error in the orientation of this coast is that its northernmost point is plotted at Cape Hermaia (Cape Bon in Tunisia, the southern entrance to the Bay of Utica), when there are several points farther west that are more to the north. The protrusion of Cape Hermaia, however, causes an emphasis on the channel of the Sicilian Strait, perhaps intentional for the service of navigators. Nevertheless it is peculiar an area so well known, that around Carthage and Utica, should be so poorly oriented.

The Roman province of Africa was a result of the collapse of Carthage in the second century BC. The heart of the province was the Carthaginian homeland, encircling the city of Carthage itself. To the south and west was the territory of Numidia, the ancestral seat of Juba II, the eventual king of Mauretania.[20] Numidia had emerged as an independent kingdom by the third century BC, and its rulers—Juba's ancestors—skillfully played the Romans and Carthaginians against each other for many generations.

With the Roman defeat of Carthage in 146 BC, its territory became the Roman province of Africa. Although the Numidian kingdom profited by the collapse of Carthage, a century of contentiousness between Numidia and Rome, and eventual entanglement of the kingdom in the Roman civil war, resulted in its elimination and the defeat of the last Numidian king, Juba I, by Julius Caesar in 46 BC. The territory was incorporated into the Roman province of Africa, with the infant son of Juba I removed to Rome to become, as Juba II, king of Mauretania 20 years later. Although there were some boundary adjustments over the years, the expanded province was the basis of Ptolemy's map.

The emphasis on the map of Africa is the urban demography, with over 150 settlements recorded. As expected, the density is greatest near the coast and around and south of the vicinity of Carthage (Karchedon) and Utica (Ptolemy's Ityke), the provincial capital. Many of the settlements were ancient, going back to Phoenician times, although some had been founded or refounded by Greeks or Romans, such as Carthage itself. For example, Kirta, or Cirta (modern Constantine), in the northwest part of the province, was the ancient Numidian royal seat, but was known to Ptolemy as Kirta Julia and was a Roman *colonia*, having received that designation from Julius Caesar.[21] At the other end of the province, on the coast, another ancient city, Leptis, was also a Phoenician foundation and often called Leptis Megale or Magna to distinguish it from a lesser Leptis farther west (*GG* 4.3.10, 13). By the fourth century BC it had received the alternate name of Neapolis, perhaps when Greeks from Kyrene moved into the region. It became a *colonia* under Trajan.[22] Ptolemy did not record its colonial status but provided both the Greek and indigenous names. Yet the Roman organization is documented frequently on the map, with a number of *coloniae* and occasional Latin toponyms such as Kisterna (Latin Cisternae, "Cisterns"), at an unknown location on the coast east of Leptis Magna, where, presumably, there were significant water storage facilities. A place named Hedaphtha (or Gidaphtha), between Carthage and Leptis, was said to be "at the boundary" and may reflect the limits of the province of Africa.

As expected, the density of settlements lessens as one goes inland, and many of the toponyms probably represent oases in the desert. Beyond the coastal strip (extending at most through 2 degrees of latitude) almost all the names are indigenous and generally otherwise unknown; the only exception is the spring or oasis of Hydata Thermae (perhaps a translation of Aquae Calidae) south of Carthage. A Roman presence is also documented by the presence of the Legio III Augusta (Sebaste to Ptolemy) at a camp in central Numidia. It had been posted in the region during the reign of Juba II as support for his campaigns against the indigenous populations, which became especially active during the transfer of power from him to his son Ptolemy in AD 24, although little is known about the disposition of the legion during the rest of the first century AD. Ptolemy's citation of it suggests that it remained in the province for some time.[23]

There are 34 ethnic groups named within the province. Expectedly, they are more common at the extremities, generally in an arc from Numidia south and around to the east. Of particular interest are the Libyphoenicians, the hellenized version of how the Carthaginians described themselves. This had expanded into a general ethnym for the indigenous populations in the former Carthaginian regions which had been assimilated into Carthaginian culture.[24] Yet by Ptolemy's day they had been theoretically reduced to a small area south of Carthage, although Ptolemy's inclusion of them may merely be a reflection of the ethnic heritage of the region.

Another ethnic curiosity are the Lotophagoi, located west of the Kinyps River and south of Leptis. This was originally a Homeric ethnym, describing the ideal society encountered by Odysseus, which subsisted on the lotus plant.[25] Ptolemy's Lotophagoi are some distance from the island of the Lotus Eaters, Lotophagitis, or Meninx (modern Djerba off the Tunisian coast), but it is probable that even in Ptolemy's day there was an inland group that believed itself to be descendants of the Homeric Lotophagoi.

Mountains, rivers, and lakes are a feature of the map, and are data gathered from the trade routes to and from the interior of Africa. Ten mountains are located, either as single points or by their endpoints. Although none can be certainly identified, one (perhaps Ptolemy's Buzara or Thammes) is probably Mt. Tabat, at 2908 m. the highest point in the region covered by the map and located in southern modern Algeria. The Sittaphion Plain, at the southern edge of the western portion of the map, is probably one of the many *ergs* of the northern Sahara, and the only such feature on the map, but cannot be identified.

An unusual feature is the several rivers whose origins are at or south of the southern margin of the map. These include the Ampsaga (modern el-Kebir), Bagradas (modern Mejerda), and Kinyps (modern Oued Caam). Since none of the rivers entering the Mediterranean in this region have their sources far above their outlets—the Mejerda, one of the longest, is only 460 km. in length—these rivers are the way Ptolemy represented trade routes into the interior, rather than actual watercourses. The Bagradas, in particular, was said to have an origin well into central Africa (approximately somewhere in modern Mali), near the great rivers of western Africa, as depicted on Map 4 of Africa; thus it provided a connection with that part of the world. Routes into central Africa had long been known: Greeks were aware of them by the fifth century BC.[26] Map 2 also shows several lakes in the interior, some of which have Greek names (e.g., Triton, Pallas, and Libya). These are all on the Triton River, presumably a Greek record of a route to the south. In fact, despite popular opinion, the Sahara is well watered, which perhaps explains why a certain Carthaginian, Mago, claimed to have crossed the desert three times without water.[27]

Islands are also a feature of the map. Some cannot be identified or only tentatively. In the west are Hydras and Drakontios, the latter perhaps known for its snakes. Between them was Galata (modern Galita), a volcanic outcrop in an isolated position 40 km. from the coast and which is a Tunisian bird sanctuary today. The greatest density of islands is in the vicinity of the Sicilian Strait, extending between Sicily and the Carthaginian region and which was also the narrowest part of the Mediterranean. Some of the islands are closer to Sicily than Africa, and were administratively attached to Sicily. The greatest details are provided for Melite (modern Malta), where temples to Hera and Herakles were recorded (perhaps a Greco-Roman interpretation of the famous megalithic structures), and its companion Gaulos (modern Gozo). Kossyra island (modern Pantelleria) was also attached to Sicily. In addition, there is a series of islands south of Malta, most of which can easily be identified even if not accurately placed (largely due to issues of the orientation of the African coast), including Aigimios (more commonly Aigimouros, modern Zembra), Aithousa (modern Linosa), Kerkina (modern Kerkenna), and the Lotophagitis island or Mininx (modern Gerba). Thus there is a solid repertory of islands from south and west of Sicily to the eastern Tunisian coast. Three islands lie in the sea southeast of Leptis (Misynos, Pontia, and Gaia), and are probably the Tre Scogli off the west coast of modern Libya south of Benghazi; if so, they would be in Cyrenaica, not Africa.

Map 3 of Libya

The third map of Libya covers the rest of the territories along the Mediterranean coast (which Ptolemy called the Libyan Sea in the west and the Egyptian Sea in the

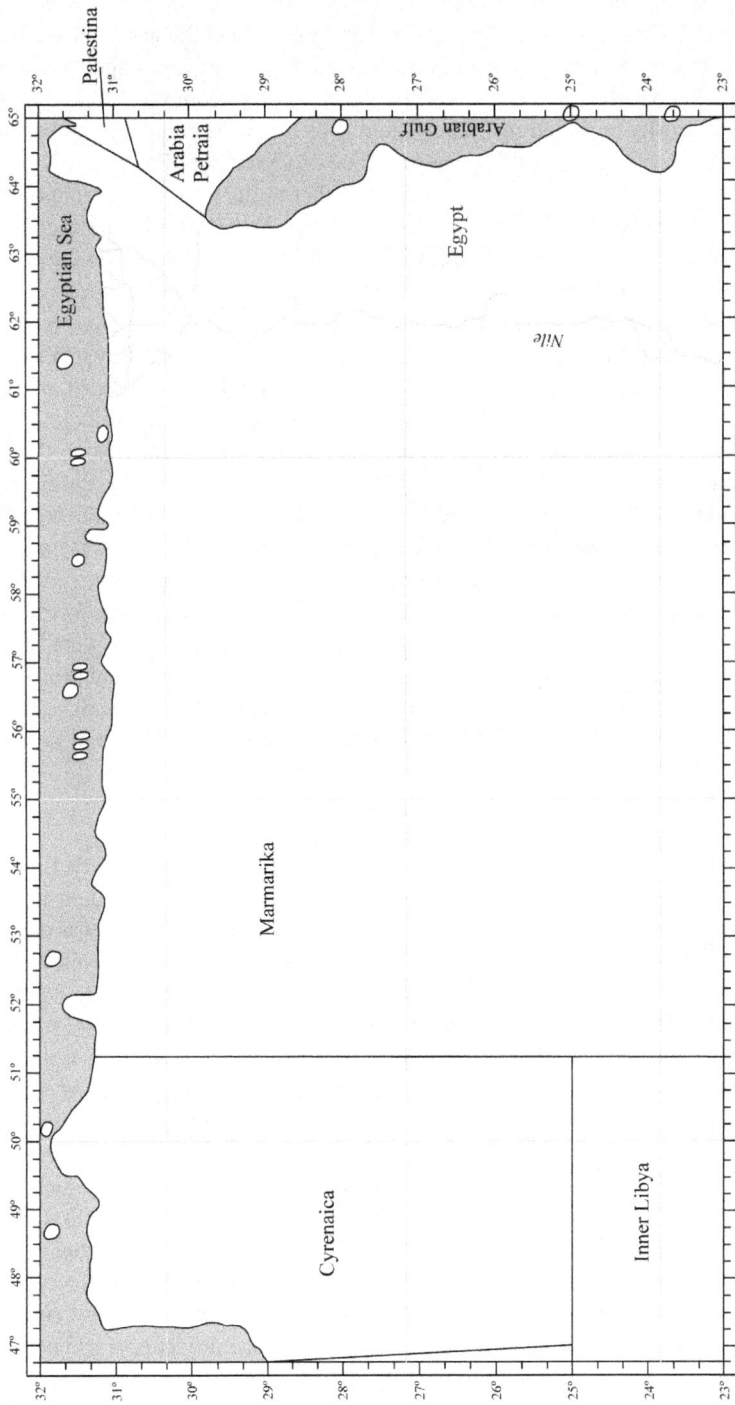

Map 5.3 Map 3 of Libya

east), from the eastern end of the province of Africa at the Altars of the Philainoi to the southern Levantine coast at Anthedon, beyond the Nile Delta (*GG* 4.4.1–77). The southern margin of the map is a continuation of the northern edge of Inner Libya, as depicted on Map 4 of Libya, a line extending east from the Atlas Mountains and the Atlantic, recorded on Maps 1 and 2 of Libya. It comes to its end at the presumed eastern edge of the Roman province of Cyrenaica (Ptolemy's Kyrenaika), since he considered Egypt as extending farther to the south. This boundary lies at 23 degrees north latitude and continues to the Arabian Gulf, the modern Red Sea. From that point the eastern limit of Map 3 runs north along the western coast of the gulf (although including some islands) to its head at Heroonpolis (*GG* 4.5.13), and then to the Mediterranean at Anthedon.

Map 3 covers the territory of two Roman provinces, Cyrenaica and Egypt (Ptolemy's Aigyptos). Although they are divided by a north-south line just east of 51 degrees longitude, running south of Darnis (modern Derna in Libya), the boundary between the two provinces was more arbitrary than usual, since there was no topographical feature that could be used, and Darnis is simply the easternmost significant town in coastal Cyrenaica. The recognized division was generally well to the east at the feature called the Katabathmos (the "Descent," at modern Sollum), where the coastal road drops several hundred meters from the uplands.[28] It remains the modern border between Egypt and Libya. By the first century BC there was a border fortification at this point called the Four Towers.[29] But frontiers in remote country were fluid and could easily move according to the contemporary political situation, and Ptolemy seems to have chosen a point just east of the populated area of the Cyrenaica. This also allowed better coordination with the interior regions depicted on Map 4 of Libya.

Cyrenaica

The Roman province of Cyrenaica approximated the eastern half of modern Libya (*GG* 4.4.1–14). The coastal area was a region of intensive Greek settlement from the late seventh century BC; as described in vivid detail by Herodotos,[30] it was early thought to be the location of the mythical Garden of the Hesperides. After the time of Alexander the Great the area generally became a dependency of Ptolemaic Egypt, until its last king, Ptolemy Apion, willed it to Rome in 74 BC, although due to the activities of Kleopatra VII and Marcus Antonius its status as a province was not settled until after their deaths in 30 BC.[31] Ptolemy's account of the Cyrenaica is more anachronistic than many portions of the *Geographical Guide*: there is no evidence of the Roman organization, which was completed over a century before he wrote.

There is a rich variety of toponyms, including Carthaginian vestiges such as Barke and famous Greek cities such as Kyrene and Berenike (*GG* 4.4.4, 11). Citation of the city of Ptolemais indicates that Ptolemy's data, at least, came from after 323 BC. Many of the names in the interior are Greek—Kainopolis, Echinos, Neapolis, and Hydra, among others—demonstrating that intensity of Greek movement away from the coast, in part due to the cultivation of the primary export project of the region, the spice known as silphium. The southern part of the Cyrenaica was called the Silphiophoros ("Silphium Producing") territory. Silphium was a plant like fennel that had been known to the Greek world since their arrival in the region; it was a prosperous resource, yet became extinct by the late first century AD.[32] Thus the plant was no longer available when Ptolemy wrote, but the southern Cyrenaica was still famed as the silphium region.

There is an unusual quality to Ptolemy's map of the Cyrenaica, perhaps because of its intensive and early Greek settlement in a region far from the Greek homeland. Well into the Roman period, Libya (or Africa, as it was then called) was the most exotic of the three continents, unique in its topography, flora, and fauna. In the latter part of the first century BC Vitruvius could speak of it as "the parent and nurse of wild beasts, especially snakes,"[33] an attitude reflected in Ptolemy's "land infested with wild beasts" (*GG* 4.4.10), a phrase used toponymically. Needless to say, the Cyrenaica was the jumping-off point for explorations into the even more exotic interior of the continent; the routes south had been known to Greeks since at least the fifth century BC, some of which were described by Ptolemy himself, with his reports of travelers reaching the hippopotamus territory and beyond, as well as the upper Nile and the Selene Mountains. Beyond these extremities there was nothing but "unknown land" as far as the south pole.

Other unusual toponyms in the Cyrenaica included the Caves of the Laganikoi, located in the south-central portion and unexplained. Less enigmatic are the Sands (Thines) of Herakles. Although only a point on the map, the toponym presumably refers to the great sandy regions such as the Rebiana or Calanshu; the term *thines* was applied to this region by early Hellenistic times.[34] Ptolemy actually considered them to be a mountain, a conclusion not easily explained. Near the sands was the probable source of the only river Ptolemy documented in the Cyrenaica, the Lathon, which he plotted flowing slightly west of north through a lagoon (*limenothalassa*) and into the ocean west of Ptolemais (*GG* 4.4.8). This is probably the river called the Lethe or Lathon by earlier authors;[35] the name Lethe indicates its strange properties, for it flows through a karst region (the modern Bu Shatin) with numerous sinkholes and underground watercourses.[36] Ptolemy seemingly placed its mouth too close to Ptolemais; it was more to the southwest at Berenike. Other than the river, the only other topographical features noted are two mountains in the southern portion of the Cyrenaica, Ouelpa (perhaps Velpa) to the west and Baikolikon to the east. The former may be modern Bikku Betti, at 2266 m. the highest point in modern Libya and located near the border with Chad.

There are two islands recorded off the coast, Myrmex and Laia. Neither can be identified with certainty: the only Myrmex on this coast is far to the east in Egypt, which means that Ptolemy's island is badly misplaced or not the same, but its descriptive name ("ant") may be generic. Laia, also known as the island of Aphrodite, may be Gezerit Chersa near Darnis. As always, coastal islands tend in time to become submerged or joined to the mainland.

The Boundaries of Marmarike, Libya, and All of Egypt

The remainder of Map 3 of Libya covers the areas east of the Cyrenaica. This includes not only Egypt proper, but the territory between the Cyrenaica and the

Nile valley and its immediate environs. Since the traditional western boundary of Egypt was at the Katabathmos, extending from there (in theory) to the south and passing west of the famous oasis of Ammon, in Ptolemy's scheme this left a region of about three degrees of latitude (at the Mediterranean) that was neither in Egypt nor the Cyrenaica. Except for the coastal strip and the oasis of Augila, this was desolate country, usually administered as part of Cyrenaica, although control varied between that region and Egypt.[37] The historical name for this territory was Marmarika, or Marmarike, a name documented (as an ethnym) from at least the fourth century BC.[38] As Ptolemy presented it, it was a strip from the coast south to a line at 23 degrees north latitude which corresponded to the southern boundary of Egypt. There were few toponyms in Marmarika, even along the coast. Inland the locations inevitably had indigenous names—whether villages or ethnic groups—indicating that Greco-Roman penetration was far more limited than in the Cyrenaica. The only place of importance was the oasis of Augila (*GG* 4.5.30), actually a collection of oases at a junction of ancient trading routes, situated about 400 km. south of the coast and known to the Greek world since the fifth century BC.[39] In the first century BC camel caravans took four days to reach Augila from the coast, and it was especially famous for its dates.

East of Marmarika was Egypt. From the Katabathmos, on the coast of Ptolemy's Egyptian Sea and the traditional western boundary of the territory, it was nearly 500 km. along the coast to Alexandria, with population increasing as one headed east. In fact, whether this vast region—stretching south to the northern boundary of Aithiopia at 23 degrees north latitude—could properly be called Egypt was a matter of dispute, because it all lay west of the Nile Delta and the valley of the upper river: Egypt was properly limited to the territory watered by the Nile.[40] But administratively the Roman province extended west to the Katabathmos and south to the First Cataract at Syene, which was essentially the definition of Egypt that Ptolemy used, although for regularity his Egypt went south beyond the cataract to the line at 23 degrees north latitude that went across the continent.

The region from the western edge of the Nile valley to the Katabathmos, and to the south, was a wild and remote area except for the coastal strip. The most famous site was the oracle and oasis of Ammon (modern Siwa) (*GG* 4.5.33), 475 km. west-southwest of Alexandria. It was of ancient origin, but gained prominence when Alexander the Great visited it in 331 BC.[41] The site was also of geologic interest, lying far from the ocean but below sea level, with seashells conspicuous in the region; thus it was important in understanding the formative processes of the earth. A short distance to the east was the Camp of Alexander, presumably a location still pointed out to tourists hundreds of years after the king was there.

Scattered through these western regions were a number of lakes and springs, often with Greek names, such as the Helios Spring or Lykomedes Lake. A number of oases were also plotted. These reflect points along various trade and caravan routes across the desert, connecting the ethnic groups that lived in this wilderness, of which nearly 20 are named. The exceedingly few settlements—probably themselves oases—all have indigenous names. But if one went too far south, a "sandy and rainless region" was reached, believed to have no inhabitants (*GG* 4.5.26).

Along the coast, from Darnis to Alexandria, and a short distance inland, was a more populated agricultural zone. There were relatively few towns on the coast itself—although this well-traveled sailing route was carefully plotted with numerous topographical features and several offshore islands—but there was a relatively dense strip of settlements extending a number of kilometers inland. Most have indigenous names but ones such as Leukon, Menelaos, Bibliaphorion, Klimax, and Glaukon demonstrate Greek penetration, perhaps as early as the seventh century BC.

After traveling hundreds of kilometers east from Darnis, the voyager would reach Alexandria and the Nile, and Egypt proper. Ptolemy separated the western desert regions from the Nile valley by a lengthy mountain range, running north-south west of the river, which he called the Libyan Mountains. It ran from west of the north end of the Delta to the latitude of the First Cataract—theoretically a distance of as much as 700 km.—but was probably little more than a conceived dividing line to mark the western edge of the Nile valley. To the east of this range was Egypt itself.

Greeks had been in Egypt since the opening of the international trade emporium of Naukratis just west of the Delta in the seventh century BC, if not earlier.[42] Closer relations between Egypt and the Greek world developed after that time, leading to the founding of the Ptolemaic dynasty in the late fourth century BC and the Roman province in 30 BC. Ptolemy's map of Egypt reflects the status of the province in his own era.

The Nile valley itself was one of the most carefully surveyed regions of the ancient world. From the Mediterranean (here Ptolemy's Egyptian Sea) to the limit of Map 3 of Libya south of Syene and the First Cataract there were dozens of towns, villages, and religious centers that Ptolemy plotted, with the density in the Delta especially high. Most of these were exceedingly ancient, although the Hellenistic organization of the region was also apparent, not only with Alexandria, but locations such as Ptolemais (the name of two towns) and Arsinoë. Many toponyms, especially those reflecting the name of a divinity, were Greek translations of indigenous theophoric names, such as Hermoupolis, Aphroditopolis, or Apollonopolis. The toponyms also show the diverse ethnic history of Egypt, with, in addition to the Greek names, a place called Babylon at the head of the Delta (*GG* 4.5.54), and Arabians scattered along the eastern side of the Nile (*GG* 4.5.53, 74). Moreover, Ptolemy was aware of the historic organization of Egypt into nomes (*nomoi*), a word originally meaning "pasture" but which became the Greek word for Egyptian administrative districts.[43] The number of them varied: Ptolemy listed 47, comparable to the 45 of Pliny somewhat over half a century earlier.[44]

The traditional southern limit of the Roman province (and, historically, Egypt itself) was the First Cataract at Syene. But to provide a sense of cartographic regularity, Ptolemy extended the territory to the alignment of 23 degrees north latitude that he had generally used as the southern boundary of Map 3 of Libya, thereby adding a few towns south of the cataract to Egypt proper.

East of the Nile was a rugged area between the river and the Arabian Gulf (modern Red Sea). There was practically no permanent population in this remote region, and it was considered ethnically Arabian although part of the Roman province of Egypt. There were various access routes from the Nile to the Arabian Gulf across this territory, and ports were developed along the coast, some of which had had their origins in dynastic times but which became especially prominent from the early Ptolemaic period. From early in the third century BC the kings had an increasing need for war elephants, which could be obtained by sailing south on the Red Sea to the elephant territory in east Africa.[45] In the late second century BC, the utility of the Red Sea routes was enhanced by the development of trade with India, and by the Roman period as many as 120 ships made the journey to India regularly.[46] Ptolemy listed the major Red Sea ports, from Arsinoë in the north to two named Berenike farther south. The most important was the more northern Berenike; extensive excavations have revealed a rich and complex international emporium where evidence for over a dozen languages has been found.[47]

The lands between the Nile and the Red Sea were also a region of mineral resources: Ptolemy documented several "mountains"—more likely prominent outcrops—including the Alabastrinos Mountain, and, more importantly, the Smaragdos Mountain. The latter indicated the location of emerald mines; the gem was becoming more common in Roman times.[48] Ptolemy's material on the Red Sea was collected after AD 117, since he included what he called the "Trajan River" (*GG* 4.5.54), actually the canal that connected the Nile at the head of the delta with Heroonpolis on the Red Sea, an improved version of a Ptolemaic watercourse.[49]

The final section of Map 3 of Libya is a small anomalous region east of the Nile Delta. The boundary between the continents of Libya and Asia had always been problematic, unlike the ones between Europe and Libya or Europe and Asia, where distinct geographical features gave clarity to the division. The original determination of the boundary between Libya and Asia was at the Nile, documented by the fifth century BC,[50] but this became increasingly unreasonable, especially by Hellenistic times, as Ptolemaic interests spread far to the east of the river into the Levant and Syria. Ptolemy, probably influenced by Roman administrative practice, simply drew a line northeast from the head of the Red Sea at Heroonpolis to the Mediterranean, making this his boundary. But any division produced difficulties, and Ptolemy's choice had the unfortunate effect of placing the ancient Levantine city of Anthedon in Egypt (*GG* 4.5.12). Yet at the same time he realized the difficulty and repeated Anthedon in his description of Palestine and Judaea (*GG* 5.16.2).

Map 4 of Libya

Ptolemy's final map of Libya covers a vast area south of the Roman territory and a line from the Atlantic at just south of 27 degrees north latitude (the southern edge of Mauretania Tingitana) to the Red Sea at 23 degrees north latitude (Ptolemy's southern edge of Egypt). The southern boundary of the map is around 16 degrees north latitude in "unknown territory." Thus the map includes all the rest of the continent insofar as it was known. This was not Roman territory, so Ptolemy had to rely on sources other than Roman administrative details, but Carthaginians, Greeks, and Roman explorers and traders had long penetrated the region.

Ptolemy divided the map into three portions. The northwest he called Interior Libya (*GG* 4.6.1–34), located south of the Mauretanian provinces, the province of Africa, and Cyrenaica. To its east was Aithiopia Below Egypt (4.7.1–41), as far as

Map 4 of Asia

Red Sea

Sea of Hippalos

Barbaric Gulf

Arabian Gulf

Syrian Sea

Egyptian Sea

Egypt

Nile

Aithiopia South
of Egypt

Libyan Sea

Marmarica

Cyrenaica

Province of Africa

Inner Libya

Inner Aithiopia

African Sea

Sardian Sea

Mauretania
Caesarienses

Iberian Sea

Mauretania
Tingitana

Strait of
Herakles

Western
Ocean

Outer Sea

Hesperian Gulf

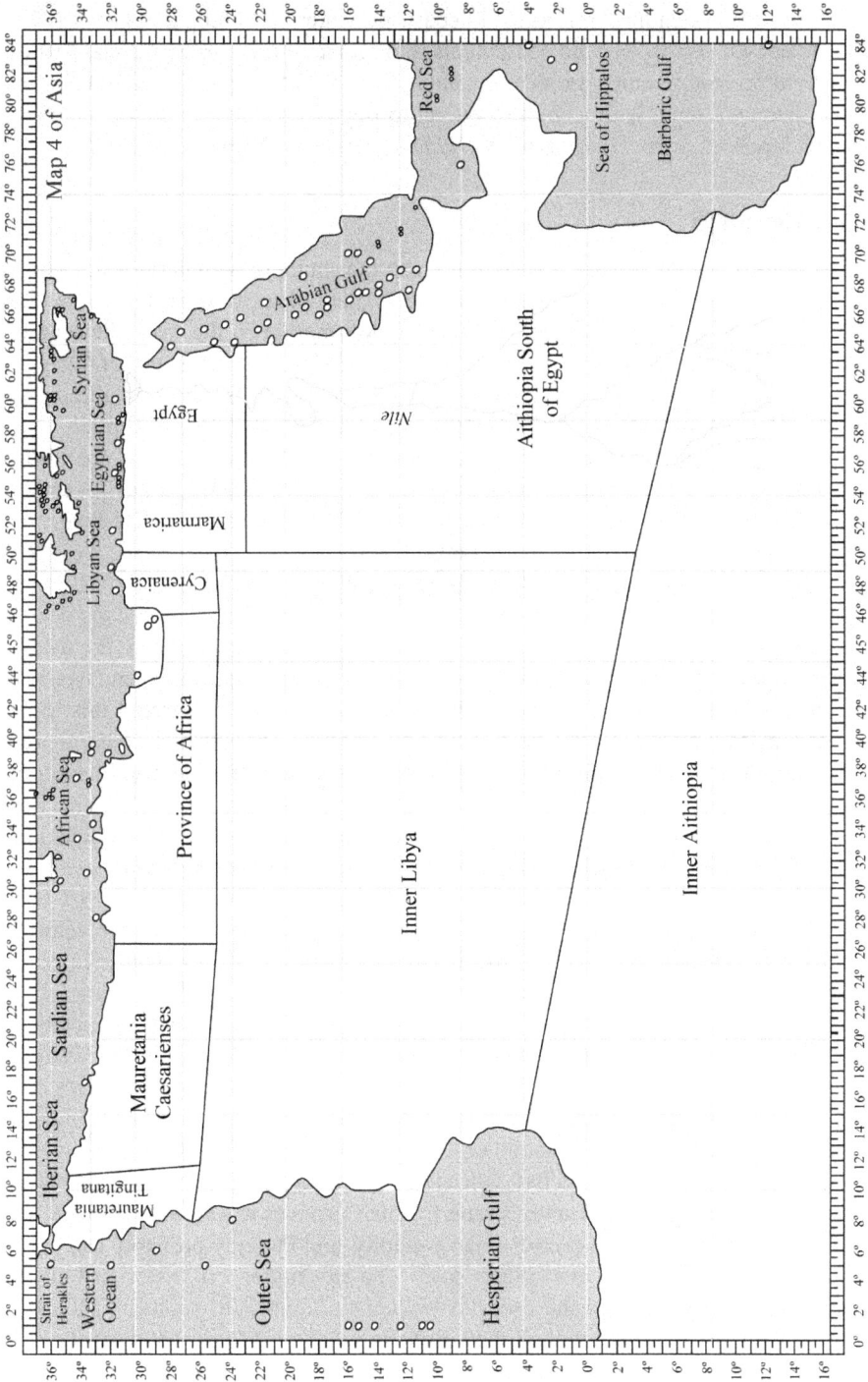

Map 5.4 Map 4 of Libya

the Red Sea and the western extremity of the modern Indian Ocean. South of these two regions, extending from coast to coast, was Interior Aithiopia (*GG* 4.8.1–7), a little-known region with no settlements, only a handful of topographical features and ethnic groups, and, significantly, unusual animals. The southern boundary was the unknown land, with Cape Prason, at 15 degrees south latitude, the southernmost toponym in the *Geographical Guide*.

Interior Libya

Interior Libya was roughly a rectangle, with the southern boundary of the Roman provinces at its north, the External Ocean (Atlantic) to its west, and a southern extension of the line dividing Cyrenaica and Marmarika on the east. The southern boundary was an arbitrary line from the External Ocean at just south of 4 degrees north latitude to the join with the eastern boundary at 8 degrees south latitude.

The orientation of the region is skewed. Ptolemy had little conception of the long east-west alignment of the African coast along the north side of the Gulf of Guinea, and the Atlantic coast was presented as running essentially north-south, so that Theon Ochema (Mt. Cameroon), which lies on the longitude of Carthage, was placed near the longitude of Mauretanian Caesarea, which is far to the west. This error is an obvious product of the inability to calculate longitude.

The density of toponyms in this remote region is astonishing, perhaps demonstrative of the numerous ancient trade routes across the Sahara. There are over 50 settlements—probably mostly oases—a dozen rivers, 11 mountains, and nearly 60 ethnic groups. People from the Mediterranean world had penetrated this region since at least the sixth century BC, but, except along the Atlantic coast, there is practically no evidence that they left any mark on the topographic map: the interior names are indigenous. Some of the rivers are probably trade routes rather than watercourses: the Bagrades (modern Mejerda) extends for 14 degrees of latitude into the interior yet its source is only 460 km. from the Mediterranean.

The major river of the region was the Nigeir, presumably the modern Niger (although the name is generic for "river"), running westward across the center of the region through 28 degrees of longitude. This is presumably the unnamed central African river mentioned by Herodotos, but due to the problems of longitude and the orientation of the coast its direction has become reversed.[51] Other rivers drain into the Atlantic, presumably the major rivers of West Africa such as the Senegal and Gambia.

It is along the Atlantic coast that the greatest density of toponyms occurs. This had been explored extensively almost as far as the equator since at least the sixth century BC. Around 500 BC Hanno of Carthage mounted an expedition that went to equatorial regions and the active volcano Theon Ochema, the Chariot of the Gods, just north of the equator. Hanno's report, which survives in a Greek summary, was well known throughout Greco-Roman antiquity, and Ptolemy probably had access to a fuller account than available today.[52] The adventurer and historian Polybios and others also explored the coast in Hellenistic times.[53] In addition, Juba II of Mauretania wrote an account of Hanno's voyage.[54] Thus Ptolemy had ample material about this portion of the West African coast, and the scattered Greek toponyms, such as the Ophiodes River, the Megas Harbor, and the Hesperou Keras all presumably go back to these explorers.

Inland there was a large number of ethnic groups. Some, like the Garamantes, had been known through trade contacts since early times.[55] Others were identified as various groups of Aithiopians, including the Pyrrhaian, Nigritian, or Odrangidan. These were indigenous groups that had had the name Aithiopian attached to them because of the persistent idea that Aithiopians (whose traditional territory was south of Egypt) actually spread across the entire continent. This belief was derived from the statement in the *Iliad* that Aithiopians lived both in traditional Aithiopia and on the External Ocean.[56] By the fourth century BC they were thought to occupy the entire southwestern quadrant of the inhabited world.[57] There is no evidence for any criteria other than geographical that identified certain ethnic groups as Aithiopian, but from earliest times they were said to be found in many places.

Several islands were documented in the External Ocean. At the north edge of the map, offshore from the Atlas Mountains, was Kerne (*GG* 4.6.33). This was a Carthaginian outpost founded by Hanno; its exact location has not been determined.[58] It became the best-known place on this coast, although by Ptolemy's time it had probably been abandoned.[59] It was on an island in the recess of a bay, but Ptolemy placed it over three degrees of longitude out to sea, so much had its actual location been forgotten.

Of particular importance for Ptolemy's cartographic theory were the six Makaroi Nesos, the Fortunate or Blessed Islands (*GG* 4.6.34). They were placed on a north-south line from 16 to 10 degrees north longitude and as much as 8 degrees of longitude west of the coast. These are the modern Canary Islands, which had been visited by the Carthaginians but were examined, surveyed, and named by Juba II of Mauretania.[60] Again there is the problem of longitude, since Ptolemy placed the islands on a north-south line when their orientation is more east-west. In addition, one island, Hera, was cited twice, both within the group and to the north in an isolated position. But the islands were of notable significance to Ptolemy since they were his westernmost toponyms and thus determined his prime meridian (*GG* 1.11.2), a calculation actually made by Marinos. For clarity, Ptolemy or Marinos placed the islands at 1 degree east longitude, leaving the prime meridian just to the west and untouched by any land.

Aithiopia Below Egypt

To the east of Interior Libya was Aithiopia Below Egypt, the traditional Aithiopian territory (*GG* 4.7.1–41). It extended east from Interior Libya to the Red Sea and Indian Ocean, and south to the eastern extension of the line that marked the south edge of Interior Libya. Aithiopia in its more limited sense was the region along the Nile south of the First Cataract, but this portion of the map included regions far up the Nile and its tributaries through the kingdom of Meroë and beyond, and east of the Nile including modern Somalia, the Horn of Africa, and points south into modern coastal Kenya and Tanzania. Many islands in the Red Sea and northwestern Indian Ocean were also plotted.

Despite the remoteness, the unexpected density of toponyms demonstrates the large amount of exploration and commercial penetration far to the south. Greeks had gone up the Nile beyond the First Cataract as early as the fifth century BC and had made contact with the famous kingdom of Meroë.[61] At that time reports were received from as much as two months farther upstream, to the point where the White and Blue Nile join at modern Khartoum. With the establishment of the Ptolemaic government there was a vigorous attempt to explore and exploit the resources of the world of the upper Nile. Expeditions were sent far beyond Meroë; there are few details but they obtained information from well into the tropics. The Red Sea was also explored in detail. The primary goal of these explorers was to support the economic basis of the Ptolemaic world, features reflected in Map 4 of Libya, but they also showed an interest in the unusual celestial phenomena and flora and fauna of low latitudes.

There was further exploration after the Romans acquired Egypt. The source of the Nile had long been an issue of eternal fascination: Julius Caesar wanted to find it[62] and Juba II thought that he had, locating it in his own kingdom.[63] The emperor Nero sent an expedition that probably went as far as modern Uganda,[64] and Ptolemy himself in the *Geographical Guide* reported on expeditions, probably of the late first century AD, that reached the mountains at the source of the river (*GG* 1.17.6, 4.7.26, 4.8.3). The data from all these explorations became incorporated into Map 4 of Libya.

The greatest density of toponyms on the map, as might be expected, is along the Nile from the First Cataract to Meroë. To the west of the river, and extending south of the equator, were the Aithiopian Mountains; as with the Libyan Mountains of Map 3 of Libya, these merely marked the western boundary of the Nile valley, with only a few ethnyms beyond. Upstream from Meroë the map shows the various branches of the Nile, beginning with the Astaboras and the Astapous (*GG* 4.7.20–4). Additional branches are noted to the south but not named. Throughout this territory of the upper Nile there were numerous ethnic groups, some with indigenous names but others with Greek descriptive ones, defined by diet, such as the Root Eaters or the Elephant Eaters (*GG* 4.7.29, 34).

The natural resources exploited by the Ptolemies are apparent on the map. There are two localities named Elephant Mountain (*GG* 4.7.10, 26), as well as a region producing myrrh (the Smyrnophoros territory) and one producing cinnamon (the Kinnanomophoros territory) (*GG* 4.7.31, 34). Although myrrh did come from the modern Somali region of east Africa, cinnamon, whose origin Ptolemy placed on the upper Nile south of the equator, was not produced locally but was imported to the Mediterranean world from southern Asia or beyond.[65] This is a solid example of confusing the home region of a trade product with points along the trade route from little-known to better-known areas. Greek and Roman merchants obtained cinnamon (and other south Asian spices and aromatics) from eastern merchants, with the transfer taking place at the eastern extremities of Africa in modern Somalia. The aptly titled Aromata (Spice Port, at modern Cape Guardafui), which Ptolemy identified as one of several trading emporia in the region (*GG* 4.7.10), demonstrates by its name the local role in acquiring eastern aromatics and transshipping them to vessels heading up the Red Sea.[66]

Ptolemy also recorded localities along the coast south of Aromata, a route along which trade had also been established in Hellenistic times.[67] The final port to the south was Rhapta (*GG* 4.7.12), an important city generally identified with modern Dar es Salaam.[68] Rhapta had been known to the Mediterranean world at least since the early first century AD, and it was here at some time later in the century that Diogenes learned about the source of the Nile (*GG* 1.9).

From Aromata and the other emporia in extreme east Africa ships carrying exotic commodities headed into and up the Red Sea to the various ports on its west side. By Ptolemy's time the sea had been exhaustively surveyed, essentially for navigational purposes, and nearly 20 islands had been plotted on the west side of the southern portion of the sea, with many others on the east side (*GG* 6.7.43–5). Most of these have Greek names, often descriptive (Chelonides, or Turtle Islands) or named after Greco-Roman divinities (Athena, Demeter, Bakchos). The primary interest in locating these islands was for the safety of shipping routes; the more detailed description of them in the *Periplous of the Erythraian Sea* demonstrates that they offered little if any economic benefit.[69]

Interior Aithiopia

The final segment of Map 4 of Libya extends across its southern portion from the Atlantic to the Indian Ocean, which Ptolemy called the Barbarian Gulf or Shallow (Bracheia) Sea (*GG* 4.8.1–7). It is clear that he knew little about this region, which was included on the map merely to account for certain explorations that went farther south than his Inner Libya or Aithiopia South of Egypt. The material on this map comes from the journeys of Septimius Flaccus and Julius Maternus to Agisymba, and those of Diogenes, Theophilos, and Dioskouros along the East African coast, and, in the case of Diogenes, perhaps inland to the lakes that were the source of the Nile (*GG* 1.8–9). All these expeditions took place in the late first century AD.

Agisymba is a vague toponym located south of the equator, defined as a "land" (*choros*), with no further specifics, but which must lie far to the south of Leptis Magna, from which it was accessed. Yet its location was so uncertain that no coordinates were provided. In this poorly defined region were several mountains, located between 8 and 13 degrees south latitude and 10 and 45 degrees east longitude, having indigenous toponyms, information presumably acquired by the Agisymba expeditions. Several groups of Aithiopians were also nearby. Of particular interest were the fauna: white elephants, rhinoceroses, and tigers, again probably reported by Flaccus or Maternus. The rhinoceros is the African variety (*Diceros bicornis*), which in antiquity was found in central Africa and thus the region penetrated by the Roman explorers.[70] There is no species of white elephant; if the text of the *Geographical Guide* is correct, the expeditions may have encountered albinos. Tigers are not indigenous to Africa, but had been known to the Greco-Roman world since the time of Alexander the Great. Yet the expeditions may have seen another type of feline, perhaps a cheetah or leopard.[71]

The eastern portions of Interior Aithiopia are also poorly documented, but include the famous Selene Mountains (*GG* 4.8.6), probably the modern Ruwenzori, believed by Ptolemy to contain the source of the Nile, which flowed from them into the lakes that had been reported by Diogenes. There were also the inevitable groups of Aithiopians—in this case Fish Eaters and Cannibals—in this region.

On the coast of the Indian Ocean in Interior Aithiopia there were two toponyms, Cape Rhapton and Cape Prason. The former was just south of the Rhapta region, and may have been reached (or at least seen) by Diogenes. Almost certainly it is the prominent Ras Kanzi at the southern end of the bay on which Dar es Salaam lies. Farther along the coast to the south was Cape Prason, reached by Dioskouros in the late first century BC (*GG* 1.9.4, 1.17.6). The name may be Greek (*prason*, a type of seaweed, that was believed to infest the Indian Ocean).[72] Nevertheless it was the most southern toponym in the *Geographical Guide*, at 15 degrees south latitude. Ptolemy believed that the Indian Ocean was an enclosed sea, an idea that probably originated with Hipparchos in the second century BC,[73] and Cape Prason was believed to be where the Libyan coast turned toward the east eventually to join with southeast Asia. Yet Ptolemy provided no coordinates for this assumed coast but only mentioned its existence in passing, indicating that he had no specific data to support Hipparchos' theory, hardly unexpected. To some extent such an idea was counterintuitive, because Phoenicians had circumnavigated Africa around 600 BC,[74] but this report had been largely dismissed in subsequent years,[75] and thus the concept of an enclosed Indian Ocean may have seemed more plausible. Nevertheless it was a restraint on later exploration until the time of the Renaissance and Vasco da Gama.[76]

One island was plotted off the eastern coast of Interior Aithiopia, Menouthias, which Ptolemy located northeast of Cape Prason. The *Periplous of the Erythraian Sea* described it in greater detail, commenting on its topography, flora, and fauna, and the fishing habits of the locals.[77] The obvious candidate is one of the three islands near Dar es Salaam: Zanzibar, Pemba, or Mafia, but Ptolemy's location (certainly not totally reliable) places it 3 degrees of latitude northeast of Cape Prason, whereas the known islands are close to the shore. Nevertheless, modern opinion has settled on Pemba or Zanzibar although the plotted location may suggest the Seychelles.[78]

To end the map of Interior Aithiopia, and indeed the four maps of Libya, Ptolemy told his readers that he had reached the southern limit of the inhabited world, and that beyond were only unknown (*agnostoi*) lands for the remaining 74 degrees to the south pole. Needless to say, the south pole was purely a theoretical concept, based on the idea of terrestrial zones developed probably by Parmenides in the fifth century BC.[79] Yet it would have been visually represented on the globe of the earth that Krates of Mallos constructed in the second century BC.[80]

Notes

1 Homer, *Odyssey* 4.85 etc.

2 Klaus Geus and Florian Mittenhüber, "Die Länderkarten Africas," in *Klaudios Ptolemaios: Handbuch der Geographie, Ergänzungsband* (ed. Alfred Stückelberger and Florian Mittenhuber, Basel 2009) 282–9.
3 Sallust, *Jugurtha* 17.3; Vitruvius 2.9.13; Strabo, *Geography* 2.3.8.
4 Duane W. Roller, *The World of Juba II and Cleopatra Selene: Royal Scholarship on Rome's African Frontier* (New York 2003) 46–58.
5 Suetonius, *Divine Julius* 52.1.
6 Roller, *Three Ancient Geographical Treatises* 14–15.
7 Polybios 34.15.7–9.
8 Strabo, *Geography* 17.3.3.
9 Strabo, *Geography* 3.1.8; Dio 48.45.1–3.
10 Pliny, *Natural History* 5.2.
11 Hanno 6–7.
12 Pliny, *Natural History* 5.14–15; Dio 60.9.
13 Dikaiarchos F125; Eratosthenes, *Geography* F106; Strabo, *Geography* 3.5.5; Pliny, *Natural History* 3.4.
14 Juba F43; Pliny, *Natural History* 6.201; Roller, *World of Juba* 115–16.
15 Herodotos 4.8; Strabo, *Geography* 3.5.4; Pliny, *Natural History* 4.120.
16 Roller, *Through the Pillars* 45–7.
17 Pomponius Mela 1.30; J. Lassus, "Portus Magnus," *PECS* 732–3.
18 Philippe Leveau, "Recherches historiques sur une région montagneuse de Maurétanie Césarienne: des Tigava Castra à la mer," *MEFR* (1977) 296–7.
19 Stückelberger and Graßhoff, *Klaudios Ptolemaios* 395.
20 Roller, *World of Juba* 11–38.
21 Appian, *Civil War* 4.54.
22 Pseudo-Skylax 109–10; Strabo, *Geography* 17.3.18; Pliny, *Natural History* 5.27; Werner Huss, "Leptis Magna," *BNP* 7 (2005) 419–25.
23 Pollard and Berry, *Complete Roman Legions* 114–19.
24 Hanno 1; *King Nikomedes Periplous* 196.
25 Homer, *Odyssey* 9.82–104; Strabo, *Geography* 1.2.17; Roller, *Historical and Topographical Guide* 24.
26 Herodotos 2.30–1.
27 Athenaios 2.44d; Klaus Geus, *Prosopographie der Literarisch bezeugten Karthager* (Leuven 1994) 179–80.
28 Polybios 31.18.9.
29 Strabo, *Geography* 17.3.22.
30 Herodotos 4.150–8.
31 Günther Hölbl, *A History of the Ptolemaic Empire* (tr. Tina Saavedra, London 2001) 210; Duane W. Roller, *Cleopatra: A Biography* (Oxford 2010) 94–5, 100.
32 Pliny, *Natural History* 19.38–45; 22.100–6; Andrew Dalby, *Food in the Ancient World From A to Z* (London 2003) 303–4.
33 Vitruvius 8.3.24.
34 Apollonios 4.1384.
35 Strabo, *Geography* 17.3.20; Lucan 9.355; Pliny, *Natural History* 5.31.
36 G. D. B. Jones and J. H. Little, "Coastal Settlement in Cyrenaica," *JRS* 61 (1971) 78.
37 *BNP Historical Atlas of the Ancient World* 163.
38 Pseudo-Skylax 108.
39 Herodotos 4.172, 183; Strabo, *Geography* 17.3.23.
40 Strabo, *Geography* 17.1.4–5.
41 Strabo, *Geography* 1.3.4, 17.1.43; Plutarch, *Alexander* 26–7; Arrian, *Anabasis* 3.2.3–4; see also Herodotos 1.46, 2.32, 55.
42 Herodotos 2.178.
43 Herodotos 2.4.
44 Pliny, *Natural History* 5.49.

45 Steven E. Sidebotham, *Berenike and the Ancient Maritime Spice Route* (Berkeley 2011) 41–53.
46 Poseidonios F49; Strabo, *Geography* 2.3.4, 2.5.12.
47 Sidebotham, *Berenike* 74–5.
48 Lisbet Thoresen, "Archaeogemmology and Ancient Literary Sources on Gems and Their Origins," in *Gemstones in the First Millenium* AD*: Mines, Trade, Workshops and Symbolism* (ed. Alexandra Hilgner et al., Mainz 2017) 162, 182–3.
49 Sidebotham, *Berenike* 180.
50 Herodotos 4.45.
51 Herodotos 2.32.
52 Roller, *Three Ancient Geographical Treatises* 12–15.
53 Polybios 34.15.7–9; Pliny, *Natural History* 5.8–11; 6.199; 18.22–3.
54 Juba F6; Athenaios 3.83b.
55 Herodotos 4.174, 183.
56 Homer, *Iliad* 1.423 etc.
57 Ephoros F30a; Strabo, *Geography* 1.2.28.
58 Hanno 8.
59 Roller, *Three Ancient Geographical Treatises* 29–30.
60 Juba F43–4; Roller, *Through the Pillars* 47–8.
61 Herodotos 2.29–30.
62 Lucan 10.191–2.
63 Pliny, *Natural History* 5.51–4.
64 Seneca, *Natural Questions* 4a.2.7; 6.7.2–3.
65 Dalby, *Food* 87–8.
66 *Periplous of the Erythraian Sea* 12.
67 *Periplous of the Erythraian Sea* 13–18.
68 Casson, *Periplus* 141–2.
69 *Periplous of the Erythraian Sea* 1–12.
70 Kitchell, *Animals* 161–3.
71 Kitchell, *Animals* 183–5.
72 Theophrastos, *Research on Plants* 4.6.2; *GG* 7.2.1.
73 Hipparchos, *Against the Geography of Eratosthenes* F4; *GG* 7.5.2, 8.1.4; Berggren and Jones, *Ptolemy's Geography* 22.
74 Herodotos 4.42.
75 Roller, *Ancient Geography* 24–5.
76 Thomson, *History of Ancient Geography* 277–8.
77 *Periplous of the Erythraian Sea* 15.
78 Casson, *Periplus* 140.
79 Roller, *Ancient Geography* 73–4.
80 Strabo, *Geography* 2.5.10.

6   Asia

Introduction

Books 5, 6, and the first half of Book 7 are devoted to the continent of Asia, which was displayed on 12 maps.[1] Although the true extent of the continent was hardly known, it had a certain vastness that made it different from Europe and Asia. According to Ptolemy, it extended two thirds of the way around the earth, from 55 degrees east longitude at Cape Sigeion, the western edge of Asia Minor (modern Kumkale in Turkey) to Sinai and other uncertain points at 180 degrees east longitude. This is an extent of 125 degrees of longitude, but from the Aegean coast of Turkey to the Chinese coast at Shanghai is only about 100 degrees, demonstrating Ptolemy's east-west elongation of the inhabited world. Nevertheless this vast distance gave Asia an almost mythic quality not apparent in the other two continents. Moreover, it was realized that much of Asia remained unknown: the author of the *Periplous of the Erythraian Sea*, writing a century before Ptolemy, wrote of the stormy cold land with rugged terrain that existed at the far eastern extremity of Asia, but which had "not been discovered," an ancient hint of the wilds of Siberia.[2] It was believed that the perimeter of Europe was known, and all of Libya except the extreme south, which seemed of little importance, but eastern Asia was a strange region unlike anywhere else on earth. Traders had gone far to the east, with the expedition commissioned by Maes Titianos the prime example, whose agents had reached the Silk People (*GG* 1.11–12). Others had gone to the land of the Sinai (*GG* 7.3.1–3) and the great city known as Thina, whose inhabitants traded with the Indians.[3] That these may be names associated with the Chinese adds little to understanding the ancient knowledge of these regions. Unlike the perimeters of the other continents, there seemed to be no obvious eastern or even northern edge to Asia: somewhere out there was the External Ocean (with the portion east of the continent theoretically called the Eastern Ocean), but it never seemed to be found. Just as explorers in Libya had searched for the source of the Nile, those heading into eastern Asia expected to reach the Ocean. In the early third century BC Demodamas of Miletos went far into central Asia but never found the Ocean.[4] There was no report of the Ocean from Maes Titianos' expedition. The author of the *Periplous of the Erythraian Sea* assumed that the Silk People were near the Ocean but had no proof. These frustrations meant that Asia could not be defined in the same way that

DOI: 10.4324/9781003248590-7

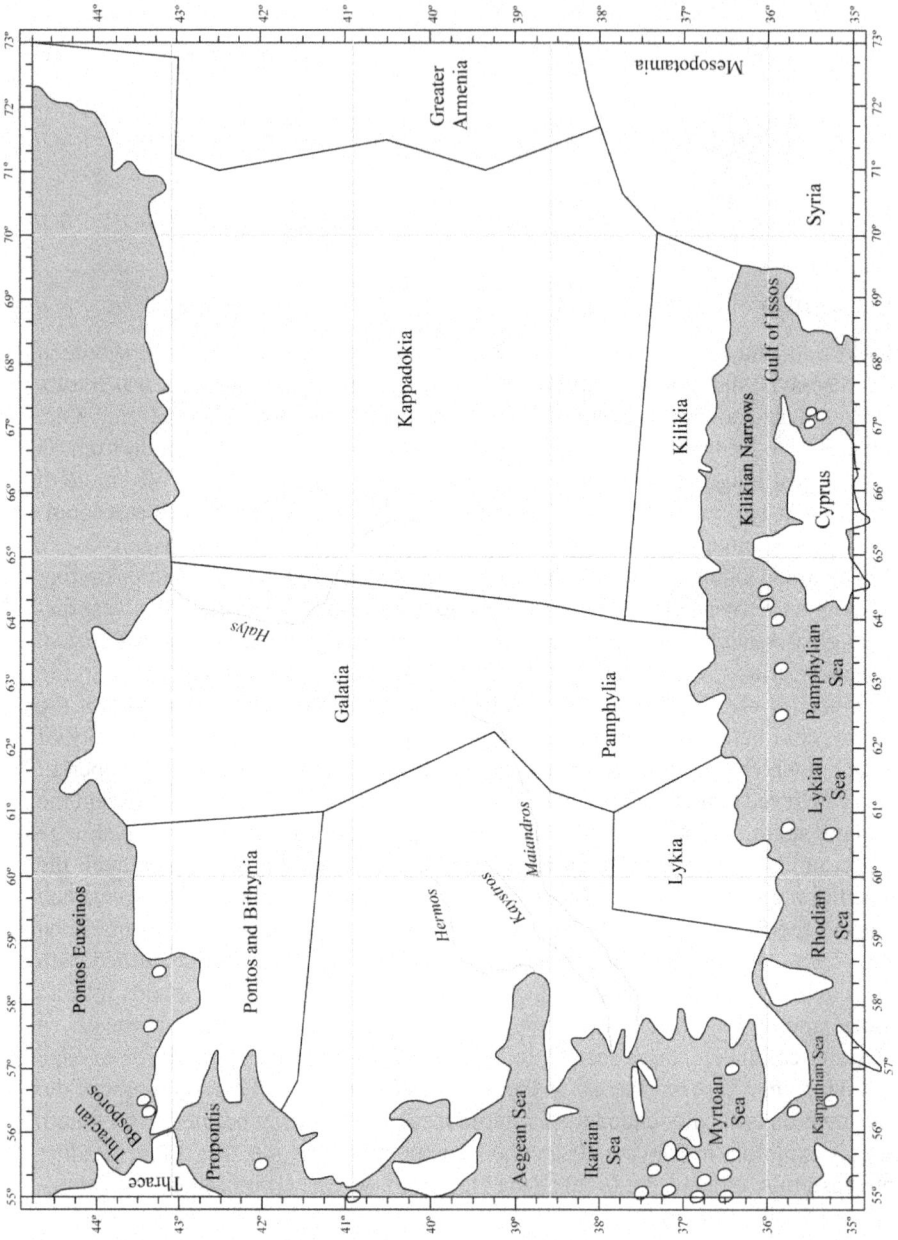

Map 6.1 Map 1 of Asia

the other two continents were, and Ptolemy's maps of the far east and north (Maps 7, 8, and 10 of Asia) have no definable eastern and northern boundary.

To be sure, western Asia was well known, a region of a complex mixture of cultures that extended from the Aegean to India. Here Ptolemy plotted hundreds of toponyms, especially in the ancient regions of Anatolia, Syria, the Levant, and Mesopotamia. But much like Africa in the nineteenth century, Asia remained the unknown continent.

The three books of the *Geographical Guide* about Asia describe it from west to east. Book 5 (the first four maps of Asia) covers Anatolia, the territory northeast of the Black Sea and east of Map 8 of Europe, and east to the Caspian Sea, as well as Levant and Mesopotamia. Book 6 (Maps 5–9 of Asia) catalogues the lands south of the Caspian and east to India, and include the Arabian Peninsula and points east and north of the Caspian. Book 7 (Maps 10–12) includes the Indian peninsula and east and north to the limits of the inhabited world, as well as Taprobane (modern Sri Lanka).

Map 1 of Asia

Ptolemy did not have a general name for the territory covered by Map 1 of Asia, and defined it in terms of the several Roman provinces in the region (Pontos and Bithynia, Asia, Lykia, Galatia, Pamphylia, Cappadocia, and Cilicia) (*GG* 5.1.1–7.12). A separate section was devoted to Lesser Armenia. His lack of an overall name may have been to avoid compounding the confusion implicit in the varying meanings of the name "Asia," which could mean the entire continent, the region covered by Map 1 (generally called Anatolia or Asia Minor today, names from a later era than Ptolemy), and the Roman province on the Aegean (which Ptolemy called *Idios Asia*, or "Specific Asia." By avoiding any use of "Asia" for the region that is modern Turkey, he did his best to clarify these overlapping toponyms.

All the boundaries of Map 1 of Asia except the eastern were defined by coastlines: the Pontos Euxeinos (Black Sea) in the north, the Aegean and (to its south) the Ikarian and Myrtoan Seas in the west, and the Mediterranean in the south. Needless to say this last was not a term Ptolemy used, but defined it by a series of local seas (Karpathian, Rhodian, Lykian, and Pamphylian), from the west to the Kilikian Channel between the mainland and Cyprus and, to its east, the Gulf of Issos, the northeastern corner of the Mediterranean.

The eastern boundary of the map was more problematic. The Euphrates River defined much of it, but was not available at the northern and southern extremities. In the north a line followed the theoretical eastern boundary of the province of Cappadocia, running from Sinibra (modern Erzurum) on the upper Euphrates along mountains to the Black Sea at Sebastopolis. In the south, the boundary also followed mountains from a point on the Euphrates at the northern edge of Mesopotamia to the northeast corner of the Mediterranean, along the theoretical northern edge of Syria.

Pontos and Bithynia

The northwestern corner of Anatolia was the province of Pontos and Bithynia (or Bithynia and Pontos: the name varied) (*GG* 5.1.1–15). This had been Roman territory since 74 BC, when its last king, Nikomedes IV—the final representative of a dynasty that had been established at the beginning of the third century BC—willed the kingdom to Rome. The dual name of the province was due to the lengthy and complex entanglement of the region with the kingdom of Pontos just to the east of Bithynia; that kingdom had ended with the death of Mithridates VI the Great in 63 BC, whose dynasty may have been founded as early as the late fifth century BC and which dominated different regions of northern Anatolia. After 63 BC control of Pontos varied between the Romans and local rulers, with the territory passing permanently to the Romans in AD 64, but as part of the province of Galatia. Ptolemy's map reflects this status, but he chose to use the older name for the region of Bithynia—Pontos and Bithynia—even though it was essentially obsolete by the second century AD,[5] and Ptolemy's province includes only the western portion of historic Pontos.

Ptolemy's Pontos and Bithynia is bounded on the north by the Black Sea, and the Bosporos and Propontis on the west. On the east a north-south line marked the edge of Galatia, running from the city of Kytoros (at modern Kidros) south to a point at the northeast corner of the province of Asia. Then the southern boundary headed slightly north of west to the Propontis, oriented so that it would pass over Mt. Olympos (here called "Mysian" to distinguish it from the more famous mountain in the northern Greek peninsula), reaching the coast at the mouth of the Rhyndakos River.

Within Ptolemy's province there were about 30 towns and a few topographical features. Most of the towns reflect the period of the Bithynian kingdom, whose urban centers, such as Prousias and Nikomedeia, survived into the Roman period. But there is also some reflection of the Roman presence, with towns such as Iuliopolis, Klaudiopolis, and Flaviopolis, the latest identifiable toponym, at modern Gerede in the northeastern portion of the province.

Several coastal points, rivers, ethnic groups, mountains, and islands are also plotted, but the toponymic detail of the province is relatively sparse. Ptolemy did note the highest mountain, Olympos (modern Ulu Dağ, at 2543 m.), and the longest river, the Sangarios (modern Sakarya, 824 km. long). One toponym of curious interest is Libyssa, near modern Tavşancil on the Propontic coast), the estate of the former Carthaginian commander Hannibal, given to him by King Prousias I of Bithynia and where he retired and eventually committed suicide in 183 BC.[6] The fact that the place was still remembered 300 years later may have meant that it had particular local significance as the probable burial place of the great leader.

Specific Asia

South of Pontos and Bithynia was "specific" (*idios*) Asia, the Roman province (*GG* 5.2.1–34). This was a region of immense antiquity, including the famous districts of the Troad, Lydia, Ionia, and Karia, and the cities of Pergamon, Ephesos, Sardis, Miletos, Halikarnassos, and many others. The original province was established after the end of the Pergamene dynasty in 133 BC. Over the following years it was expanded south to the Mediterranean and included the coastal islands from Tenedos in the north to Rhodes in the south, in particular the large eastern Aegean islands of Lesbos, Samos, and Chios. As with much of Anatolia, the provincial boundaries were constantly changing and evolving, but Ptolemy showed the province at its largest extent. Dozens of toponyms were plotted, but these are mostly ancient; the Roman presence was hardly acknowledged. Other features include the great rivers of the region (the Kaikos, Hermos, Kaystros, and Maiandros), and the famous mountains, from Ida in the north to Sipylos, Tmolos, and Mykale in the south.

One might expect that this region would be among the most accurately described in the *Geographical Guide*, since it had been intensively studied geographically for hundreds of years before Ptolemy, and was the very area where Greek mapmaking originated, at Miletos in the sixth century BC.[7] But oddly Ptolemy's map contains major geographical errors in and around Lydia, and it is astonishing that famous Lydian locales should be so poorly placed. The problems center on Sardis, certainly one of the best-known places in the region (*GG* 5.2.16–17). The ancient Lydian capital is placed south of the Kaystros River, with Mt. Tmolos north of the city and near Smyrna. In actuality, Sardis is north of the Kaystros, and the mountain lies between the river and the city. Sardis itself lies on the Paktolos River, which is placed nearly a degree of latitude north of the city. Mt. Sipylos is located east of the Paktolos (it is actually well to the west), and far away from the nearby town that takes its name, Magnesia at Sipylos.

It is not unusual for there to be problems of orientation in portions of the maps, especially in remote areas. But the errors in Lydia are unusually egregious, all the more so because it is a well-known region. It is possible there was some major manuscript error in this portion of the *Geographical Guide*, but in fact these problems are impossible to explain.

Lykia

Southeast of the province of Asia, on the Mediterranean coast, was Lykia (Roman Lycia) (*GG* 5.3.1–9). Its status varied from independence to domination by the major powers, but in AD 43, after over a century of autonomy, it became a Roman province. At that time, or somewhat later (perhaps in AD 74) it was combined with Pamphylia, the territory immediately to its east, forming a joint province. This was its status when Ptolemy wrote, but he chose to treat the two regions as separate entities.

Ptolemy's Lykia was a small region with the province of Asia to its west and north, and Pamphylia on the east, which was separated from Lykia by the Masikytos Mountains (modern Alaca Dağı), running north from the coast. There is no evidence of a Roman presence on the map; the 36 towns and villages represent the ancient Lykian demography as first established in the Bronze Age.[8] The major stream is the Xanthos River (modern Esen Çayi), which, although short (120 km.), produces the only significant floodplain in the territory, with Xanthos (modern Kinik) at its mouth. It was already prominent in early times and was the largest city in Lykia, notable for its impressive remains; Xanthos and its fertile surroundings were the heartland of ancient Lykia. Several offshore islands were also considered part of the territory.

Galatia

East of Pontos and Bithynia and Asia was Galatia (*GG* 5.4.1–12), which Ptolemy defined in terms of the Roman province established in 25 BC with the death of its last king, Amyntas. Although Anatolia generally was a region of ancient peoples whose origins were in prehistoric times, the Galatians were newcomers and the latest ethnic group (other than the Romans) to occupy the territory. In 278 BC migratory groups of Kelts, some of which had attacked Rome a century previously, arrived from western Europe and settled in north central Anatolia, a region that historically had been Phrygia and western Cappadocia. They came to be called "Galatoi," or "Galatians," the Greek version of their indigenous name, usually Galloi (Gauls) or a variant thereof.[9]

The Roman province formed after the death of Augustus originally comprised the territory of the Galatians, but over the next century it was expanded: north into Paphlagonia and the western part of Pontos, a result of the lengthy reorganization of the region in the century after the death of Mithridates VI, and south and east into Cappadocia and northern Pamphylia. This expanded province is what Ptolemy based his map on; he does not seem to have incorporated some territorial adjustments of the late first and early second centuries BC.[10]

Thus Ptolemy's province is a north-south strip between Pontos and Bithynia and Asia on the west and the province of Cappadocia on the east. A number of ethnic groups are located, including the three Galatian divisions (Tektosagai, Trokmoi, and Tolistobogoi), who may have retained some individual identity even into the second century AD. Otherwise there is an abundance of towns—nearly 70 in all—on the map, some of which are ancient, and others that represent Hellenistic (Antiocheia, Laodikeia) and Roman (Pompeiopolis, Germanikopolis, Klaudiopolis) foundations. None of these is later than the mid-first century BC. The major cities of Galatia were Pessinous (at modern Balıhisar) in the west, an ancient cult center, Sinope, the capital of the Mithridatic kingdom of Pontos, and, in Roman times, Ancyra.

Morphological features are few, but an unusual one of particular interest is what Ptolemy called "Burned Laodikeia" (*GG* 5.4.10). "Burned" (Katakeukamene) was originally a local toponym applied here to a city name, a recognition of the intense volcanic activity of the southwestern part of the province.[11] The region, where vulcanism is still apparent today, included frequent and devastating earthquakes.

The longest river of Anatolia crossed Galatia in its lower course. This was the Halys (modern Kızıl İrmak), flowing for 1355 km. and emptying into the Black Sea about 110 km. east of Sinope. It had an important cultural profile, believed to be the division between the Aegean and eastern worlds, a view established by the sixth century BC.[12]

To the west of Sinope was another important topographical location, Cape Karambis (modern Kerembu Burnu), the northernmost point on the Black Sea coast of Anatolia and the sailing place for crossing the sea to its northern shore, which was possible without being out of sight of land. Otherwise Ptolemy paid little attention to the topography of Galatia, naming only two mountains (Oligas in the north and Dindymos in the west), with little sense of their extent. But an odd toponym is the

Ridge (Lophos) of Kelainai in the southwest, presumably one of the volcanic summits in the vicinity of the ancient center of Kelainai (at modern Dinar). Although the city was important from at least Persian times,[13] there is no obvious reason why this seemingly minor topographical feature should have been singled out by Ptolemy.

Pamphylia

To the east of Lykia and south of Galatia was Pamphylia (*GG* 5.5.1–10). It was more of a region of various ethnic groups that a cohesive territory, as demonstrated by its name, "All Peoples." It tended to be appropriated by the major powers. In Roman times it was originally part of the province of Asia, but was attached to other provinces until becoming part of Lycia in AD 43. Ptolemy, however, chose to consider it a distinct region.

With little independent history its demographics reflected the presence of the surrounding powers, although the map shows no particular indication of Roman involvement. Ptolemy listed over 40 towns; the latest are Seleukid foundations (Antiocheia and Seleukeia). Except for four short rivers flowing south into the Mediterranean (Katarrhaktes, Kestros, Eurymedon, and Melas), which created a triangular coastal plain where the ancient cities of Perge and Aspendos were located (both preserving extensive remains), Ptolemy indicated no morphological features in this small territory.

Cappadocia

East of Galatia and Pamphylia was the region Ptolemy described as Kappadokia, or Cappadocia (*GG* 5.6.1–18). This was an ancient district of Anatolia, which originally included everything from the Black Sea almost to the Mediterranean, but in Hellenistic times and with the rise of the Mithridatic dynasty of Pontos it was restricted to the southern portions. The last king of an independent Cappadocia was the scholarly Archelaos I, who died around AD 17, after which his territory was provincialized. Further boundary adjustments and the end of the dynasties of Lesser Armenia and Pontos meant that by the end of the first century AD the Roman province stretched north to the Black Sea and east to the Euphrates. Ptolemy's map shows the province at its largest extent, south to the border of Kilikia (Cilicia), and then a boundary along the Amanos Mountains to the Euphrates at the northern edge of Mesopotamia. Its eastern boundary ran north along the river to the vicinity of Sinibra (at modern Erzurum), and northeast to the Black Sea on its east coast, an extremity of the province more theoretical than real.

The Roman province as defined by Ptolemy was a particularly diverse collection of ethnicities and petty states which circulated in and out of Roman control. In the north was the heartland of ancient Pontos, which became Roman after the death of Mithridates VI in 63 BC, but was returned to an independent dynasty by the triumvir Marcus Antonius in the early 30s BC. Yet it became Roman again in AD 64, but was attached to Galatia.[14] In time the two regions were separated. Lesser Armenia also varied between independence and provincial organization. Eventually, during the reign of Trajan, the entire region was consolidated into a sprawling Cappadocian province.

It was divided into a number of administrative districts (*strategiai*), which to some extent replicated the ancient historical and ethnic divisions (*GG* 5.6.12–18). There seem to have been 10 of them, created under the indigenous monarchy, probably in or before the first century BC.[15] The Romans inherited this organiza- tion (although Ptolemy listed only seven), and made few if any changes, using the ancient city of Mazaka—renamed Kaisareia by Archelaos in the first century BC—as their provincial administrative center.

Greater Cappadocia, as one of the most eastern regions of the Roman empire, had an unusual history, since it was more connected to the world of the east than that of the Mediterranean. Much of its history had been established in Hittite and Assyrian times, and significant remnants of both cultures remained into the Roman period. In particular, there were a number of religious states that were ruled by hereditary priest-kings or priest-queens with royal powers, which often still func- tioned within the Roman province. Ptolemy noted a number of these, such as the two named Komana, one in Pontos and the other in traditional Cappadocia. Oth- erwise the province consisted of a large number of towns and villages and exten- sive rural areas. The urbanization reflected the efforts of the indigenous dynasts: Amaseia in Pontos was founded by Mithridates I in the early third century BC and remained an important city. At times there was recognition of the Roman presence, but this resulted in few changes to the historic demography and was largely lim- ited to renaming of ancient locations. Kabeira in Pontos, one of the temple states (at modern Niksar) became Sebasteia or Sebastopolis in the early first century AD through the efforts of the Pontic queen Pythodoris: the town continued as a temple state, Greco-Roman city, and royal capital. As was often the case, the imposition of a new culture had hardly any effect on the old established one.

Ptolemy paid little attention to the complex topography of provincial Cappado- cia. The Halys River had its source in the region, and some of the other major rivers (the Iris, Thermodon, Melas, and Kydnos) were also plotted, with greater or lesser degrees of accuracy, although one of the longest and most important, the Lykos, did not appear on the map. The many and complex mountain ranges of northern Cappadocia were mostly ignored; oddly the only one located was the Skordiskos, probably noted merely because it was believed to be the source of the Iris and Ther- modon, not true in either case, and in fact the Skordiskos is only a minor ridge in southeastern Pontos. The major range of the province, duly recorded by Ptolemy, is the Antitauros, plotted somewhat accurately across the southern half of the region and heading east into Armenia.

The map has the inevitable inaccuracies. Amaseia, which lies in a spectacular location in the deep gorge of the Iris River, is placed over a degree of longitude to its west. Yet this type of error is not egregious, and the only region where Ptolemy seems to have had serious difficulty was in the far northeast of the province, where the Black Sea makes its turn to the north. Here there are a number of towns on the coast and several rivers, including one named the Lykos, which, unless it is a second Lykos, is badly misplaced far to the east of the more famous river with that name. Moreover, the northern limit of the province is at the mouth of the Apsorros River, the modern Tchorokhi estuary in Georgia.[16] This is reasonable, but Ptolemy's map adds one additional name, just to the north, Sebastopolis. There is a remote possibility that an otherwise unknown town of this name was in the region, but localities named after the emperor (in this case Augustus) tended to be significant places, and, moreover, there is a famous Sebastopolis farther north along the coast at modern Sukhumi (*GG* 5.10.2), so named by queen Pythodoris of Pontos and Colchis at the location of the ancient trading center of Dioskourias.[17] Ptolemy knew of the dual name, but it would seem that the toponym Sebastopolis has been transferred by error to the margin of Map 1 of Asia, duplicating erroneously its proper location on Map 3 of Asia. Even though Sebastopolis was outside Roman territory, any Roman presence there would have been administered from Cappadocia, the probable reason for the error.[18]

Lesser Armenia

Although in Ptolemy's day Lesser Armenia was part of the province of Cappadocia—its last independent ruler, Aristoboulos, had died in AD 72—Ptolemy considered it as a separate unit (*GG* 5.7.1–12). Moreover, his view of Lesser Armenia went far beyond the traditional boundaries of that territory and included several minor states surrounding it, especially to its south, which had been incorporated into Cappadocia. The northernmost of these was Orbisene, with Orsene to the south, and then Melitene, Kataonia, Mourimene, Laviansene, and Aravene. The last four evolved into Roman administrative districts, but all were minor states and had little association with Lesser Armenia. Topographically, Ptolemy's Lesser Armenia was bisected by the Antitauros, heading east into Armenia proper, and the Melas River (modern Tohma Su), which flows east across the center of the region, emptying into the Euphrates. There is no obvious reason why Ptolemy attached them to Lesser Armenia, or, why, in fact, he considered Lesser Armenia, which was a portion of Cappadocia, as a different district. It may be nothing more than that he thought the province of Cappadocia was too sprawling and complex to be considered in a single chapter.

Cilicia

The final region on Map 1 of Asia is Cilicia (Ptolemy's Kilikia), the southeastern portion of Anatolia, lying between Pamphylia, Cappadocia, Syria, and the Mediterranean (*GG* 5.8.1–7). This was some of the roughest territory in Anatolia, especially the western portion where deep and rugged canyons bring rivers down from the interior highlands to the sea. In fact, this region was called Kilikia Tracheia, or Rough Kilikia (*GG* 5.8.5). To the east the country was more benign, with a large coastal plain formed by the Pyramos River (modern Ceyhan Nehri); this was called Kilikia Pedias, or Level Kilikia.

The region was settled by Greeks at an early date, and in the Hellenistic period tended to be disputed between the Ptolemies and Seleukids, but it generally lacked

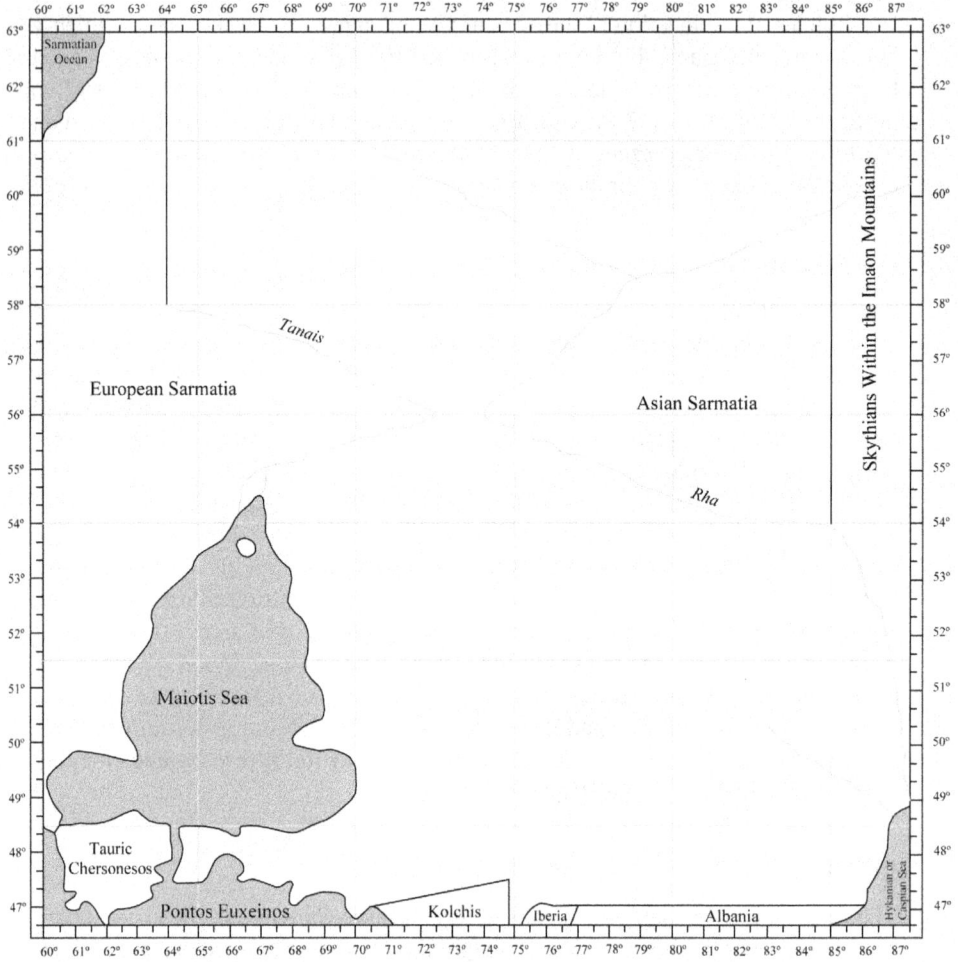

Map 6.2 Map 2 of Asia

any central government, although the temple state of Olbe retained power and influence until the late first century BC. In the early part of that century the Romans, recognizing the uncertain nature of the territory, and also its role as a haven for piracy, began to exert a presence, eventually with provincialization. Nevertheless there were various boundary adjustments and a shifting nature of Roman control alternating with periods of local dynasties, with no permanent province until late in the first century AD.[19]

Over 30 towns were plotted in Cilicia, from the ancient settlements of Tarsos and Adana—whose origins were in Hittite and Assyrian times—to a large number of Hellenistic foundations and a few instances of Roman renaming (Olbe became Diokaisareia in the early first century AD). The two most important rivers were the Pyramos in the east (the modern Ceyhan, 509 km. long), and the Kydnos (modern Berdan, 123 km. long) in the west, less important for its hydraulic features than its association with Kleopatra VII. There is a high density of navigational points plotted on the Cilician coast, necessary for sailing along this popular shipping lane from the Levant to points west. With his description of Cilicia, Ptolemy has completed Map 1 of Asia.

Map 2 of Asia

Sarmatia in Asia

Map 2 of Asia is devoted entirely to the Asian part of Sarmatia (*GG* 5.9.1–32), the European portion having been examined in Map 8 of Europe. Sarmatia was a general toponym used by Ptolemy for the territory north of the Black Sea, named for the nomadic peoples who had been known to the Greek world since at least the fifth century BC.[20] The division between the European and Asian Sarmatians was, as expected, that between the two continents, which had been established at the Tanais River (modern Don).[21] The source of the Tanais was not known (the river is 1870 km. long with headwaters southeast of Moscow), but by the seventh century BC traders had gone at least 400 km. upstream, since Greek pottery of that period has been found near Krivoroshie, which is probably roughly where Ptolemy placed the headwaters of the river.[22] From there he established a line due north, continuing to divide Asian and European Sarmatia. This terminates at 63 degrees north latitude, Ptolemy's standard for the northern limit of civilization. The Baltic (Ptolemy's Sarmatian Ocean) lies just to the west, but there is a gap of about two degrees of longitude between the line and the ocean that is not reconciled. To the north are "unknown" lands.

Thus the western boundary of Map 2 of Asia (and thus the division between Europe and Asia) extends south from this arbitrary point to the perceived source of the Tanais, and then along the river to its mouth at the city of Tanais. The boundary then passes through the Maiotic Lake (modern Sea of Azov) and the Kimmerian Bosporos (modern Strait of Kerch) into the Black Sea. This is also the eastern limit of Map 8 of Europe.

The southern boundary of Map 2 of Asia follows the north coast of the Black Sea to the mouth of the Korax River (perhaps the modern Byzb' but this is not certain) and then inland on a line across the Caucasus Mountains to the Caspian Sea. The line is at approximately along 47 degrees north latitude but is not straight, and is affected by the northern boundaries of the territories represented to its south on Map 3 of Asia (Kolchis, Iberia, and Albania). It joins the Caspian at the mouth of the Soanas River (perhaps the modern Sundhza). The eastern boundary of Map 2 of Asia follows the western shore of the Caspian to the mouth of the Rha River (modern Volga), and then continues up the river to 54 degrees north latitude, where Ptolemy assumed the river turned sharply west, although such a point cannot be identified. From this point the eastern boundary is a line running due north to 63 degrees north latitude, the assumed limit of habitation. East of this boundary—whether along the lower Volga or the line to the north—were the Skythians, represented on Map 7 of Asia.

The southern portions of the map, along and near the Black Sea, reveal a heavily populated region, especially on the peninsula immediately east of the Kimmerian Bosporos between the Black Sea and the Maiotic Lake. This had been an area of Greek settlement since early times, and in Ptolemy's day it was the Roman allied kingdom of Bosporos, centered at Pantikapaion west of the Bosporos. Its sphere of influence included most of the littoral of the Maiotic Sea and east along the Black Sea for some distance; the western (European) portions of the kingdom are represented on Map 8 of Europe. The rest of the territory on Map 2 of Asia—north of the Greek settlements—was wild and remote, occupied by various ethnic groups that traded with the Mediterranean world but seem to have had no urbanization beyond a short distance up the Tanais, as far as Exapolis, perhaps a hellenized confederation of six indigenous villages. Beyond Exapolis only indigenous peoples, rivers, and mountains were known.

There are 15 towns, a number of topographical points, and several rivers represented on the peninsula east of the Bosporos. The most important city was Phanagoreia, on an inlet of the Bosporos and founded in the sixth century BC.[23] A sanctuary of Achilles at the east side of the northern exit of the Kimmerian Bosporos is another demonstration of the prominence of the hero in this region, such as his racecourse west of the strait.[24] Ptolemy's sanctuary has not been located.

Extending north from the peninsula is the Maiotic Lake, which spread from the Bosporos to the mouth of the Tanais River. The lake is oriented due north-south when actually it is tilted west to east (Tanais city is 3 degrees east of the longitude of the Bosporos, not on essentially the same longitude as Ptolemy would have it). The lake is also too large on the map, a fine example of the tendency to elongate cumulative distances. Along its east side are a succession of towns, topographic points, and river mouths, almost evenly spaced, the record of a sailing route to the city of Tanais at the head of the lake and the mouth of its homonymous river. The city (at modern Nedvigavka) was probably founded in the third century BC as an outpost of the Bosporanian kingdom and as the trading access point to the vast northern hinterland.[25]

Several mountain ranges were located in Asian Sarmatia. Across its southern edge was the massive Caucasus system, spreading from the Black Sea to the Caspian Sea and rising to 5642 m., the highest mountains in the Mediterranean world. Greeks had known about them as early as the sixth century BC and they had been popularized by Aeschylus in his *Prometheus Bound*.[26] Although traders and merchants had probably probed the range for a long time, the first systematic

Greco-Roman penetration was by Pompey the Great and his chronicler Theophanes in 63 BC.[27] This expedition included identification of some of the passes, one of which was probably the Sarmatian Pass of Ptolemy, not easily identified today.

Four other mountain ranges appear on the map. The Korax Mountains are a western extension of the Caucasus, near the Black Sea. To the north of the Caucasus are the Keraunia and Hippika. The generic name of the former ("Thunderer") indicates that it was identified by Greeks, perhaps by Theophanes, and may be the Dagestan Mountains at the northeast end of the Caucasus, rising to 4470 m. The Hippika are more problematic but perhaps are the low hills east of the upper Maiotis Lake.[28]

The fourth range on the map is the Hyperboreian. The Greek descriptive name ("Beyond the North") shows its mythical status. The name is usually an ethnym, referring to far northern peoples who seem to have been known to Homer and Hesiod.[29] Perhaps connected conceptually were the Rhipaian ("Tossing") Mountains, also in the far north. Their existence was predicated on a cosmological theory that mountains were essential in the north to hide the path of the sun.[30] But no such mountains exist, regardless of name, and there is no reason to see them as anything other than something mythical that became institutionalized in geographical theory.

The two major rivers of the northern portion of Map 2 of Asia were the Tanais (modern Don) and Rha (modern Volga). Although there was little detail about their courses, enough was known to realize that they approached each other, the Tanais coming from the southwest and the Rha from the southeast, before sharply turning (as one goes upstream) to the northwest and northeast respectively. The two rivers are only about 75 km. apart (at modern Volgograd), a fact clearly depicted on the map. Someone going upstream on the Rha from the Caspian would pass through the Land of Mithridates, perhaps more likely a region claimed at some time by a local dynast than any association with the famous dynasty of Pontos; the name Mithridates became common throughout the world east of the Mediterranean in Roman times. The upper Rha was believed to be composed of two large tributaries, both of which had their sources in the Hyperboreian range. Ptolemy placed their join at 58.5 degrees north latitude; the Rha is the western one, and the eastern one is probably the modern Kama, which joins the Volga south of modern Kazan. This is probably the most northern point in Asian Sarmatia reached by trade goods from the Mediterranean.

A final and curious point on the map is the existence of the Columns of Alexander, lying north of the Caucasus between the Rha and the Maiotic Lake. Alexander the Great came nowhere close to this region, but wanted to believe that he had crossed the Caucasus, and his people manipulated topography to make it seem that he had done so.[31] The Columns of Alexander—perhaps a topographical feature—may reflect a local feeling that Alexander had indeed crossed the mountains. It was a feature with a similar history as the Altars of Alexander in European Sarmatia (*GG* 3.5.26).

Map 3 of Asia

This map covers a broad swath from the Asian Sarmatians of Map 2 of Asia south to the northern limit of Mesopotamia and Assyria (*GG* 5.10.1–13.22). Much of it

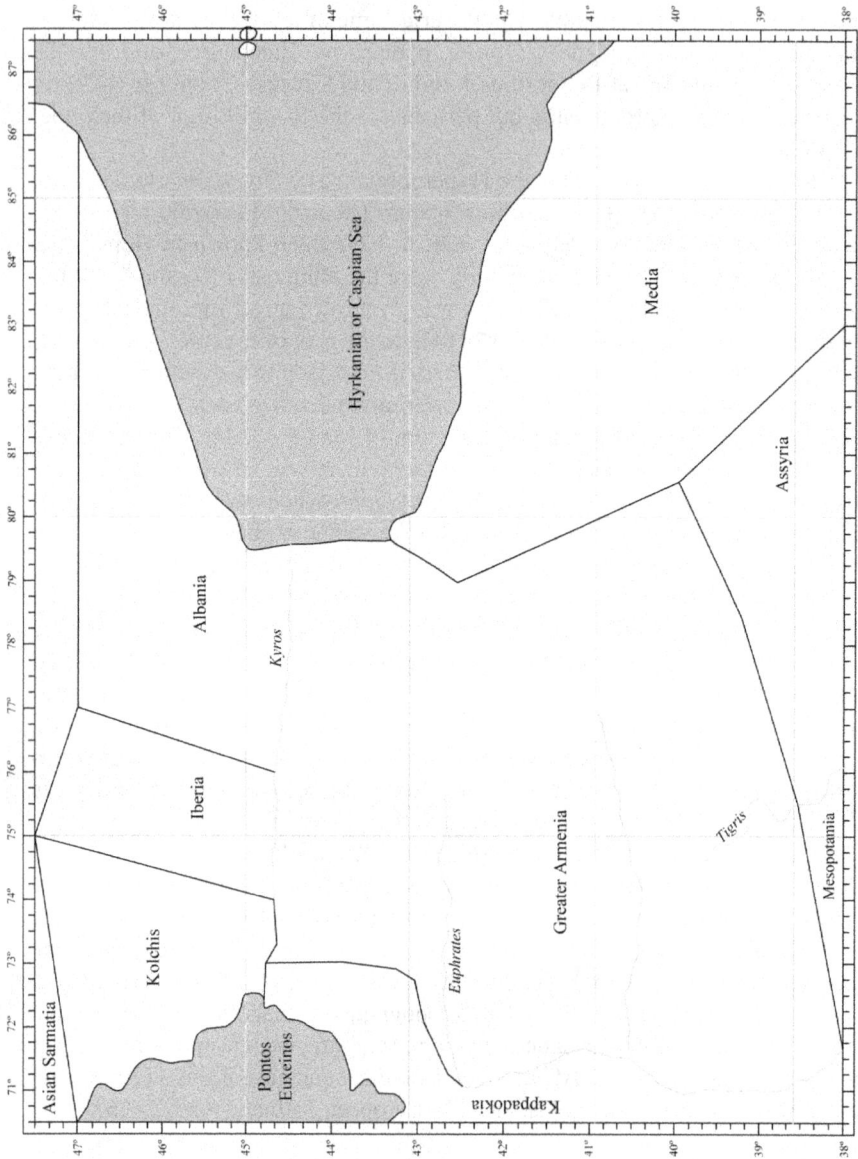

Map 6.3 Map 3 of Asia

is the region between the Black and Caspian (or Hyrkanian) Seas. Between them, from west to east, are the three districts of Kolchis, Iberia, and Albania, and to their south the large region of Greater Armenia.

The northern boundary of the map is the southern edge of Map 2 of Asia, an arbitrary line at 47 degrees north latitude from the mouth of the Korax River (where it empties into the Black Sea) across the Caucasus Mountains to the Caspian Sea at the mouth of the Soanas River. The western boundary of the map follows the coast of the Black Sea south from the Korax to the northern limit of Map 1 of Asia near Phasis and the mouth of its homonymous river (the modern Rioni). The boundary then follows the eastern boundary of Map 1 until reaching the upper Euphrates at Sinibra, and then runs south along the river, which is still the eastern boundary of Map 1, as far as the Tauros Mountains and the northern limit of Mesopotamia at Samosata (modern Samsat). From there it runs east and then north along various mountain ranges: first the Tauros, and then (east of the Tigris) the Niphates and Caspian Mountains, all part of the complex mountain systems in the borderlands of modern Turkey, Armenia, Iraq, and Iran. The line turns sharply north at the transition between the Niphates and Caspian ranges (also seen as the border between Armenia and Media). This is to account for the erroneous placement of the southern edge of the Caspian Sea at roughly the same latitude at the southern limit of the Black Sea: the former is actually 4 degrees south of the latter.

From where the boundary joins the Caspian Sea (a short distance south of the mouth of the Araxes—modern Aras—River), it follows the coast of the sea to the mouth of the Soanas River at the northeast corner of the map. But this line should be almost due south to north; rather it runs easterly with the mouth of the Soanas 7 degrees of longitude east of that of the Araxes. The Caspian, as shown on Map 7 of Asia, has its long dimension west to east, when it should be south to north. Although Ptolemy knew its complete coast, this failure to orient the sea properly may be due to an inability to collate information from several different sources of the Seleukid into the Roman periods.

Kolchis

Kolchis (*GG* 5.10.1–6), famous in Greek myth for the story of Jason, the Argonauts, and Medea, occupies the southeastern corner of the Black Sea coast, lying between the Phasis and Korax Rivers and inland to the Caucasus Mountains (essentially the western part of modern Georgia). Historically part of the Mithridatic kingdom, which at its peak in the early first century BC controlled most of the Black Sea coast, the weakening of that kingdom caused the Romans in the first century AD to attach Colchis to the province of Cappadocia.[32] This may be the reason some of the coastal data on Map 3 of Asia duplicates that on the Cappadocian section of Map 1 of Asia, perhaps not only reflecting two different sources but the fact that Kolchis was part of the province of Cappadocia when Ptolemy wrote. The Kolchian coast was dominated by the ancient and major trading centers of Phasis and Dioskourias (the latter known as Sebastopolis in Ptolemy's day, although both names were still in use), which was at modern Sukhumi. The Phasis River emptied into the sea at the homonymous city, and traders went upriver into the Caucasus. Few from the Mediterranean world had penetrated them except for the problematic expedition of Gnaeus Pompeius in the 60s BC, and some of Ptolemy's toponyms in interior Kolchis and the Caucasus may have come from the report of Ptolemy's chronicler, Theophanes of Mytilene,[33] although this is not acknowledged and Ptolemy's information for Kolchis beyond the coast is scant, limited to several rivers, a few interior settlements, and some ethnic groups.

Iberia

This was a land-locked territory east of Kolchis (*GG* 5.11.1–3), today the interior portion of modern Georgia. To Ptolemy its southern boundary was the Kyros (modern Kura) River, with Armenia to the south, but Iberia actually included the entire floodplain. The western and northern boundaries were part of the Caucasus, and the east an arbitrary north-south line between 75 and 77 degrees east longitude that separated Iberia and Albania. The territory of these prosperous agricultural peoples was more oriented on the Kyros than Ptolemy realized.[34] After minimal Seleukid control faded away, an indigenous dynasty rose to power. Pompeius penetrated the region in the 60s BC, but it remained at the fringes of the Roman world. Ptolemy knew of no geographical features beyond the bordering Kyros River and the Caucasus, and a handful of towns, including the centers of Artanissa and Armaktika.

Albania

Extending from the eastern edge of Iberia to the Caspian Sea, and north of the Kyros River, was Albania (*GG* 5.12.1–8). Its interior was the mountainous eastern summits of the Caucasus, which dropped to a coastal plain of varying width along the Caspian Sea east and north of the Kyros River. Albania extended north to the arbitrary boundary at 47 degrees north latitude that marked the southern edge of the Sarmatian territory, a line running west from the mouth of the Soanas River. The Albanians were outside the Greco-Roman world, and like many remote societies were presented in ideal pastoral terms. They may have been visited by the Seleukid explorer Patroklos in the 280s BC,[35] but were more extensively contacted by Pompeius in the 60s BC; a detailed report of his reconnaissance was provided by Theophanes and summarized by Strabo;[36] the former may have been Ptolemy's major source, although there is no explicit evidence of this.

Ptolemy's account of Albania included several rivers (the Albanos, Kasios, and Gerrhos) draining into the Caspian Sea between the mouths of the Kyros and the Soanas; none of these is specifically identified among the many rivers along this coast. As usual, these river mouths were access points to the interior. Over 20 villages are named, with two on the coast (Gangara to the south and Albana to the north), probably the major Albanian urban centers and trading emporia. None of the towns can be located with any precision, and in fact Ptolemy's map of Albania is problematic, showing no knowledge of the sharp southward turn of the Kyros River, which he presented as running due east between 44 and 45 degrees north latitude from its source in southwestern Iberia. It actually runs southeast from central Iberia and then heads southerly through 2 degrees of latitude before entering the Caspian Sea. The uncertain plotting of Albania is also reflected in the single other topographical feature cited, the Albanian Gates, placed on the border with Sarmatia some 6 degrees inland but also at the east end of the Caucasus. The Gates are certainly one of the passes through the Caucasus, and identification of them is probably due to Pompeius and Theophanes,[37] but the toponym Albanian Gates is otherwise unknown and has become confused with the better-known Caspian Gates, generally located where the easternmost extension of the Caucasus comes within the few kilometers of the Caspian Sea near modern Derbend.[38] The nomenclature and location of these "gates" through the Caucasus have been confused since the first century BC.

Greater Armenia

The remainder of Map 3 of Asia, over half its area, covers Greater Armenia (*GG* 5.13.1–22). The term "Greater" (Megale) refers to the heartland of Armenia, distinguishing it from the smaller territory to its west, known as Lesser (Mikra) Armenia (*GG* 5.7.1–12), a division created as part of the Seleukid organization of the region. In Ptolemy's time Lesser Armenia was part of the Roman province of Cappadocia, whereas Greater Armenia was an independent kingdom. Ptolemy defined the territory as everything south of the Kyros River as far as the northern limits of Mesopotamia and Assyria, situated between Cappadocia to the west and Media to the east, but including a small portion on the Caspian seacoast just south of the Kyros. This was an expansive conception of the territories, and within it there were a number of localized districts and ethnic groups, such as Sirakene, Sophene, and Gordyene, whose relations with the Armenian central government varied over time.

Armenia was a rugged territory that included the famous Mt. Ararat, at 5165 m. the highest point in Asia Minor and known from earliest times. The region was the home of several indigenous populations that developed in the Bronze Age, and the biblical mention of Mt. Ararat demonstrates that the district was well known to far distant peoples from an early date.[39] It eventually came under Seleukid control, and then was ruled by a prominent dynasty that had imperial ambitions, especially under Tigranes II (95–55 BC), who extended Armenian power to the Mediterranean. This dynasty, with varying amounts of Roman involvement, was still in power over a much-reduced territory when Ptolemy wrote.

Ptolemy's map is surprisingly devoid of topographical features in this rugged area; it is possible that the extreme nature of the landscape defied description. Only four mountains are listed: Gordyaia in the south central part, Abos to its northeast, Paryadres in the center of the territory, and the eastern end of the Anti-Tauros in the west, which continues across the Euphrates into Cappadocia. These mountains are well placed, but the others are problematic. Mt. Paryadres is an enigma, because its conventional location is part of the Pontic mountains along the south shore of the Black Sea, rather than in Armenia. It should be on Map 1 of Asia, although there may be another range of the same name in Armenia. The location of the Armenian Paryadres suggests Mt. Ararat, but that peak could also be Mt. Abos to the south, which is actually a better position for the famous mountain. To the southwest was Mt Gordyaia; its location just east of Lake Thospitis suggests one of the high summits on the modern Turkish-Iranian border.

Lake Thospitis (modern Lake Van), one of three lakes in Armenia recorded by Ptolemy, is the largest lake in Turkey, and Ptolemy believed that it was the source of the Tigris River, a misconception from as early as the time of Eratosthenes.[40] The river is actually within a few kilometers of the lake, but not in the same watershed. To the northeast of the lake was Lake Arsessa, perhaps modern Erçek Gölü. The third lake was Lychnitis, in the northeast, probably modern Lake Seren just south of the Lesser Caucasus.

As usual, mountains seem to serve the function of providing river sources: the major river of northern Armenia, the Araxes (modern Aras) flows out of the Paryadres Mountains into either the Kyros River or directly into the Caspian Sea: Ptolemy had it going both places, which may reflect two different reports or varying courses due to changes in the flow. The Euphrates was also said to originate in the Paryadres. Regardless of how the Paryadres in Armenia should be identified, Ptolemy has placed the sources of the two great rivers near each other.[41] Ptolemy also plotted an unnamed river flowing west from Mt. Abos to the Euphrates which may be a doublet for that river.

The human presence in Armenia was well documented, with over 80 towns and villages and a number of localized ethnic groups. There are few details on most of these places, but the density of toponyms reflects that despite its ruggedness and remoteness, Armenia was well known to the Greco-Roman world.[42] The most important locality was Artaxata, near modern Aralik in northeastern Armenia, founded by King Artaxias I as his capital in the early second century BC in a bend of the Araxes River.[43] Also of importance were Arsamosata—the capital of Sophene—as well as Tigranokerta in southern Armenia, not precisely located but probably closer to the Tigris than Ptolemy placed it.[44] This was the expansive capital founded by Tigranes II in the early first century BC, but which was never finished and was largely destroyed by the Romans in 69 BC. In Ptolemy's day it survived merely as a village.[45]

Map 4 of Asia

This map covers a sprawling region from the eastern Mediterranean coast to the Tigris River. Cyprus is also included. Because this was a region of ancient civilizations, the density of toponyms is great, especially in the western portion. The map adjoins Maps 2 and 3 of Asia on its north at the borders of Cappadocia and Armenia, and Map 5 of Asia along the Tigris River. To the south is the head of the Persian Gulf, and then the southern boundary of Map 4 is plotted across undefined mountains that also mark the northern edge of Map 6 of Asia. The western edge is the Egyptian border and Map 4 of Libya in the south and the Mediterranean to the north, with the addition of Cyprus. Seven regions are defined: Cyprus, Syria, Palestina Judaea, Arabia Petraia, Mesopotamia, Arabia Eremos, and Babylonia.

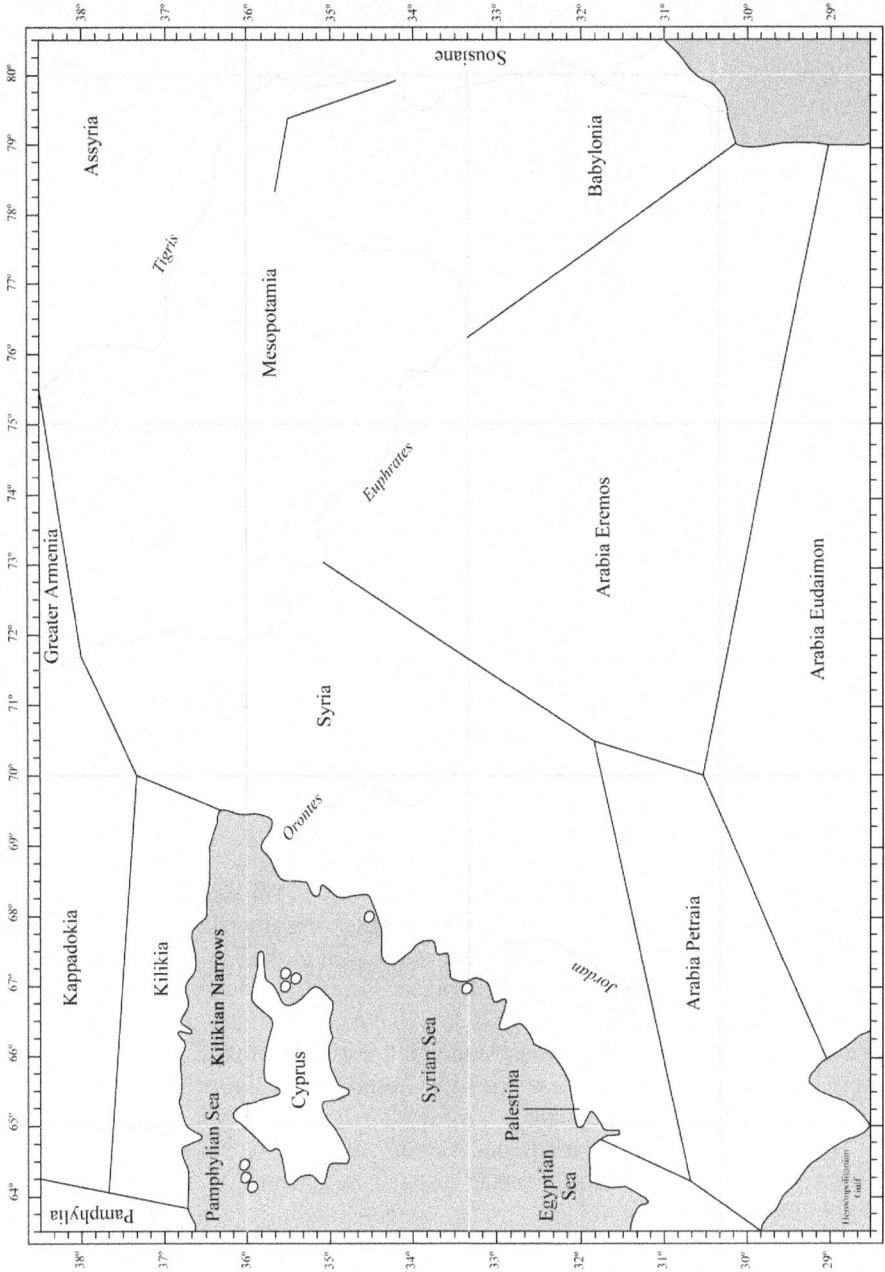

Map 6.4 Map 4 of Asia

Cyprus

The island of Cyprus (Kypros) had been a Roman province since the first century BC, and previously the seat of various indigenous kingdoms (*GG* 5.14.1–7). Its shape and location were accurately plotted by Ptolemy. Because it was an important locality on the sailing routes of the eastern Mediterranean, its coast is well defined, marked by 16 promontories and river mouths, most of which have Greek names. In addition there are several offshore islands. This is certainly information from a sailing manual, perhaps that of Timosthenes of Rhodes, who surveyed the Mediterranean for Ptolemy II in the early third century BC and whose work was known to Ptolemy.[46]

The four rivers appearing on the map—the Tetios, Pediaios, Lapethos, and Lykos—all originate from a central mountain known as Olympos, another example of geographical regularity. Cypriot Mt. Olympos, the high point of the island (1952 m.) is actually at the western end, but Ptolemy has moved it to the center of the island to provide a common source for the four rivers.[47] Cypriot rivers are short and low in volume; the longest, Pediaios, is only 98 km. Inclusion of them on the map is not so much to record their course but to note their mouths, something of importance to sailors, and their actual sources are somewhat more scattered than Ptolemy would have it.

There are 15 villages and towns plotted on the island, including ancient centers such as Salamis, Amathous, and Old and New Paphos. There seems to have been little if any change to the demography of the island in the Roman period.

Syria

The Roman province of Syria was created in 63 BC out of the remnants of the Seleukid empire (*GG* 5.15.1–27). It included not only the Syrian heartland, but an extensive array of ancient independent cities and local dynasties. Ptolemy defined this territory as extending east from the Mediterranean coast to the Euphrates and as far as its great turn to the east (the site of Alamatha) and the western edge of Mesopotamia (which lay across the river). The boundary then follows an arbitrary line southwest and then more westerly to the vicinity of Philadelphia (modern Amman), following the northern edge of the various Arabian territories. Then it curves northwest and west to the Mediterranean, reaching the sea at the mouth of the Chorseas River (modern Nahal Daliya), just north of Caesarea Maritima. The northern boundary of Ptolemy's Syria leaves the coast at the Amanos Gates at the northeast corner of the Mediterranean and the eastern limit of Cilicia, and then curves along the Amanos Mountains to Samosata on the Euphrates, separating Syria from Cilicia and Cappadocia. Given the ethnic and political diversity of this territory, much of the inland boundaries are speculative, and Ptolemy has included within his Syria such ancient states as Palmyra, Batanaia, and the Phoenician cities, where Roman control was limited to non-existent. Famous rivers were also depicted, most notably the Orontes (modern el-Asi) and the upper Iordanos (Jordan), whose origin at the ancient shrine of Caesarea Panias (modern Banias) is accurately depicted; in an unusual editorial note, Ptolemy commented that this was in fact the source of the river (*GG* 5.15.21).

In depicting rivers and mountains, even in this well-known area, Ptolemy again fell victim to the need for geographical regularity, especially the desire to have the origin of rivers in mountains: the source of the Jordan is a rare exception. The large cave at Banias from which the river flows was a major topographical point in southern Levantine culture, and this was probably Ptolemy's reason for emphasizing it. The Jordan then flows due south to Lake Gennesarit (Sea of Galilee) and, upon leaving the lake, crosses into Palestina for the rest of its course.

But the other rivers in Syria are prime examples of the dangers of forcing them into the theory that rivers generally originate in mountains. The source of the Orontes is relatively correctly placed in the district of Laodikene, but also at the perceived eastern end of the Lebanon Mountains, which requires those mountains, which run southwest to northeast, to extend southeast to northwest. Ptolemy was aware that the other major range of southern Syria, the Anti-Lebanon, runs parallel to the Lebanon, so they too are wrongly oriented. Damascus, which in actuality lies south of the southern end of the Anti-Lebanon, is placed between the two ranges in Koile (Hollow) Syria, near the source of the Orontes, which is nowhere close to Damascus. It is difficult to understand such massive confusion in such a known region, with the result that here Ptolemy's map becomes almost incomprehensible. Other mountains, such as Karmelos (Carmel) and Kassios, are more accurately placed, perhaps because they have no associated rivers, and the Chrysoras River (modern Nahr Barada) is essentially properly located, flowing from between the Lebanon and Anti-Lebanon past Damascus, and then losing itself in the desert.

Two other rivers appear on the map. The Eleutheros is a minor stream (78 km., the modern Nahr el-Kebir and the frontier between modern Syria and Lebanon), flowing from the north end of the Lebanon Mountains west into the Mediterranean; it was probably noted because of its historic role as a boundary between Seleukid and Ptolemaic territory. The other river is the Singas, flowing northerly from the Piera Mountains near Antioch and eventually reaching the Euphrates just below Samosata. Its lower portion (modern Göksu) is properly depicted, but the lengthy stream depicted on the map (in fact the longest river Ptolemy recorded in Syria) has no actual parallel, and its upper course is probably confused with the Labotas (modern Kara Su), which was not mentioned by Ptolemy but is the major northern affluent of the Orontes and which flows southwest across Syria.[48] The depiction of mouths of rivers again shows the indebtedness to sailing manuals, as well as coastal promontories; these are all spaced rather evenly along the Syrian coast and would provide important navigational markers, from the mouth of the Chorseas River in the south to the Rhosikos Promontory (modern Hinzir Barun) in the north.

Numerous towns and cities appear on the map (over 100 in all), many of which are ancient pre-Greek centers, but others, such as Antioch-on-the-Orontes, Alexandria-on-the-Issos, and more than one place named Laodikeia, reflect the Hellenistic demography. There are generally Seleukid names in the north and Ptolemaic ones in the south, since the territory tended to be divided (usually at the Eleutheros River) between the two kingdoms. Evidence for the Roman presence is represented by the camp of Legio XVI Flavia near Samosata and Legio III Gallica in the Kassiotis district near Rhaphaneai (modern Rafniye); the former legion had been established by Vespasian in the AD 70s and spent many years near Samosata, and the latter was in the region from at least the AD 130s; noting its presence may be one of the later entries in the *Geographical Guide*.[49]

Palestina Judaea

Ptolemy saw the terms Palestina and Judaea as synonyms (*GG* 5.16.1–10), which is anomalous to standard usage; normally, Palestina was the larger area, including not only Judaea but the ancient districts of Samaria and Galilee, all three of which, rather contradictorily, are shown as portions of the map of Palestina Judaea. Ptolemy's region is bounded on the north and east by Syria, on the south by Arabia Petraia, and on the west by Egypt and the Mediterranean. When the *Geographical Guide* was written, the area was largely the Roman administrative district of Syria-Palestina, ever since the end of the reign of the last indigenous king, Agrippa II, around AD 100.

In addition to the historic districts of the region—Galilee, Samaria, Judaea, and Idumaea—Ptolemy located approximately 40 towns and villages, most notably the Caesarea, or Kaisareia, of Herod the Great on the coast, and the ancient center of Jerusalem (Ptolemy's Hierosolymna) inland. Ptolemy called Caesarea, "Kaisareia of Straton," and thus it still retained a vestige of its pre-Herodian nomenclature, when it had been a modest anchorage founded by Straton I of Sidon in the first half of the fourth century BC, a demonstration of the survival of Hellenistic names even when theoretically anachronistic. On the other hand, Jerusalem was said to be "now called Aelia Capitolina," reflecting the renaming of the city by Hadrian during the disturbances of AD 132–135, and another late entry in the *Geographical Guide*.

Coastal features are rare on this map, largely because the shore is unusually smooth and straight, with few capes or promontories. The mouth of the Chorseas River north of Caesarea was noted, as were two harbors, that of Gaza at the south and Iamneia just to its north. Strangely Gaza itself, for many years the greatest trading center of the southern Levant, is not mentioned; in Ptolemy's world the port of Gaza was more important that the city itself, since it was here that camel caravans from the southern Arabian peninsula, Petra, and even Babylonia, would unload their wares—especially the valuable and exotic Arabian aromatics—for transfer to ships and to be sent all over the Mediterranean.[50]

Other than the Chorseas, the only other river depicted is the Iordanes (Jordan), which continues from Lake Gennesarit (the Sea of Galilee) in southern Syria south to Lake Asphaltitis, or the Dead Sea, which, as a strange phenomenon, had intrigued Greeks since the time of Alexander the Great.[51] Despite the ruggedness of the region and its many historic mountains, there are none on the map of Palestina Judaea.

Arabia Petraia

To the south of Palestina Judaea and Syria was Arabia Petraia (*GG* 5.17.1–7); the name is not so much "Rocky Arabia"—although this would be appropriate given the nature of its topography—but designating the portion of Arabia around the famous city of Petra. Ptolemy oriented his region from northeast to southwest—from Arabia Eremos in the east to the head of the Red Sea in the west—but the territory is actually a north-south strip from Bostra (usually considered part of Syria) running east of the Jordan, the Dead Sea, and the rift valley to its south, to the Elantic or Ailantic Gulf (modern Gulf of Aqaba). In addition much of the Sinai Peninsula was included. Part of this confusion is that Ptolemy has moved beyond the heavily populated regions near the Mediterranean coast into the desert territory across the Jordan. Even though the strip immediately east of the river was well known—Trajan had built a north-south highway through the area—farther to the east information tended to coalesce around the camel routes to southern Arabia and Babylonia. Yet even after the Roman arrival in the second century AD, Petra retained its status as a major trading emporium and crossroads.

Ptolemy recorded few features; there were no rivers in Arabia Petraia and the only mountain range was the Melana (Black) west of the Elantic Gulf, probably Mt. Sinai. Ptolemy also noted Cape Pharan (modern Ras Muhammed), the southern promontory of the Sinai Peninsula. The southern border of Arabia Petraia was marked by unnamed mountains that separated it from Arabia Eudaimon, the major part of the Arabian peninsula.

Within Arabia Petraia about 30 towns and villages were listed, most notably Bostra, Medaba, Elana, and Petra itself. Because of the error in orientation these are strangely placed: they are generally accurate relative to one another but in an east-west rather than north-south line. The data are probably based on trade routes, with the toponyms, probably stopping places, located along their corridors.[52] Bostra (modern Bosra), the Roman administrative capital, was noted as the home of Legio III Cyrenaica, which seems to have been placed there when the region became Roman territory at the beginning of the second century AD.[53] The remaining locations on the map are towns and villages scattered throughout the district. Except for the placement of the legion at Bostra, there is no record of any Roman presence; all the toponyms are indigenous. Again there is evidence of confusion: Gerasa (modern Jerash) is placed near the head of the Gulf of Aqaba and as such becomes one of the most southernmost towns in Arabia Petraia, but it is actually at the other end, between Bostra and the Jordan River. Mesada, if this is the famous Masada, west of the Dead Sea, is located near Bostra; it should be in Palestina Judaea. Needless to say, there is always the possibility that these towns that seem to be so badly placed are localities with similar names as the more famous ones, but this is unlikely.

Mesopotamia

Mesopotamia is the region immediately south of Armenia (*GG* 5.18.1–7).[54] The name derives from its position between the Tigris and Euphrates Rivers, and the toponym probably dates to the time of Alexander the Great.[55] The river system of Mesopotamia is unusually complex. Ptolemy believed that the Euphrates came to an end north of Babylon, where a canal—the Royal River—led to the Tigris, which is only a few kilometers away. Another stream, unnamed, headed south through Babylon and then terminated in several unnamed lakes near the ancient city of Orchoe (Warka). This is actually the Euphrates, but Ptolemy did not recognize it as such, ignoring the overwhelming evidence from throughout history that the river flowed through Babylon. Yet he may be excused for his anomalous interpretations in a region where extensive changes in the rivers, canalization, and siltation have long played a role and have made the river courses difficult to determine from any period. Nevertheless, joining the two streams via the Royal River gave Ptolemy a neat southern limit to Mesopotamia.

As noted Mesopotamia lay south of Armenia. On its easterly side (the left bank of the Tigris) was Assyria, and on the western side (the right bank of the Euphrates) were Syria in the north, and, downstream, Arabia Eremos and then Babylonia, which is on the right bank of both the lower Euphrates and the Royal River.

There are two rivers (in addition to the Tigris and Euphrates) and two mountains in Mesopotamia. The Chaboras (or Aborrhas, modern Khabur) is the largest tributary of the Euphrates in Mesopotamia (456 km.). To Ptolemy it originated in the Masion Mountains, which were uplands of northern Mesopotamia that lie south of the Tigris, although Ptolemy presented them as an isolated summit. To the east is the other river of Mesopotamia, the Saokoras, which cannot be identified. East of this river is another mountain, Singaras (modern Sinjar), which is an isolated anticline rising to 1463 m. in northern Iraq. It is a rare uplift in the plains of Mesopotamia, and thus has long been a defensive location.

Over 60 towns and several ethnic groups are depicted, from exceedingly old sites such as Sipphara (ancient Sippar, at modern Tell Abu Habba southwest of Baghdad) and Nisibis (modern Nusaybin on the Syrian-Turkish border), to Hellenistic foundations such as Nikephorion (modern Raqqah) and Seleukeia, normally Seleukeia-on-the-Tigris (at Tell Umar), the most important Hellenistic city in the region. In Roman times Mesopotamia was Parthian territory, and thus there is no evidence recorded of a Roman demographic presence, although one may note Karrhai (ancient and modern Harran), famous not only as the home of Abraham but where M. Licinius Crassus was soundly defeated by the Parthians in 53 BC.

Arabia Eremos

The large wedge between the Euphrates and Syria was called Arabia Eremos (*GG* 5.19.1–7), best translated as Desolate Arabia. This was a triangular and lightly populated region that lay north of Arabia Eudaimon, or the Arabian Peninsula. Its western boundary was the line marking the eastern extremity of Syria and Arabia Petraia, which is arbitrary and probably determined by the eastern limit of the more populated areas of Syria and the Levant. The northeastern boundary was the Euphrates, until Ptolemy had the river swing to the northeast toward the Tigris, whereupon the boundary became a line of unnamed mountains extending as far as the Persian Gulf. This probably marked where the land begins to rise west of the Tigris-Euphrates floodplain, or the western limit of Babylonia. The southern limit was a line separating Arabia Eremos from Arabia Eudaimon, running southeast to northwest between 29 and 30 degrees north latitude.

The phrase Arabia Eremos seems to have come into use by at least the third century BC.[56] The word *eremos* can have a variety of related meanings, all with the connotation of desolation or loneliness.[57] It can also mean "deserted" but that cannot be the case here, since the region was populated by various groups of Tent Dwellers

(Skenitai), the nomads who still inhabit the area. These were said to be aggressive in the north and more benign in the south, where it was probably to their advantage not to harass the caravans that traveled between Gaza, Petra, and Babylonia.[58]

There are no physical features recorded in Arabia Eremos, merely a number of localities that are probably oases or resting points on trade routes.[59] Their distribution shows the routes across the territory: there is a series of 11 in a row along the right bank of the Euphrates, all with indigenous names. Most of these are obscure watering holes, but the northernmost, Thapsakos, was at an ancient crossing of the river and where routes to Assyria and Babylonia divided. It is not located with certainty today, but is near the modern Syria-Iraqi border. The crossing was used by Alexander the Great but had been abandoned by the third century BC, surviving only as a historic name and a major topographical point in Eratosthenes' grid of the inhabited world.[60]

Most of the remaining toponyms in Arabia Eremos define routes across the southern part of the region, those between Petra or Gaza and Babylonia. There are about 20 localities on these routes and three ethnic groups, which indicates that this region was somewhat better known than the northern route. One of the western places, Artemita, suggests a Greek foundation, but no details are known.[61]

Babylonia

The remainder of Map 5 covers Babylonia (*GG* 5.20.1–8). To Ptolemy this was the lower (southern) portion of the plain of the Tigris and Euphrates, bounded on the north by Mesopotamia, on the east by Sousiane, on the south by the Persian Gulf, and on the west by the unnamed mountains—the western edge of the floodplain—that marked the beginning of Arabia Eremos. The layout of this region is peculiar because Ptolemy believed that the Euphrates (below the mouth of the Saokoras River) swung to the north and then around to the east, coming close to the Tigris. It then separated into three channels: first the Naarsares and then an unnamed stream that flowed through Babylon, which split from the Royal River. The Naarsares is the western branch of the Euphrates, rejoining the main stream between Babylon and Borsippa. The unnamed stream is the main channel of the Euphrates, but rather than reach the lower Tigris or the Persian Gulf, it disappeared into lakes in the Amardokaia district near Orchoe. The Royal River is the ancient connection to the Tigris, located where it and the Euphrates come to within a few kilometers of each other near modern Baghdad. Moreover, the lower Tigris splits into two streams which enter the Persian Gulf separately.

Obviously Babylonia has always been a region of intense hydraulic complexity, including the rivers themselves, canals that have been constructed over thousands of years, extensive marshlands into which rivers may disappear and emerge from, and siltation so intense that from the time of Alexander the Great to the mid-first century AD the coast had extended itself nearly 200 km.[62] Exactly what period is represented by Ptolemy's data for the head of the Persian Gulf cannot be determined.

Nearly 30 towns are shown in Babylonia, from the ancient cities of Babylon, Orchoe, and Borsippa, to the Parthian settlement of Vologaisia (at Abu Halafiya), a foundation of King Vologaesus I (ruled AD 51–76/80). The Greek and Roman presence was slight to non-existent, and Trajan's ephemeral province of Mesopotamia (AD 114–117) was too limited chronologically to be recorded by Ptolemy.

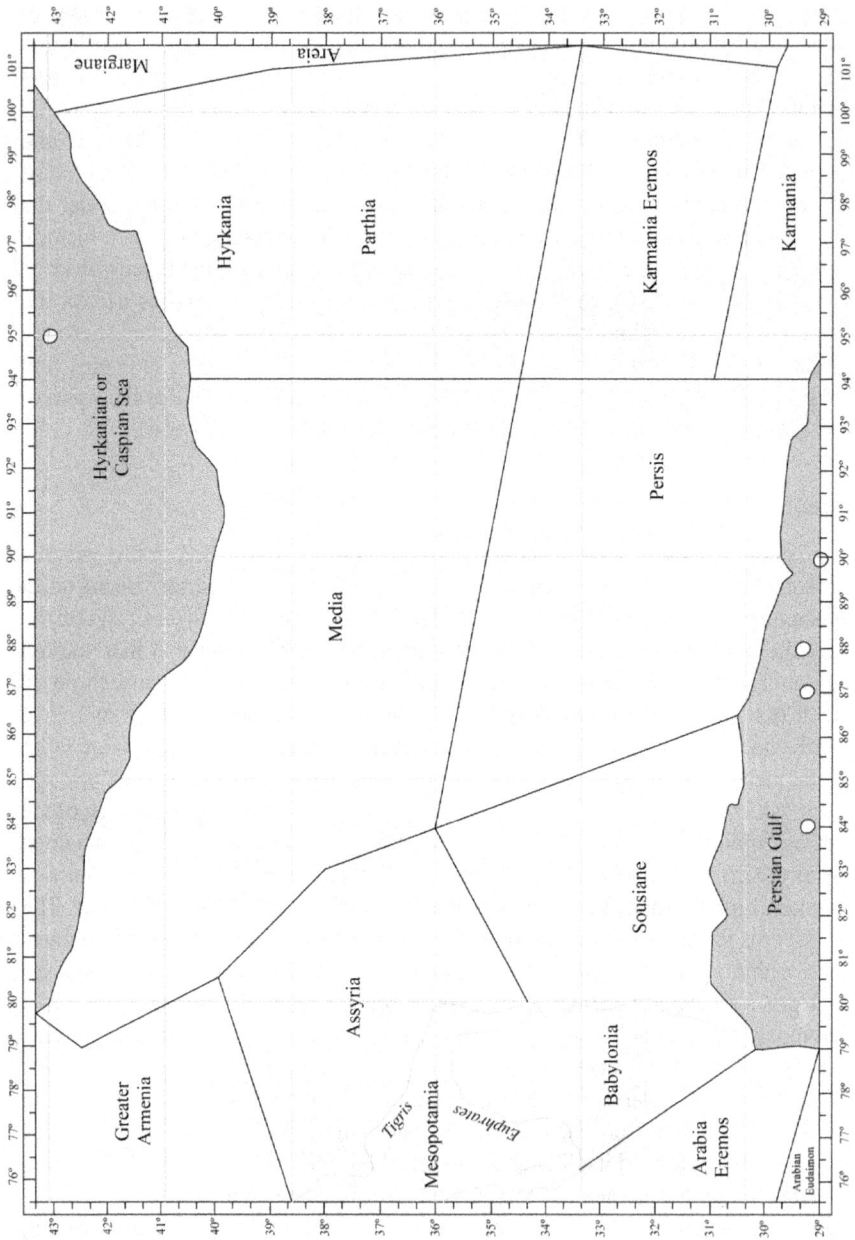

Map 6.5 Map 5 of Asia

Map 5 of Asia

This map covers the region east of the Tigris River as far as Areia, or essentially most of modern Iran. The southern boundary is the Persian Gulf (except that coastal Karmania was placed in Map 6 of Asia) and the northern the Caspian Sea, except for Hyrkania (placed in Map 7 of Asia). The territory depicted on the map is elongated east-west, so that the Caspian Sea stretched through 20 degrees of longitude (it is actually 5 degrees), another of Ptolemy's instances of extending the east-west length of the inhabited world. Except for the ephemeral Seleukid presence as far as India, Map 5 and the successive ones lie outside any Greek or Roman territory. Six regions are represented on the map: in the west is Assyria, with Sousiane to its south. To their east are Media to the north and Persis to its south. East of Media is Parthia, and to its south, east of Persis, is Karmania Eremos, the inland portion of that district. Most of the internal boundaries of these regions are arbitrary lines or ephemeral mountains.

Assyria

The ancient territory of Assyria (*GG* 6.1.1–7) lay east of Mesopotamia and south of Armenia and the Niphates River. To its east was Media, and Sousiane was to the south. Several rivers flow south across the territory into the Tigris, which separates Assyria from Mesopotamia. The Lykos and Kapros, the northernmost river and the next to the south, are the modern Greater and Lesser Zab, the major eastern tributaries of the Tigris; their Greek names suggest they were so designated at the time of Alexander the Great. The third river, the Gorgos (perhaps the modern Adheim), joining the Tigris far downstream, may also have a Greek name for the same reason, but it may come from Persian *gorg*, meaning "wolf," perhaps a doublet with the Lykos River.[63] The Choathras Mountains run across northeastern Assyria; their position suggests the Zagros range, although there is an actual Zagros Mountain to the east in Media (*GG* 6.2.6), which is doubtless the same range under a different name, yet shown by Ptolemy as a separate mountain.

The towns in Assyria (over 30 in all) cover the broad extent of the demographic history of the region, including ancient Nineveh (or Ninos, at modern Kuyunjik on the Tigris). The Assyrian city had been destroyed in the late seventh century BC[64] but survived as a village into Ptolemy's time.[65] Other sites of importance include Arbela (modern Erbil), an important local center, and Ktesiphon, of uncertain origin but with a Greek—or at least hellenized—name and which became a Parthian royal center.[66] Additional towns shown are probably indigenous villages. At the southwestern corner of Assyria, where it joins Babylonia, Mesopotamia, and Sousiane, are the Altars of Herakles, but no further details are available.

Media

Media (*GG* 6.2.1–18) lies to the east of Assyria and Armenia, with its northern border the shore of the Caspian Sea. It extended east to Parthia and Hyrkania and south to Persis. The boundaries of Media tend to be arbitrary, except for the Caspian Sea and the Parchoathras Mountains on the south, separating it from

Persis; these mountains are unidentified, lying in the rugged territory of southern modern Iran. Because of the east-west elongation of the Caspian Sea, Media is also oriented too much in this direction rather than its proper northwest-southeast alignment, another example of lengthening the inhabited world. Media was an ancient territory that was under Parthian control at the time of the production of the *Geographical Guide.*

There are five rivers that flowed north across Media into the Caspian. The two westernmost, the Kambysos and Kyros, which are hellenized Persian royal names, replicate river names farther to the north in Albania and Iberia, although the northern Kambysos was not recorded by Ptolemy. The duplication may be a transference of those names into Media. East of the Median Kyros was the Amardos, perhaps also a hellenized Persian name and probably the modern Safid Ruz, the longest of the Median rivers emptying into the Caspian (670 km.). Next to the east is another river with a Greek name, the Straton (modern Chalus), and then, near the Hyrkanian border, the Charindas. These five rivers show the ethnic diversity of Median history with names from the Median and Persian period into the Greek (probably from the Seleukid era), but presented by Ptolemy in hellenized form, perhaps from a Seleukid topographical report.

Only one river flows south from Media, the Eulaios (modern Karun), the longest river in the region (950 km.). It passes through the Parchoathras Mountains into Persis and then Sousiane, emptying, according to Ptolemy, into the Persian Gulf (*GG* 6.3.2). Its source is strangely not in the isolated Zagros Mountains but just one degree of longitude to their east; similarly the source of the Amardos is half a degree to the northwest of the mountains. For rivers to have their sources so close to but not in mountains is anomalous and may be a transmission error.

On its course to the Caspian the Amardos passes through Lake Matiane (modern Urmia), the largest lake in southwestern Asia. It actually lies within its own watershed and the actual course of the Amardos is far to the east. There are two isolated mountains in Media, the Orontes and Iasonion. The former is probably Mt. Alvand just south of Akbatana (or Ektabana, modern Hamadan), the ancient Median capital, so vividly described by Herodotos.[67] In the southeast of Media was Mt. Iasonion, not precisely located but an example of the long association of Jason and Medea with Media.[68]

The ruggedness of the territory meant that mountain passes were an important topographical feature. The Zagros Gates were the access through those mountains into Assyria and Mesopotamia. In the northeast were the Caspian Gates, the route from the Median highlands to the lowlands around the eastern Caspian Sea. This was used by Alexander the Great and seen as the boundary between the southern and northern parts of the inhabited world. They are not located with certainty today, but may be the modern Sar Darreh.[69]

Many towns and villages were located in Media, most of which, like Agbatana, were indigenous centers. Other than the geographical features already noted, there is no evidence of the scant Greek presence in the region except for the renaming of Rhagai (as Europos) by Seleukos I in the early third century BC,[70] and perhaps the river names that seem hellenized.

Sousiane

Sousiane (*GG* 6.3.1–6) is the region around the ancient city of Sousa, at the northern edge of the Persian Gulf. Arbitrary lines separate it from Assyria and Persis; Babylonia is west of the Tigris. Here the elongation of the east-west dimension of the inhabited world is quite apparent, since Sousa is placed 4 degrees of longitude east of the Tigris when it is actually only 1 1/2. Sousa was the ancient Elamite capital, and then a Persian winter royal city. In Ptolemy's time it was an unusually cosmopolitan city where many ethnic groups came together.

The major river of Sousiane was the Eulaios, flowing south from its source in Media through Persis. According to Ptolemy it entered the Persian Gulf well east of the mouth of the Tigris, although in earlier times it seems to have actually emptied into that river.[71] West of the mouth of the Eulaios was the major trading center of Charax, which Ptolemy placed on the shore of the Persian Gulf just east of the mouth of the Tigris, but in his day it was actually more than 80 km. inland due to siltation, indicating that at least in this locality his information was quite out of date.[72] Probably Charax had not been on the coast since the third century BC; Ptolemy may have been using Seleukid maritime sources.

The other river in Sousiane was the Mosaios, not identified among the complex river and canal patterns around the head of the Persian Gulf. Other than Sousa and Charax, the demography of Ptolemy's Sousiane cannot be interpreted beyond the assumption of a number of towns and villages in a much more compact area than displayed. An island south of the Eulaios mouth, Taxiana, may have been lost to siltation.

Persis

The heartland of the Persian Empire was Persis (*GG* 6.4.1–8), lying between Sousiane, Media, Karmania, and the Persian Gulf. In the center of the territory was the famous city of Persopolis (a more correct spelling than Persepolis), the ancient Persian capital. Most of the localities depicted are indigenous towns and villages, but the town of Tanagra may have been where settlers from Central Greece were located in the fifth century BC. The towns in Persis are evenly spaced, and there is no indication of the higher density of population that existed around Persopolis.

In addition to the Eulaios, three rivers—the Oroatis, Rhogomanis, and Brisoanas— cross Persis from north to south, as well as the Bagradas, the boundary with Karmania. None of these can be identified other than speculatively, but the high density of rivers that entered the Persian Gulf indicates that their mouths were important navigational markers in this well-traveled region. The same can be said for the positioning of several islands, of which Arakia (probably modern Kharg), which has the alternate name of the Island of Alexander, is a remnant of the king's explorers who were the first to survey in detail the Persian Gulf.

Parthia

The Parthians were the dominant power in the region from the Euphrates to the east in Ptolemy's time, and their original homeland was Parthia (*GG* 6.5.1–4), a toponym that he chose to present in its hellenized Latin form rather than the usual Greek form, Parthyaia (although Parthyene was a district of Parthia). Parthia lay east of Media and between Hyrkania and Karmania. To its east was Areia. Despite the importance of the region, Ptolemy's documentation is scant. There are no geographical features except for speculative bordering mountains, and only one river, the Maxeras, which is not easily identified, but which flows north from the northern boundary of Parthia through Hyrkania to the Caspian Sea.

There is a scattering of towns and villages. The most intriguing is Hekatonpylos (perhaps at modern Shahr-I Qumis), a foundation of either Alexander the Great or the Seleukids, and allegedly so named because many roads came together at it.[73] It became a Parthian royal seat. Additional Seleukid foundations were Apameia in the west, which was actually in Media, and perhaps Mysia in the northeast.[74] The other towns seem to be indigenous.

Karmania Eremos

South of Parthia and between it and Karmania proper was Karmania Eremos, Desolate Karmania (*GG* 6.6.1–2). As previously noted, the word *eremos* was used by Ptolemy to indicate a region of exceedingly light population and little if any urbanization. In the case of Karmania Eremos it truly was empty space, from a demographic point of view, probably identified by Ptolemy to avoid a gap on his map. The Parchoathras Mountains, which have extended across Map 5 of Asia since Assyria, marked the northern boundary and the Bagradas River the western. There are no geographical features or population centers; part of the region was called Modomastike, otherwise unknown, and there were three ethnic groups, the Isatichai, Chouthoi, and Gadanopydres, also unique toponyms.

Map 6 of Asia

The territories around the lower Persian Gulf are the subject of Map 6 of Asia, divided into two parts, with Arabia Eudaimon to the west and Karmania to the east. The islands in the gulf, the coastal parts of the Indian Ocean, and the eastern Red Sea are also included. The fact that Karmania and Arabia Eudaimon appear on the same map but are separated by the Persian Gulf demonstrates that despite many interior locations in both regions, Ptolemy's thoughts were oriented toward coastal navigation, shown by the inclusion of essentially the entire Persian Gulf on the map, excepting only its extreme north, which is depicted on Maps 4 and 5 of Asia.

Arabia Eudaimon

The term Arabia Eudaimon (*GG* 6.7.1–47) (Arabia Felix in Latin, or Fortunate Arabia), the modern Arabian Peninsula, had been known to the Greek world since at least the fifth century BC;[75] the navigator Skylax of Karyanda may have passed along its southern coast around 500 BC and given the Greek world some awareness of the peninsula.[76]

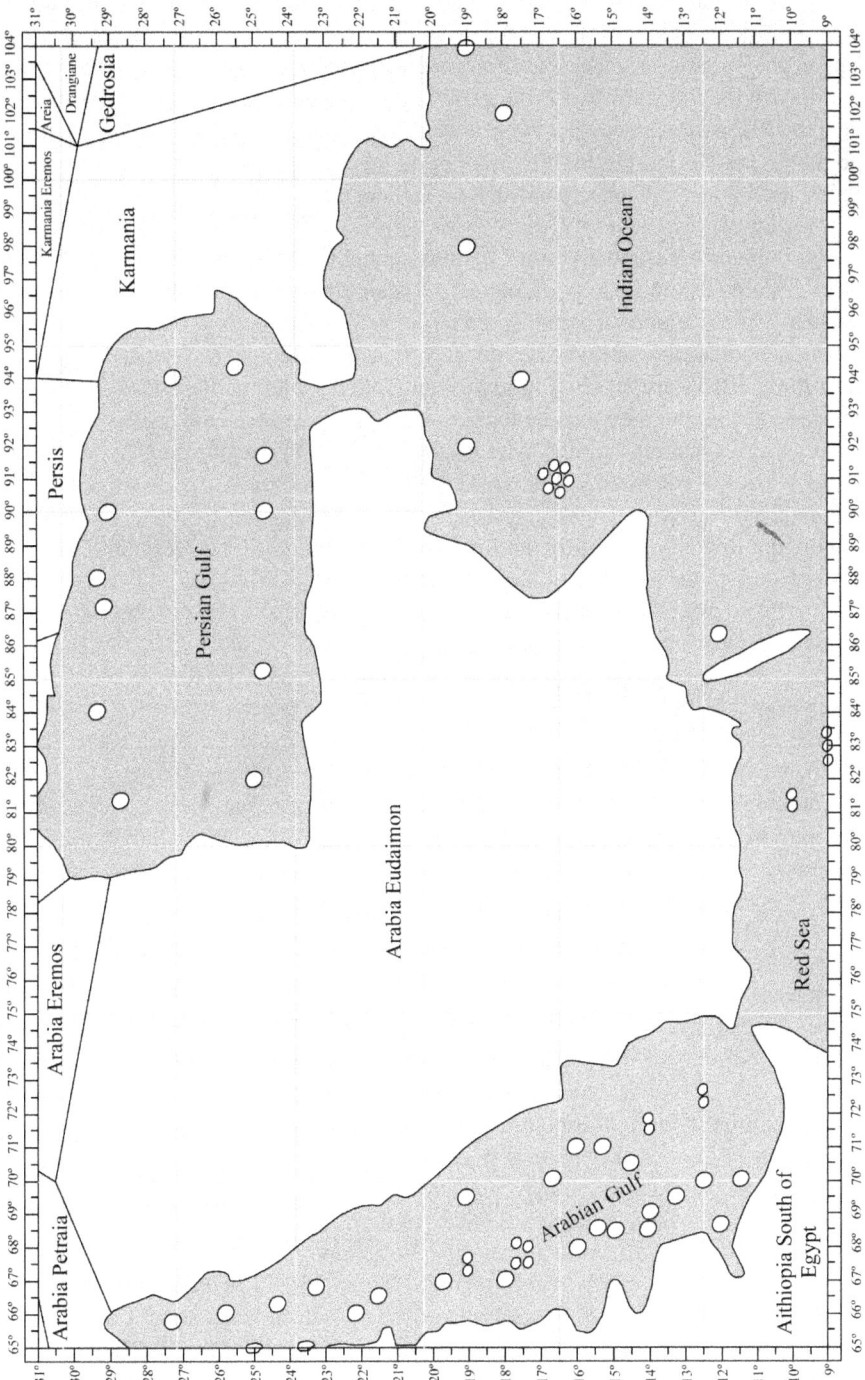

Map 6.6 Map 6 of Asia

But it was the aromatics trade that brought Arabia into Greek and then Roman knowledge: camel caravans went from the frankincense- and myrrh-producing regions of the southwest to the Mediterranean at Gaza in 65 days; this trade was established by the first century BC but probably had long existed.[77] Aromatics had reached the Macedonian court by the mid-fourth century BC, which probably stimulated the interest of Alexander the Great in the peninsula, and his last months he commissioned a number of expeditions to explore its coast.[78] Later the Romans made some attempts to penetrate the interior; although these efforts were problematic, they gathered a certain amount of ethnographic and topographic information that was collated in the treatise of Juba II of Mauretania, *On Arabia*, which was probably a primary source in Ptolemy's day.

The density of toponyms and ethnyms in Arabia in this remote and rugged region is astonishing. There are nearly 150 towns and villages, and, perhaps less surprisingly, over 50 ethnyms. Few can be located today; in most cases the toponyms are probably oases on caravan routes. The best-known region was the southwest, the aromatic district in and around modern Yemen, with its major city at Sabbatha (or Sabata, modern Shabwa).

Although a large portion of the aromatic trade headed north to Petra and Gaza, there were important and ancient emporia on the coasts. Okelis, or Akila, was at the southwest corner of the peninsula, and was not primarily a trade center but the last mainland location on the route to India, since ships would head from there across the northern Indian Ocean to the western Indian coast.[79] At the other end of the peninsula, somewhere in the upper Persian Gulf, was Gerrha, noted for its salt-based architecture.[80] Neither of these centers has been precisely located. Kane, on the south coast, may have been where Greek and Roman ships met Indian ones.[81]

The numerous ethnic groups generally remain enigmatic, but show the diverse populations of the region. Some have Greek names, such as the Ichthyophagoi (Fish Eaters) near the mouth of the Persian Gulf, a generic and descriptive Greek term for coastal populations, usually on the Indian Ocean. Much of the peninsula was said to be the home of the Smyrnophoroi (Myrrh Bearers) or the Libanotophoroi (Frankincense Bearers), demonstrating how the Greeks viewed the indigenous populations.

There are scattered topographical features depicted, 12 mountains and 4 rivers. Significantly, several of the mountains are described as *anonyma* ("nameless"), indicating a lack of topographical detail, even along the camel routes. Few of the mountains can be identified although Klimax, in the southwest, is probably An-Nabi Shy'ayb, at 3666 m. the highest point on the peninsula and which was close enough to the Red Sea coast to have been seen by Greek sailors; Klimax is a common Greek name for the highest point in a region.

In fact, most of the Greek names in Arabia are along the coasts, with Hippos town and village in the northwest, Thebai farther south, Cape Poseidion in the extreme southwest, the Melan (Black) Mountains on the south coast, an Oracle of Artemis (perhaps an indigenous cultic center) near the mouth of the Persian Gulf, and others. These are certainly remnants of the toponymic efforts of the Greek navigators who explored the coasts, beginning in the fourth century BC. In one of his relatively few commercial comments, Ptolemy noted the presence of pearl fisheries along the southeast coast; this may be a slightly misplaced reference to the Great Pearl Bank, which lies between Dubai and Oman.[82] Ptolemy's almost irrelevant mention of the fisheries shows that as the *Geographical Guide* moved farther east, it became more dependent on mercantile data from the eastern part of the inhabited world.

There are many islands off the Arabian coast; a number of them have Greek personal names (e.g., Timagenes, Polybios, Sokrates, and others), reflecting the Ptolemaic exploration of the Red Sea. The explorers sent out by Alexander the Great had probably limited themselves to coastal phenomena,[83] but Ptolemaic efforts, especially in the Red Sea, were more intensive and more oriented toward navigational issues. The first thorough exploration was by Agatharchides of Knidos in the mid-second century BC, who may have established many of the Greek island toponyms.[84] The personalities whose names became attached to toponyms are generally unknown, although it is intriguing that they tend to be those of well-known figures in Greek history.

All these islands are small—some essentially reefs—but were essential knowledge for the large number of seamen who traveled the Arabian coasts in the Ptolemaic and Roman periods. The exception is Dioskourides, the Island of the Dioskouroi, modern Sokotra, slightly misplaced off the southern Arabian coast when it is actually 250 km. east of the Horn of Africa, but culturally linked more to Arabia (it is Yemeni territory today). It is the largest island in the region (110 km. long) and is noted today for its biological diversity. It may have been the first major rest stop on the voyage to India.[85]

Karmania

Karmania (*GG* 6.8.1–16), the modern Kerman district of Iran, lies along the eastern Persian Gulf between Persis and Gedrosia, extending beyond the mouth of the gulf to its east. As with the Caspian Sea, the shape of the Persian Gulf is disoriented, presented as an east-west oblong extending through 17 degrees of longitude. In fact the gulf only covers 10 degrees of longitude, and thus Ptolemy's presentation of it becomes another example of elongation of the east-west dimension of the inhabited world. The north-south extent, 6 degrees of latitude, is accurate but the excessive longitude means the gulf is more rectangular than it actually is, and thus Karmania is located at the east end of the rectangle when its coast should be more east-west than north-south.

Karmania was part of the heartland of the Persian Empire, and came into Greek knowledge at the time of Alexander the Great. Although known for its wine— Alexander and his companions overindulged in the product[86]—and gold, it was not an area of particular Greco-Roman interest beyond those who sailed past it to and from India. Thus most of the toponyms recorded by Ptolemy are coastal, perhaps originating with Nearchos and Onesikritos, who in 325–324 BC sailed from the mouth of the Indos back to the Persian Gulf while Alexander took the land route, with additional information from later sailors going to and from India. There are 10 rivers, most of which have their origin in a single unnamed mountain in the center of the territory—another example of a mountain created to provide river sources—and a handful of settlements. The eastern boundary of Karmania was the Persian Mountains; this probably reflects the volcanic summits in eastern Iran which reach 4000 m. elevation.

There are only four ethnic groups; one of them, the Chelidonophagoi, or Turtle Eaters, in the southeast, were probably identified as such at the time of Alexander. At the northern border of the district were the Kameloboskoi, the Camel Herders, presumably a descriptive name reflecting the demographic realities of northern Karmania and beyond into Desolate Karmania.

The settlement names are indigenous with two interesting exceptions. In what Ptolemy depicted as the extreme west of Karmania, near the modern Strait of Hormuz (a name derived from ancient Cape Harmozan), was Mt. Semiramis, one of many toponyms scattered through the eastern part of the inhabited world named after the mythical Assyrian queen.[87] These are often early habitation mounds whose origin was unknown in Greco-Roman times; the Karmanian mountain may have its name applied during the Persian era.

To the northeast was a place named Alexandria. This is one of the most problematic of the cities of that name, and its association with Alexander the Great is dubious, but presumably there was a town of that name at least by the first century BC.[88]

Map 7 of Asia

This map covers a large region north and east of the Caspian Sea, bounded on the south by Media and other districts as far as Indike (India), the Caspian and Asian Sarmatia on the west, and what is defined as "Skythia Beyond the Imaon Mountains" on the east. To the north the map ends at 63 degrees north latitude, the traditional northern boundary of the inhabited world, with "unknown land" beyond. The territories on the map include (moving from the Caspian Sea to the east) Hyrkania, Margiane, Baktria, the land of the Sogdianians, and that of the Sakians.

Ptolemy's understanding of the region varied greatly. A major problem continues to be the orientation of the Caspian Sea, whose long dimension should be north-south but is east-west. Opinion was varied in antiquity whether the Caspian was connected to the External Ocean; Ptolemy, contrary to the more common idea of his era, saw it as an enclosed sea.[89] It had been explored by the Seleukid navigator Patrokles in the early third century BC, who seems to have been the only one to make such an expedition before the time of Ptolemy, and although he believed it was connected to the Ocean, he probably provided Ptolemy with much of his data concerning the coasts of the sea.[90]

The region immediately east of the Caspian had been known to the Greek world since the final and fatal expedition of Cyrus the Great of Persia in the 520s BC.[91] Further knowledge was obtained from the expedition of Alexander the Great and then by Patrokles. He was followed by another Seleukid explorer, Demodamas,[92] who went far to the northeast in the early third century BC, well into modern Kazakhstan.[93] These explorations east of the Caspian followed long-established

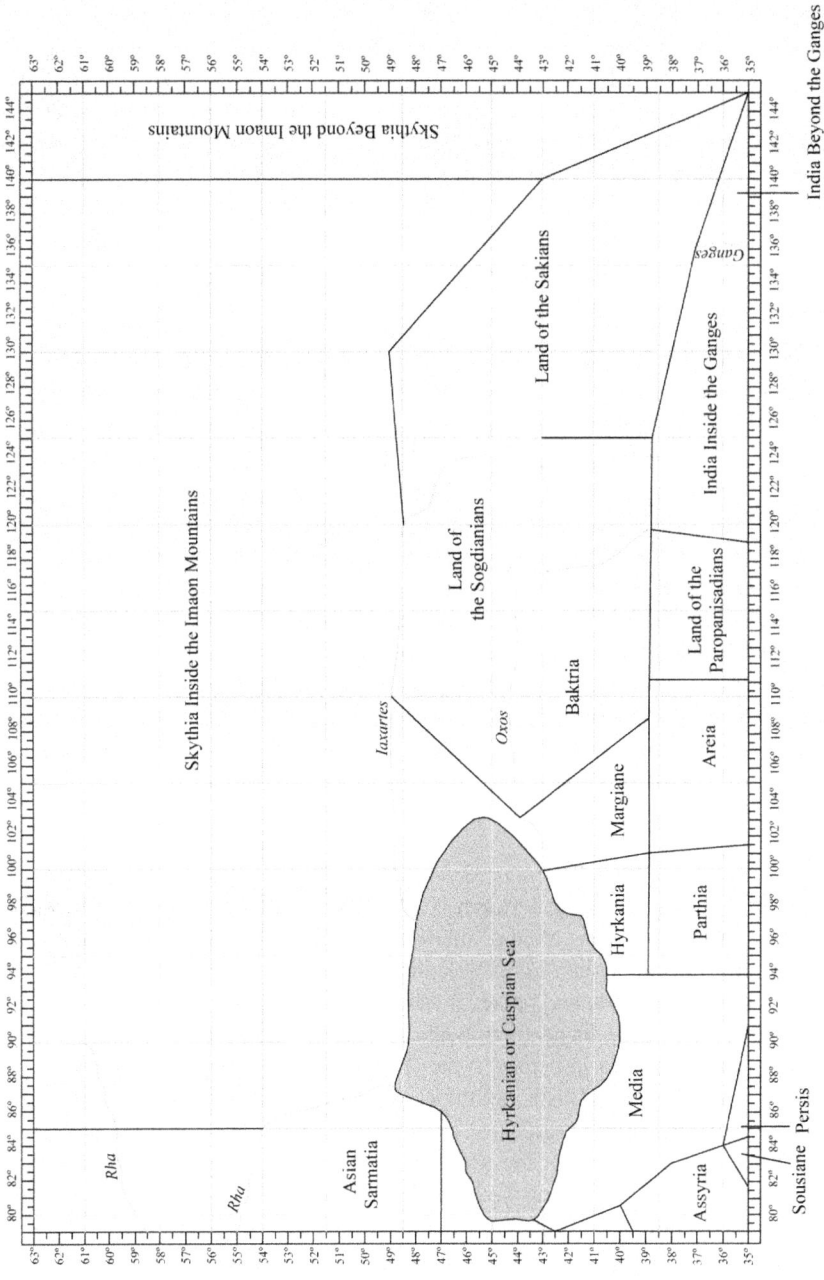

Map 6.7 Map 7 of Asia

trade routes into the little-known far eastern portions of Asia and in time coalesced into what came to be called the Silk Road. Although knowledge of Chinese silk reached the Mediterranean by the first century BC,[94] the region east of the Caspian remained poorly understood, with contradictory information available to Ptolemy from sources ranging over hundreds of years; this is best shown by the plotting of the river systems, where the two major streams of the region, the Iaxartes (modern Syr Darya) and Oxos (modern Amu Darya) have numerous intertangled tributaries that were believed to cross one another. Significant changes in the local hydrology in modern times add to the confusion of interpretation. But much of the depiction of the demography of this region is from the second century BC or earlier. Moreover, the same toponym might be plotted in two different places, due to differences in chronology or reports that were inconsistent with one another.[95]

Hyrkania

The territory at the southeastern corner of the Caspian Sea was Hyrkania (*GG* 6.9.1–80), which Ptolemy (and most other sources) also used as an alternate name for the Caspian Sea, calling it the Hyrkanian Sea. Hyrkania had been known to the Greek world since the expedition of Cyrus and historically was Persian and then Parthian territory.[96] It was largely the coastal plain of the southeastern Caspian Sea, a region noted for its unusual tropical fertility, which caused confusion for Greek climate theorists.[97] Much of the region lies in modern Turkmenistan and northeastern Iran.

Hyrkania was a small district, and Ptolemy catalogued 12 towns, 3 ethnic groups, and 2 rivers (the Maxaras and Sokandas), probably only known because Patrokles had seen their mouths where they entered the Caspian. To the south of Hyrkania was Parthia; to its east, separated by unnamed mountainous territory (presumably the uplands marking the limit of the Caspian basin) was Margiane. No evidence of a Greek presence was recorded.

Margiane

Margiane (*GG* 6.10.1–4), lying between Hyrkania and Baktria and located mostly in modern Turkmenistan, was another small district, the home of the Massagetians who had killed Cyrus the Great. Like Hyrkania, it was Persian and then Parthian, but because Alexander the Great had passed through the region it became relatively well known to the Greek world. Its northern boundary was the Oxos River (modern Amu Darya), the longest stream in central Asia (2495 km.), which Ptolemy showed as emptying into the Caspian. This is not the case today (its mouth is at the Aral Sea), but the river courses in this region were confused in antiquity and have changed greatly since then. Nevertheless the route along the Oxos, whose source is in the modern Hindu Kush, was part of the extensive east-west trade network of central Asia. Another river, the Margos (modern Marghab) flowed north across Margiane.

As part of the legacy of Alexander there was a vestigial Greek presence in Margiane. Ptolemy noted Antiocheia Margiane, which was founded by the Seleukid king Antiochos I on the abandoned site of an Alexandria; remains at modern Gyaor

Kala are probably those of the ancient city.[98] It became notorious in the mid-first century BC when prisoners from the Roman defeat at Karrhai in Mesopotamia in 53 BC were settled here by the Parthian king Orodes II.

The other important city recorded by Ptolemy was Nisaia, a Parthian royal residence that may have had its origins in Hellenistic times; extensive archaeological remains at modern Ashgabat mark the site of the ancient city.[99] Iasonion, on the Margos River and perhaps at modern Yarim Tepe, is another example of the persistent connections with Jason and Medea that are spread from Kolchis far to the east.[100]

Baktria

To the east of Margiane, across unnamed uplands, was Baktria, in antiquity the best-known region of central Asia and spread across modern Uzbekistan, Tajikistan, and northern Afghanistan (*GG* 6.11.1–9). It extended north and east to the Oxos River and south to the Paropanisos Mountains. Greeks were aware of Baktria as early as the fifth century BC, when it was part of the Persian Empire,[101] but it was Alexander the Great's extensive stay in the region—essentially two years—and his marriage to the Baktrian princess Roxane in 327 BC that established a strong Greek interest in the territory. After Alexander's death there was Seleukid control and then the establishment of the Greco-Baktrian kingdom by Diodotos I in the third century BC. Incursions from the north brought this kingdom to an end by the late second century BC.[102]

Ptolemy catalogued several rivers running across Baktria, as well as 15 towns and 13 ethnic groups; the density of the latter shows that the region was well known. The most famous town was Marakanda (modern Samarkand, with the ancient site at Afrasjab), an ancient oasis and caravan city that has retained its importance into modern times. Farther north was Eukratidia, founded by the Greco-Baktrian king Eukratides I in the second century BC, which has not been located.[103] Baktra and Zariaspa were the two components of the ancient Baktrian capital (at modern Balkh in Afghanistan); Ptolemy located the two cities nearly 2 1/2 degrees of latitude apart but they were adjoining, probably on opposite sides of the Baktros River.[104] Despite the demographic density of Baktria as Ptolemy described it, the topography of this region at the extremities of Greek geographical knowledge seems quite arbitrary, as if the towns were just placed at random. There were clearly many names but little understanding in terms of locating them.

The Sogdianians

Rather than by means of a toponym, the region north and east of Baktria was described by an ethnym, the territory of the Sogdianians (*GG* 6.12.1–6), showing that as the *Geographical Guide* moved into more remote territory, the demography was less well understood. The Sogdianians were an indigenous population beyond (essentially to the north of) the Oxos River, and had been part of the Persian Empire.[105] After its collapse the region became more disorganized, and to the Greco-Roman world the people primarily served as traders and middlemen between the west and the far east. Their primary city was Marakanda, which Ptolemy actually placed in Baktria, less an error than the perceived demographic vagueness of the region.

It is in the territory of the Sogdianians that the confusion regarding river courses is at its most extreme, with unnamed streams scattered through the region, especially between the Oxos and the Iaxartes (the latter the northern boundary of the Sogdianians). Part of this may be due to contradictory information from different sources, but in fact the "rivers" may actually be trade routes.[106]

The several towns and ethnic groups represented are generally unidentified; as with Baktria there are more of the latter than of the former. Marakanda has already been noted, and the presence of Alexander the Great was still apparent, most notably in Alexandria Eschate (Farthest Alexandria), perhaps near modern Khojend. Alexander built the city as a defense against northern raiders, believing erroneously that it would be a major Greek center at the extremities of the inhabited world. In Ptolemy's day it was probably little more than a village, or even only a historic toponym.[107] Ptolemy also located another Alexandria in the west of the Sogdianian territory, Alexandria Oxeiane, on the Oxos River; he is the only source for this city and its identification is problematic.[108]

In the west-central part of the territory was the Oxeiane Lake. It is not on the Oxos, but slightly north of it on one of its unnamed tributaries. It seems probable that this is the Aral Sea, which was hardly known in antiquity.[109] The Oxos proper (which today empties into the sea) was said by Ptolemy to flow south of the Oxeiane Lake through the Oxeia mountains and into the Caspian Sea; such a course for the river is topographically improbable, and its plotted course rather suggests a trade route. In fact, Patrokles noted that Indian goods came to the Caspian, perhaps by this means.[110]

The Sakians

To the east of the Sogdianians were the Sakians (*GG* 6.13.1–3), an ethnym linguistically related to the Skythians. Since the fourth century BC the Skythians had been said to be the nomadic population occupying the northeastern portion of the inhabited world.[111] Obviously such a geographically broad term included a variety of ethnic groups, which may not have been as closely related as Greeks ethnographers believed, and the primary location of the "Skythians" was north of the Black Sea, the peoples encountered by Dareios I of Persia in the 520s BC.[112] Yet the Greek tendency to generalize and homogenize primitive peoples at the fringes of the inhabited world meant that the Sakians were considered a related group.

The boundaries of the land of the Sakians are vague: they lived north of the Imaon Mountains (the modern Himalayas, which preserves the name), but this is more confusing than helpful. The Himalayas are so extensive that there can be little precision as to the location of the Sakians, and, moreover, Ptolemy had the Imaon Mountains turn to the north at the southeast corner of the Sakian land and continue to the end of the inhabited world. This was, needless to say, geographically imprecise but gives the Imaon Mountains, the highest and most extensive mountains known, an importance as a major boundary. To Ptolemy the Sakians were beyond the Iaxartes River (in his scheme, to its east), with the northward-trending Imaon Mountains on their east. This would roughly suggest that the population extended from Uzbekistan to Tibet, but such equivalences are highly speculative in these remote regions. Ptolemy was explicit in noting that the land of the Sakians was nomadic territory, with no cities and its peoples living in the woods and hollows.[113] The territory was a land of various ethnic groups, seven in all, unfamiliar except for the Massagetians, who, given their nomadism, may have wandered this far east. There are also two locations of note: the Lithinos Pyrgos (Stone Tower), and a pass through the north-south Imaon Mountains said to be "the trade route to the Serians." The Stone Tower was a major rendezvous point where routes from all directions came together. It was first documented by Ptolemy in his report of the eastern journey of the agents of Maes Titianos and was said to be 26,280 stadia from the Euphrates (*GG* 1.11–12).[114] A locality named Tashkurgan (Stone Tower) on the western border of China seems to preserve the name, but the toponym is common in the region; Ptolemy's Stone Tower (and perhaps others with the same name) may be constructed piles of stones used to mark routes or gathering points. Nevertheless it seems to have been an important trading center and the contact point with the Serians, or the Silk People, who were not the Chinese, as often assumed, but traders in silk who came west.

Farther east was the pass through the Imaon range, at a place called Askatankas. Ptolemy located it 2 degrees of longitude east of the Stone Tower, and it may have been the next rendezvous emporium. Since its name is indigenous rather than Greek, the cultural balance has shifted from west to east; after passing Askatankas traders headed east into the land of "the Skythians Beyond the Imaon Mountains" (*GG* 6.15.1–4). If, however, Askatankas was at a pass through the Himalayas rather than some unknown northern mountains, this would mean it was more probably on a route from the south and perhaps in the Kashmir region, or farther west, but the orientation of the mountains is too confused to be certain.

The Skythians Within the Imaon Mountains

The northern portion of Map 7 of Asia was the home of those Skythians who lived within the Imaon Mountains (*GG* 6.14.1–14). Their territory allegedly extended as far west as the Asian Sarmatians, and north to 63 degrees north latitude, the limit of human occupation and the beginning of unknown lands. The Caspian Sea was their southern boundary in the western portion of their territory, between the Rha and Oxos Rivers.

It is clear that this region is to some extent a construct to describe what would otherwise be a gap in the inhabited world, between the Asian Sarmatians and the northward trending Imaon Mountains. The failure to orient properly either the Caspian Sea or the Imaon range has made positioning questionable, but these Skythians roughly extended in a broad region from Kazakhstan to Tibet.

This was a region of obscurities: ethnic groups, mountains, and rivers and their tributaries that are hardly known. The rivers all flowed into the Caspian Sea, extending from the Rha (Volga) and Daix (Ural) in the west to the Oxos (Amu Darya) in the east (given the peculiar orientation of the sea). The source for these river mouths would be the coastal survey of Patrokles in the early third century BC, but the rivers may have been used by traders to access the interior, although

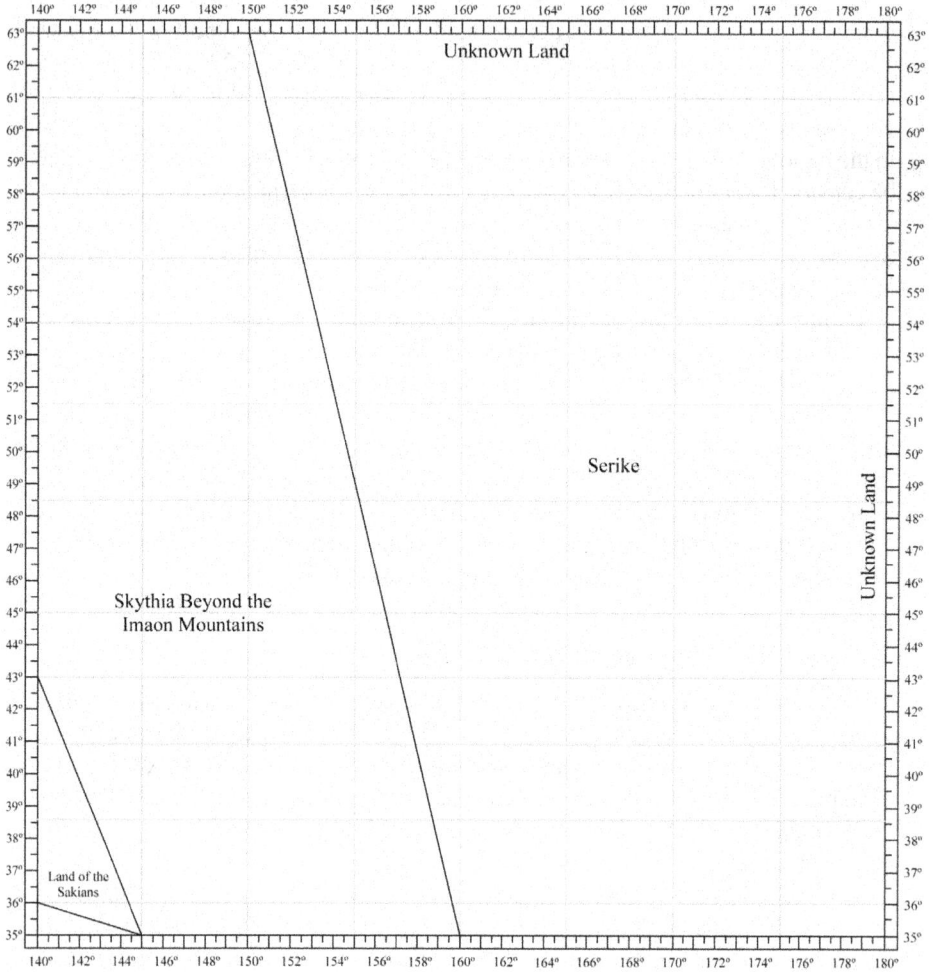

Map 6.8 Map **8** of Asia

there are no details. Ptolemy knew about the source of the Rha, which he placed 12 1/2 degrees of latitude north of the Caspian Sea, the farthest north river source documented in Asia, 3531 km. above its mouth and northwest of modern Moscow. Greek and Roman traders may not have gone all the way upstream, but far enough to know about its upper reaches.

Otherwise, except for two unlocated towns on the eastern Caspian (Aspabata and Dauaba), Map 7 of Asia is totally ethnic groups and mountains, the latter serving their usual role of providing the source of rivers. There are 30 groups, with some, such as the Galaktophagoi (Milk Eating) Skythians, having descriptive names provided by Greeks. Two have names that seem oddly out of place: the Tektosagians and the Samnitians. The former are normally one of the Galatian ethnic divisions in northwestern Asia Minor; the latter a population of central Italy (*GG* 3.1.67; 5.4.8). It seems improbable that either of these groups could have had any of their members relocate to north central Asia, and it may be the common practice of providing remote indigenous peoples with familiar names.

Map 8 of Asia

This map consists of two regions, the Skythians Beyond the Imaon Mountains and Serike. Because the Imaon Range is depicted erroneously as running north-south, it is difficult to orient this material, but presumably it is the territory north and northeast of the Himalayas, insofar as Ptolemy understood it. Unlike the other extremities of the inhabited world, the lands depicted on this map did not terminate at the External Ocean, but with unknown land to their north and west. The full extent of northeastern Asia was not known in antiquity, although it was realized that the Ocean was out there somewhere, but was never reached, because the approaches to it were through difficult terrain with inhospitable weather.[115]

The Skythians Beyond the Imaon Mountains

The western portion of Map 8 of Asia is devoted to the remainder of the Skythian territory, beyond the Imaon range (*GG* 6.15.1–4). This should be north of the Himalayas, but Ptolemy located it to the east. It was a region hardly known, only traversed by those who headed beyond the mountains to trade, or learned about the territory from traders who came west. In fact, Ptolemy had at least one source that was different from that used elsewhere in the *Geographical Guide*, since there is a reference to the Emoda Mountains, in the south of the territory which runs east into Serike; these are the same as the Imaon range, but more properly oriented. To the south of these Skythians, as expected, was Indike (India), so Ptolemy has fallen victim to two conflicting sources of information—perhaps reflecting different trade routes that originated in the west and south—with the Himalayas either running east-west and creating the border with India, or north-south and creating the border with Serike.

The territory of the Skythians Beyond the Imaon Mountains is largely bare of topographical details. Other than the mountains already mentioned, and one river, the Oichardes, there are only four towns (north to south: Auzakia, Skythian

Issedon, Chaurana, and Soita). These are certainly trading centers on the routes across the area; none can be identified. The single river, the Oichardes, which is actually mostly in Serike, is peculiar: it allegedly had three sources in three different mountain ranges, producing three streams that come together at a place called Piada, but with no connection of any of them to any other river or body of water. Again this demonstrates the deficiency and contradictory nature of the information available to Ptolemy; these alleged rivers may in fact represent trade routes.

Serike

East of the Skythians Beyond the Imaon Mountains was Serike (*GG* 6.16.1–8), with unknown land to its east and north and India to the south. Here the density of toponyms is greater than to the west, indicative of a larger population, with 14 towns and 13 ethnic groups. The rivers of the region connect with nothing else; in addition to the previously mentioned Piada, to its south is the Bautisos, which has the same characteristics of three mountain sources coming together at a single point. There are several mountain ranges but the only one that can be identified is the eastern continuation of the Emoda Mountains. As noted, these are the Himalayas, somewhat more properly oriented than the Imaon Mountains to the west.

The word Serike became the standard Greek world for silk (in the masculine, *serikos*). It was first encountered as an ethnym at the time of Alexander the Great in the phrase "the material of the Serika."[116] There was no topographical grounding, but assumedly it referred to a product brought to India by traders from the north. This was assuredly silk. By Ptolemy's day the Serike people had been located in the far northeast of the inhabited world, but there was still no reason to connect them with any specific place or ethnic group, certainly not the Chinese and China. Traders who had silk among their products came west to locations like the Stone Tower to sell their wares; a description of these people by Pliny makes it clear that they were not Chinese, because they were of large stature with blue eyes and red hair.[117] But they carried the ethnym Serike, or Sera, with them to describe their materials.

Nevertheless, the agents of Maes Titianos, traveling in the late first century AD, continued east from the Stone Tower for a distance of 36,200 stadia[118] and eventually reached Sera, seven months to the east, defined as the metropolis of the Serians. It may have been the contemporary Chinese capital of Luoyang, or, more probably, a significant provincial city to its west. Regardless, it is the most remote place in northeast Asia on Ptolemy's maps that has any chance of being identified.

Map 9 of Asia

Map 9 of Asia covers the region from Karmania and Parthia on the west to Indike (India) on the east. It was divided into five districts, with Areia, Drangiane, and Gedrosia extending north to south in the west, the Paropanisadians in the northeast, and Arachosia in the east central portion. The southern boundary was the Indian Ocean and the northern the Saripha and Parapanisos Mountains.

The area had been known to the Greek world since the expeditions commissioned by Dareios I of Persia in around 500 BC, but more intensive knowledge of

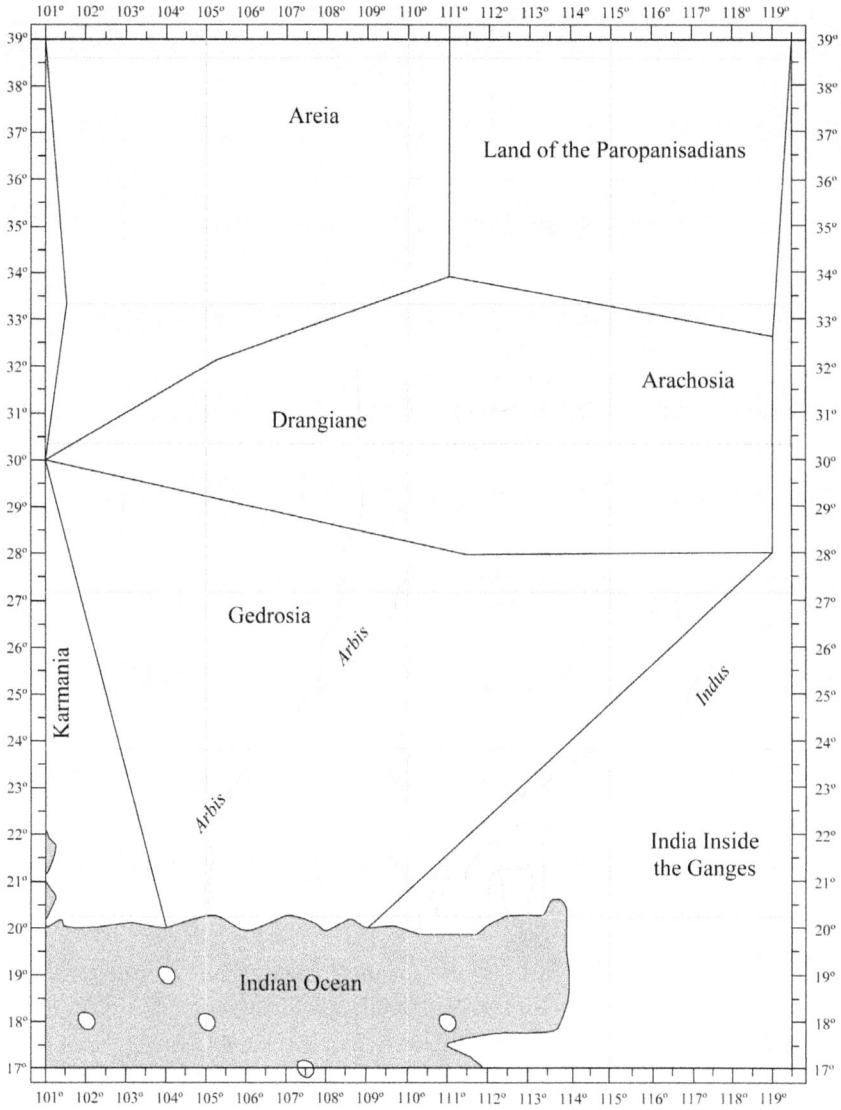

Map 6.9 Map 9 of Asia

the territory came at the time of Alexander the Great and the detailed records of those with him, as well as from the Seleukid occupation for a generation thereafter. But after the time of Alexander the region served mostly as way stations on the route to India, with relatively little Greek or Roman demographic interest. Ptolemy's record of towns—many of are probably little more than oases or stopping places—may be from the record of Baiton, the official surveyor of the expedition of Alexander the Great, but his report would only have provided relative locations.[119]

Areia

This was probably the best-known part of the region, lying south of Margiane and Baktria and east of Parthia (*GG* 6.17.1–8). The Paropanisos Mountains (generally the Hindu Kush), here running north-south, divided it from the Paropanisadian territory to the west. To the south of Areia were the Bagoon Mountains, and beyond, Drangiane. In the northwest the boundary was the Saripha Mountains (modern Barkhat Dagh). Areia is mostly in modern Afghanistan.

There was a density of towns and villages in Areia (over 30) and eight ethnic groups. The southern portion was ominously called Skorpiophoros, the Scorpion-Bearing Land. Virtually none of the towns can be identified, although Nisibis, with the same name as a famous town in Mesopotamia, may be the location of settlers from that region. Alexandria in Areia, or Alexandria Among the Areians, is generally thought to be near modern Herat, but the ancient town is unknown today.[120]

Other than the boundary mountains, the only topographical feature in Areia is the Areios River, probably the modern Hari Rud. The modern river flows west from the Hindu Kush and then turns north and loses itself in the desert west of modern Merv. Yet like so many rivers in the *Geographical Guide* that are plotted in remote locations, it is depicted in a peculiar fashion with an eastern source somewhere in the Paropanisos Mountains and a western one in the Saripha Mountains; these two streams join each other in the Areia Lake, near Alexandria in Areia. Oddly this represents the actual course of the river, with the lake probably the swampy regions near Herat, but it is not two separate streams flowing into the lake, but a single one flowing from the east and then exiting to the west and losing itself in the desert.

The Paropanisadians

The Paropanisadians (*GG* 6.18.1–5) were the inhabitants of the Paropanisos Mountains, and probably of the regions to their south around modern Kabul. They were roughly between Areia and northwest India, largely in modern eastern Afghanistan and northern Kashmir. They were mountaineers who lived in the high country and along the upper tributaries of the Indos River. Two rivers were documented in their territory: the Gorya, flowing to the southwest and part of the upper Indos system, and the Dargomanes (perhaps the modern Ab-e Safed), flowing north and connecting with the Oxos in Baktria. This shows that the locals lived, at least in part, along the divide between the drainages of the Indian Ocean and the Caspian and Aral Seas.

As expected, the territory was largely surrounded by mountains. Of particular interest is that Ptolemy called the eastern range (separating the region from India) the Caucasus, a toponym that actually refers to the mountains between the Caspian and Black Seas. This is a vestige of one of several attempts by Alexander the Great and his people to manipulate topography to his advantage. Alexander was in the region in the spring of 329 BC and named the Paropanisos the "Caucasus," an egregious attempt to enhance his reputation by saying that he had crossed the Caucasus, which, since the early fifth century BC, had been considered the limit of the inhabited world, made famous by Aeschylus.[121] Alexander never came close to the actual Caucasus, but by claiming that he did so he connected himself with mythic figures such as Prometheus and Jason. Despite attempts by explorers and geographers to rectify the situation, many of whom were outraged at the concept of two sets of Caucasus mountains, the toponym of the eastern Caucasus remained.[122] Nevertheless Ptolemy had some sense of the problem, calling the mountains in the east the "so-called" Caucasus (*GG* 6.12.1).

There are 16 towns and 5 ethnic groups in the Paropanisadian territory. Many of them seem to have names that are Persian or Parthian in origin, such as Artoarta, Ortospana, or Parsiana; the group known as the Aristophylians ("Best Tribe"), in the west, may be a Greek comment on social structure, or merely a hellenized indigenous name.

Drangiane

South of Areia was Drangiane (*GG* 6.19.1–5), with Arachosia on its east and Gedrosia on the south, the modern Iranian region of Sistan. It had a Persian history, and then Alexander the Great passed through in late 330 BC, but it retained a Persian lifestyle into the Roman period and was a region noted for its tin deposits.[123] It was lightly populated, with only 11 towns and 2 ethnic groups recorded. The Arbis River (modern Hab) flowed across the territory from north to south, and through Gedrosia to the Indian Ocean.

Arachosia

Arachosia (*GG* 6.20.1–5) lies to the east of Drangiane and north of Gedrosia. It is also south of the Paropanisadian territory and west of India, in the vicinity of modern Kandahar. After the Persian and Seleukid presence typical of this region declined, by the third century BC it became part of the Mauryan kingdom of India. Alexander had passed through in 330 BC, and a city called Alexandria in Arachosia may reflect this, but it is not mentioned until the first century BC and may be a Hellenistic foundation.[124] It would have been at modern Kandahar, which preserves the name. Arachotos, to the east, may be a doublet for Alexandria.[125] There are ten additional towns and four ethnic groups recorded by Ptolemy.

Gedrosia

Gedrosia (*GG* 6.21.1–6) is the coastal region of eastern Iran and western Pakistan, essentially modern Baluchistan. It is a rugged and treacherous territory, best known for the difficulties Alexander encountered while crossing it east to west in 325 BC.

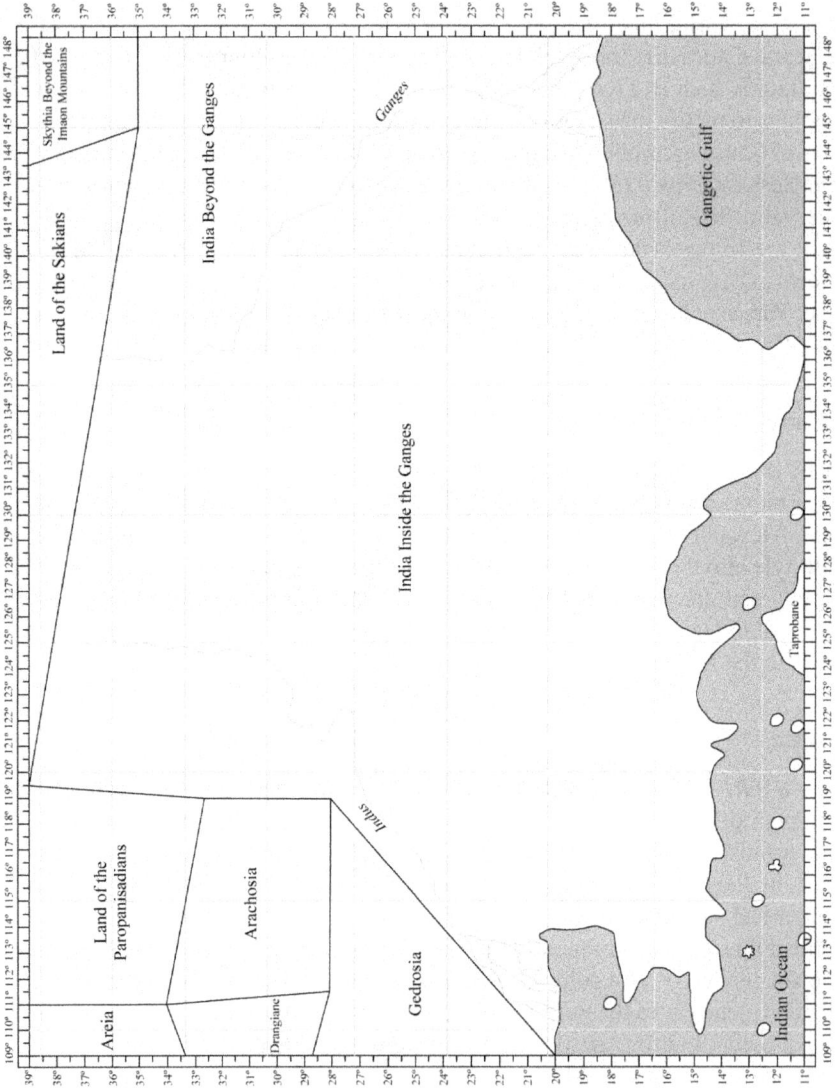

Map 6.10 Map 10 of Asia

His fleet sailed along the coast. Strabo, the earliest extant account, told in detail about the near-fatal problems that the king and his people experienced.[126]

Gedrosia lies between Drangiane and Arachosia. It was sparsely populated with only ten towns and four ethnic groups recorded by Ptolemy. The Arbis (Hab) River runs across the territory from Drangiane to the Indian Ocean, and various tributaries of the Indos are in the east. The major cities were Parsis in the north and Arbis near the coast; the former, at least, suggests a Persian or Parthian foundation. The fact that Strabo, 300 years later, had no source for Gedrosia other than the accounts of Alexander would suggest that few visited the region after the time of the king: the major routes to India either passed to the north, as described by Isidoros in his *Parthian Stations*, or made the sea voyage, as recounted by the *Periplous of the Erythraian Sea*. There is no reason to believe that Ptolemy had any information later than that of Alexander.

Map 10 of Asia

Indike Within the Ganges

This map is totally devoted to the Indian peninsula (*GG* 7.1.1–96), defined as the territory east of Gedrosia, Arachosia, and the Paropanisian territory as far as the Ganges River, with the Imaon Mountains (part of which was said to be the Caucasus) on the north, and the Indian Ocean on the south. Although there was a vast amount of detail available for the region, largely beginning at the time of Alexander the Great and continuing into the Roman period, there were serious problems of size, location, and orientation that pervaded all geographical thought about India.

There were many writers on the geography of India (Indike in Greek), and, despite its remoteness, it became one of the best-known regions of the inhabited world, although Greco-Roman understanding of Indian culture and society remained deficient in many ways. India had been known to Greeks since the late sixth century BC but it was only at the time of Alexander that intense study of the sub-continent began.[127] Alexander's expedition was limited to the Indos system, although he had heard rumors of another great river to the east, the Ganges, but was unable to pursue the matter.[128] It was only in the following generation that the entire extent of the Indian peninsula came to be known through a number of Greeks who spent time there. The most notable was Megasthenes, who visited the Mauryan court of Chandragupta at Pataliputra (modern Patna), and stayed there for several years at the end of the fourth century BC.[129] He produced his four-book *Indika* based not only on his travels with the royal entourage but also on information received at Pataliputra from all over India. He obtained a large amount of data about its size and shape, as well as a vast amount of ethnographic and cultural material. Although the *Indika* is not extant, it was used extensively by Eratosthenes, Strabo, and Pliny, and many lengthy quotations survive. Despite numerous other travelers to India who reported their experiences, it remained the definitive work on the region through the Hellenistic period.

By Roman times trade between India and the Mediterranean world had become a major endeavor, and in the Augustan period 120 ships went regularly between the

Red Sea and the sub-continent.[130] Indian envoys also came to the Roman world: ones are known from 25 and 20 BC.[131] The establishment of regular trade meant that it was important for merchants from the Mediterranean to know where to go (and how to get there) and what products were available: the best evidence is from the account of the trade routes and commodities surviving from the middle of the first century AD in the *Periplous of the Erythraian Sea*, produced by a Greek seaman based in Egypt.[132] Although one can only speculate which of these many sources, from the Alexander historians to the *Periplous*, were used by Ptolemy, there is no doubt that he had a great amount of material available to him.

Map 10 of Asia depicts the portions of the sub-continent "within" the Ganges. The river is shown running in a southeasterly direction through 19 degrees of latitude (it actually runs more easterly through only 8 degrees), and this error in orientation meant that it appeared more of a barrier and line of separation between west and east than it actually was. The Indos system is better plotted, running south-southwest across the west side of India through 16 degrees of latitude, reasonably accurate for its course from Kashmir south, but it is doubtful that its upper portions, deep in the rugged mountains, were known. India also extended west beyond the right bank of the river, bordering on Gedrosia, Arachosia, and the Paropanisadian territory.

Although there are many issues regarding the shape of India as depicted, most of which are relatively minor, the southern tip of the peninsula was erroneously plotted. Cape Kory (modern Cape Comorin), which is the actual southern point of India, was located opposite Taprobane (Sri Lanka), which is in fact over 200 km. to the east. On Ptolemy's map Cape Kory was not the southern point; 2.5 degrees farther south and 10 degrees farther east was a location called "the starting point for sailing to the Chrysos Chersonesos," which lay to the east (*GG* 7.2.12, 25). Topographical and navigational information has become confused, and the "starting point" may be farther up the coast, perhaps at the major trade emporion known today as Arikamedu, near modern Pondicherry.[133] But since this was where ships headed east, it was seen as the most remote location in India.

The topography and demography of India is recorded through a vast number of mountains, rivers, towns, and districts. The rivers of India were famous: people from the Mediterranean world with its few major streams marveled at Indian hydrology.[134] In addition to the Indos and Ganges, there are over 20 rivers recorded, most of which, as usual, have their sources in mountains, real or assumed. The navigational awareness of the Indian coast is apparent in the large number of towns and villages along it, along with the many river mouths and topographical features, so that no sailor had to go far without reaching a known point. In addition, nearly a dozen islands are shown off this coast, also valuable information for seamen.

As noted, numerous towns are plotted throughout India. These have their greatest density in the Indos drainage—the regions reached by Alexander—and along the western (Ptolemy's southern) coast and somewhat inland from there, which was the territory accessed by traders from the west. All along this coast are places marked "emporion" where ships from the Red Sea and Arabia would land and trade. A short distance inland there are a number of locations where a local dynast is named, such as Baithana, the residence of Siriptolemaios (an interesting hybrid name),

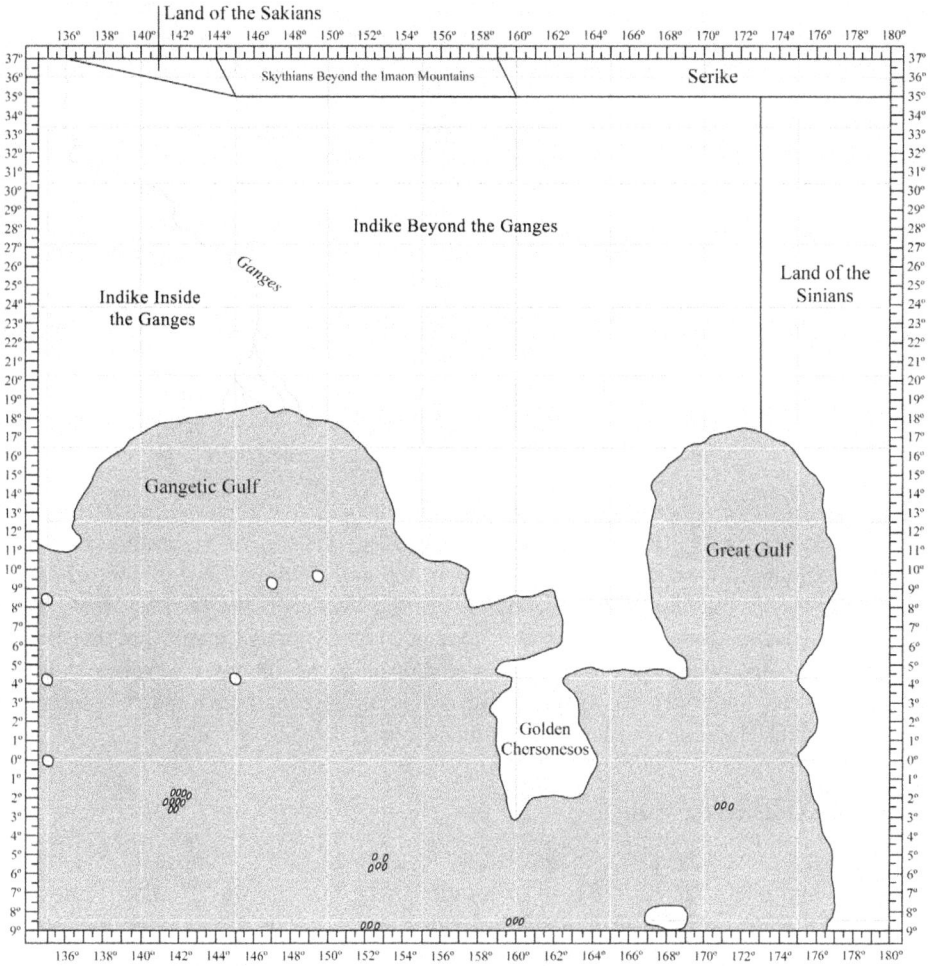

Map 6.11 **Map 11 of Asia**

or Medoura, the royal seat of Pandion. These dynastic centers were where the local rulers would have had contact with the Greek traders. But needless to say, Palimbothra (*GG* 7.1.73), on the Ganges, was also a royal residence, the seat of the famous Mauryan dynasty which dominated northern India from the fourth to the second centuries BC, ruling beyond the Indos and reaching out to the Hellenistic world. It was here that much of the Greek information on India was gathered. Nevertheless, it can be confusing, because Ptolemy had hundreds of years of data to draw upon—essentially from the time of Alexander to his own day—and he was not always sensitive about how evolving and increasing material could be contradictory: this is evident on the maps, which lack chronological consistency.[135]

There were a surprising number of Hellenistic foundations in India, although many were not documented by Ptolemy, and, as is frequently the case, they were often merely the renaming of existing towns. Nearly 40 are known, and, expectedly, almost all are in or near the Indos system. Ptolemy's citation of only a few was probably in large part due to the ephemeral nature of many of them, which, although still in existence in his time, may have reverted to local names. The most famous was perhaps Boukephala, or Boukephalos Alexandria (*GG* 7.1.46), founded by Alexander in memory of his late horse and which received a settlement of Macedonians. It was one of the few Hellenistic settlements in India that unquestionably existed under its romanized name into the Roman period, and was perhaps located near modern Jalalpur.[136]

Map 11 of Asia

This map is one of the most intriguing in the *Geographical Guide*, because it catalogues some of the most remote portions of the inhabited world. It is divided into two sections: Indike Outside the Ganges proper, which is the territory from the Ganges to the head of the Great Gulf (Megalos Kolpos), and then Sinai, the land of the Sinians, which lies to the east and runs as far as unknown territory at 180 degrees east longitude. North of the map are Skythia Beyond the Imaon Mountains and Serike. The south border is the Indian Ocean.

Indike Outside the Ganges

This region extends east from the Ganges River to an arbitrary line at 173 degrees east longitude (*GG* 7.2.1–31). The mouth of the Ganges was at what Ptolemy called the Gangetic Gulf (the modern Bay of Bengal). East of this point the coast headed southeast through the Golden Land, eventually passing a narrow isthmus (less than 2 degrees of longitude across) and ending at the Chrysos Chersonesos, or the Golden Peninsula. It is not difficult to assume that this is the coast of Bangladesh, Burma, Thailand, Malaysia, and Singapore, and that the Golden Peninsula is essentially modern Malaysia. The narrow isthmus is the Isthmus of Kra in southern Thailand. The large number of river mouths and towns (probably trading stations) plotted on this coast suggests once again a navigator's itinerary, perhaps generated by the obscure Alexandros (*GG* 1.14.1), who sailed in this region, either along the coast or directly across the Bay of Bengal from India to the Golden Peninsula.

There are also towns, rivers, and mountains in the interior that can rarely be identified today, but at the southern end of the Golden Peninsula is the emporion of Sabana, probably Singapore. More than one ethnic group in the interior is characterized as Anthropophagoi, or People Eaters, showing that those venturing inland from the coast could expect dangerously primitive country. The People Eaters were also common on the offshore islands.

Yet, unusual for the *Geographical Guide*, there is an extensive list of trading commodities for this region. There was the Land of Gold, and to its south the Land of Silver, both of which had many mines. Further north was the royal residence of Triglyphon or Trilingon (the Greek-sounding name may merely be an adaptation of an indigenous toponym or personal name). It had an economy seemingly based on avians: bearded poultry, ravens, and parrots were noted. This unusual list of products, perhaps for export to the west and both animal and mineral, along with the coastal emporia, shows Ptolemy's dependence on a trade document such as the *Periplous of the Erythraian Sea*. Moreover, there is one of Ptolemy's rare ventures into ethnography, describing the territory east of the Ganges (*GG* 7.2.15–17):[137]

> Between the Imaon Mountains and the Bepyrrhon Mountains are the Takoraians, who extend to the north. Below them are the Korankalians, and then the Passalians. Next to them, above Maiandron, are the Tiladians, who are also called the Besadians and are short, shaggy, and with flat faces, as well as white in color. Above Kirrhadia is where they say that the best *malabathron*[138] is produced, and as far as the Maiandron Mountains are the Tamerians, who are cannibals. Then there is the Silver Land, where, according to most, there are mines of high quality. Lying above the Besyngitians is the Golden Land, where there are many gold mines. Those living there have white hair, and are shaggy, short, and flat-nosed.

This is the longest ethnography in the *Geographical Guide*, and describes a territory extending roughly from north to south in modern Burma. The only similar passage is a briefer one on Taprobane (*GG* 7.4.1), perhaps from the same source. Presumably Ptolemy was directly quoting an ethnographic account, perhaps even from the mysterious Alexandros; the sense is that this remote part of the world needed more than merely a list of toponyms.

From Sabana the coast turned to the north; the itinerary has rounded the tip of the Malay Peninsula and is heading along the east coast of Malaysia and Thailand into the Great Gulf, probably the Gulf of Thailand. The first major city was Zabai, 20 days from Sabana. It is not located today but was near the dangerous Land of the Brigands.

North of Zabai there was a series of towns and rivers that are along the west side of the Great Gulf, probably additional trading stations. But at this point the topography becomes exceedingly confused, and it has been suggested that Zabai is not on the Malay Peninsula but at the southern tip of Vietnam, which is perhaps represented by the nearby Mega Akroterion (Great Promontory, perhaps modern Cape Ca-Mau).[139] This is possible, but there is no doubt that the plotting of places in this region is problematic, largely due to the remoteness of the territory as well as the complex corrections that Ptolemy made to Marinos' data (*GG* 1.14). But to locate Zabai in a totally different topographical region than the Golden Peninsula does violence to Ptolemy's positioning of toponyms.

If one is to assume that Zabai and the toponyms to its north are not on the northern Malay Peninsula but the coast of Vietnam, then the representation of the entire extremity of the inhabited world is seriously flawed, more than any other locality in the *Geographical Guide*, which is not an impossibility. East of the Golden Peninsula, Marinos provided Ptolemy with a number of places that often appear to have little relationship with one another. Moreover, as reported by Alexandros (and he was surely not the only one), there was more sailing on the open sea in this region that was customary in classical antiquity, meaning that points of reference were almost totally lacking except for a rough astronomical calculation of latitude. Alexandros' statement that Kattigara was "a few days" southeast of Zabai invites confusion: Marinos turned this into "many days," an alteration that Ptolemy ridiculed, suggesting that Alexandros actually meant "some" days (*GG* 1.14.2–3). This is of little help in locating Kattigara.

At the head of the Great Gulf—which, as Ptolemy plotted it, would be the head of the Gulf of Thailand near Bangkok—he drew an arbitrary line north along 173 degrees east longitude. This separated Indike Outside the Ganges from Sinai: the Seros River that empties at this point should be the modern Chao Phraya, the main river system of Thailand. If, however, one assumes that the Vietnamese coast is depicted, identification of the river becomes more inscrutable.

There were numerous islands in the Indian Ocean south of the mainland: many are probably various of the Andaman Islands. Several were inhabited by cannibals, but one, where the inhabitants went naked, was known for its mussels (*kolchoi*). Another group, the Maniolai, produced the Herakleian stone, a magnetic iron oxide that had been known to the Greek world since at least the fifth century BC.[140] Magnetic stones had a number of medical and homeopathic usages.[141]

Perhaps the most interesting of the islands at the south edge of Indike Beyond the Ganges is the easternmost, Iabadiou (Sabadibai, to its west, may be a doublet). Mentioned only by Ptolemy, its name meant "barley" and it also produced gold and silver, the latter assumed by the name of its major city, Argyre. It is unlikely that barley was an export product since it was common throughout the Mediterranean world, and the comment may be merely descriptive.[142] Iabadiou has long been believed to be Java, a sensible assumption based both on linguistic and topographical reasons, although its position is more suggestive of Sumatra.[143]

Sinai

The final district on the Asian mainland is Sinai, plotted east of 173 degrees east longitude (*GG* 7.3.1–6). This is the most vague of the localities on Ptolemy's maps, and may be little more than a collection of far eastern toponyms whose location or relationship to one another is not certain.[144] It is not even clear whether Sinai was a territory, ethnym, or city: Ptolemy used all these definitions. There is no doubt that the terms refer in some way to the Chinese; an alternate form, Thinai, also appears in some manuscripts of the *Geographical Guide* and in the *Periplous of the Erythraian Sea*, defined as a great inland city at the fringes of the inhabited world.[145] Ptolemy noted that the city of Sinai was noted for its metallic (*chalka*, usually meaning copper or bronze) walls; it is not clear what this means, and it may be hearsay to describe the impressive nature of a large important city.

It is the placement of the Sinai that causes confusion. There were several towns and rivers in the territory, including Sinai itself, and the mysterious Kattigara, reached by Alexandros after a southern sail from Zabai. If one assumes that Ptolemy's coordinates accurately represent the topography of this far eastern region, Kattigara would be well to the southeast (in fact, it is located near Iabadiou), but it becomes impossible to locate it other than somewhere southeast of the Malay Peninsula.

If, however, it is necessary to locate the city and land of the Sinai within historic China, this requires rejection of Ptolemy's location and the assumption that the area depicted is the east coast of Vietnam. It has even been suggested that Kattigara is modern Hanoi.[146] Supporting this view is Ptolemy's statement that Kattigara was "an anchorage of the Sinai."

But an alternative explanation fits the facts better. Sinai is so vaguely known that any topographical precision is unclear. Ptolemy's uncertainty as to just what Sinai meant supports the idea that it was merely a generalized ethnym for the people of the far east, who lived in a homonymous city. This follows the technique of Ephoros, who was writing centuries before Ptolemy and had not heard about the Sinai, but who divided the extremities of the inhabited world into four broad ethnic groups, the Skythians, Indians, Aithiopians, and Kelts.[147] The Indians were the easternmost peoples, but by Ptolemy's day the known inhabited world had gone far beyond them, and it was no longer possible to argue that they were the most remote population in that direction. Instead it was now the Sinai, who occupied the extremities as far east as was known.

Support for this point of view is provided by the Fish Eating Aithiopians, who lived around the Gulf of Sinai and near Kattigara. The Aithiopians have not been mentioned in the *Geographical Guide* since Libya, but Ptolemy saw them as adjoining the Sinai, which removes India from any extremity of the inhabited world. Furthermore, there was the idea that the Indian Ocean was an enclosed sea, a surprisingly persistent concept originating with Hipparchos and supported by Polybios and Ptolemy.[148] If this were the case, the Aithiopians spread from Libya to the Sinai, and thus, since the Aithiopians were the southernmost of peoples, the Sinai now replaced the Indians as the easternmost.

The eastern extremity of the Sinai was placed at 180 degrees east longitude (the same as that of the Serike on Map 8 of Asia). In Ptolemy's calculations, this was halfway around the earth from his zero meridian at the Canary Islands. By modern calculation, however, Singapore is at 104 degrees east longitude and Java at 110–115 degrees, so adding the 15 degrees that the Canaries are west of Greenwich, this is a maximum of 130 degrees for the inhabited world. Thus Ptolemy's 180 degrees is the culmination of the repeated east-west elongations of distances and landforms that pervade his calculations, such as with Italy and the Caspian Sea. That the world known in antiquity covered half the sphere of the earth meant

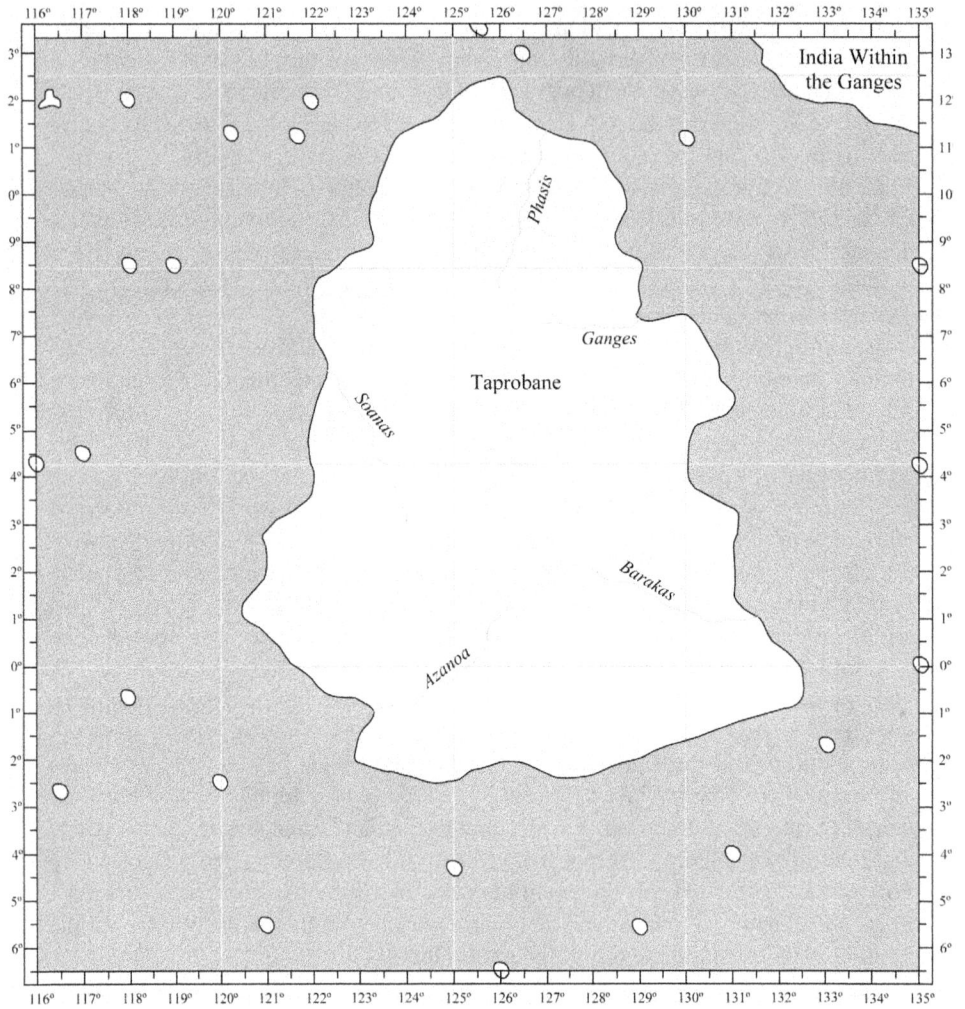

Map 6.12 Map 12 of Asia

serious miscalculations on the part of early modern explorers and was an issue not fully corrected until a method of accurate determination of longitude was established in the eighteenth century.

Map 12 of Asia

Taprobane

The island of Taprobane (modern Sri Lanka) lay to the south of India (*GG* 7.4.1–13), and is the last region discussed in the *Geographical Guide*. The island had been vaguely known since the time of Alexander the Great, but it was probably Megasthenes, a generation later, who began to gather specific information about the region and whose report was available to Eratosthenes.[149] From the very beginning there was confusion about its size and shape: early accounts had it five to eight thousand stadia in its largest dimension (perhaps 1000 to 1600 km.); this may be its circumference, measured today at 1340 km. But on Ptolemy's map it covers 15 degrees of latitude north-south, which supports the larger size (its extent is actually 4 degrees of latitude). It was also said to be a sail of 7 to 20 days from the mainland,[150] but this could not be across the narrow Palk Strait, which is only 55 km. wide, and is more probably a sailing distance from a point in northern India near the mouth of the Indos.

There was little contact between the Mediterranean world and Taprobane in Hellenistic times and it was only in the first century AD that extensive trade developed and the island became better known.[151] An expedition commissioned by a certain Annius Plocamus had gone to the island, perhaps looking for new trade markets, and this was shortly followed by Taprobanian envoys who came to Rome at the time of the emperor Claudius (reigned AD 41–54) and provided a detailed report on the island, but perhaps exaggerated its size. These expeditions may have been the basis of Ptolemy's data, but even then the island was improperly located, placed about 200 km. east of its actual position and far too large, with portions erroneously extending south of the equator.

Ptolemy's report has the usual list of coastal places, river mouths, and ethnic groups. Quite a few of the names are Greek, such as Cape Zeus, Cape Ketaion, or the Helios Harbor. This is a larger number of Greek toponyms than is usual for regions this far east, and perhaps resulted from the people of Annius Plocamus, who may have made a topographical survey of the coasts of the island. There are over 30 points of interest named, and nearly 20 islands, indicative of navigational data. In fact the coasts of Taprobane were said to have had 1378 islands.

But what is of particular interest is a detailed historical and ethnographic statement that begins Ptolemy's description of the island (*GG* 7.4.1):

Opposite the Kory Promontory of Indike lies the tip of the island of Taprobane, which was called Palai Simoundou, and now Salike.[152] Its inhabitants are collectively called the Salians, and the women completely bind up their hair. It produces rice, honey, ginger, *beryllos*, and hyakinthos, as well as gold, silver, and other metals, which are found everywhere, and also elephants and tigers.

This statement, of a type largely unique in the *Geographical Guide* (the only other example is at 7.2.15–17, in India Beyond the Ganges), was presumably taken from a report on the island, perhaps from one of the two expeditions previously mentioned, or even the obscure eastern traveler Alexandros. It is a summary of many of the major products of Taprobane, although the intrusive comment about women's hair may indicate that additional ethnographic material was removed. The island seems to have had many names, which may indicate local territories.[153]

With his description of Taprobane, Ptolemy's geographical catalogue comes to an end with a transitional paragraph (*GG* 7.4.14) letting the reader know that the previous sections provided information about the structure of the cataloguing of the places on the inhabited earth, and looking ahead to a summary of the inhabited world.

Notes

1 Gerhard Winkler and Florian Mittenhuber, "Die Länderkarten Asiens," in *Klaudios Ptolemaios: Handbuch der Geographie, Ergänzungsband* (ed. Alfred Stückelberger and Florian Mittenhuber, Basel 2009) 290–304.
2 *Periplous of the Erythraian Sea* 66.
3 *Periplous of the Erythraian Sea* 64.
4 Demodamas, *FGrHist* #428; Pliny, *Natural History* 6.49.
5 Pliny the Younger, *Letters* 10.179, etc.
6 Livy 39.51.
7 Herodotos 5.49; Agathemeros 1; Diogenes Laertios 2.1.
8 Herodotos 1.173.
9 Polybios 1.6 etc.
10 *BNP Historical Atlas of the Ancient World* 272.
11 Strabo, *Geography* 12.8.18.
12 Aeschylus, *Persians* 865; Herodotos 1.6, 75.
13 Getzel M. Cohen, *The Hellenistic Settlements in Europe, the Islands, and Asia Minor* (Berkeley 1995) 281–5.
14 *BNP Chronologies of the Ancient World* 110–11.
15 Strabo, *Geography* 12.1.4.
16 Pseudo-Skylax 81; Arrian, *Periplous of the Euxine Sea* 11.4; Graham Shipley, *Pseudo-Skylax's Periplous* (Exeter 2011) 157.
17 Duane W. Roller, *Cleopatra's Daughter and Other Royal Women of the Augustan Era* (Oxford 2018) 107.
18 David C. Braund, *Georgia in Antiquity* (Oxford 1994) 193–4.
19 Terence Bruce Mitford, "Roman Rough Cilicia," *ANRW* 2.7.2 (1980) 1243.
20 Herodotos 4.116–17.
21 Herodotos 4.45; *Airs, Waters, and Places* 13.
22 John Boardman, "Greek Archaeology on the Shores of the Black Sea," *AR* 9 (1962–1963) 41.
23 Strabo, *Geography* 11.2.10; Gocha R. Tsetskhladze, "A Survey of the Major Urban Settlements in the Kimmerian Bosporos (with a Discussion of their Status as *Poleis*)," in *Yet More Studies in the Ancient Greek Polis* (ed. Thomas Heine Nielsen, Stuttgart 1997) 51–5.
24 Strabo, *Geography* 7.3.19; Pliny, *Natural History* 4.83.
25 Strabo, *Geography* 11.2.3.
26 Hekataios F191–2.
27 Theophanes F5, 6; Strabo, *Geography* 11.4.1–8.

28 *BA* Map 85.
29 Herodotos 4.32; Bridgman, *Hyperboreans.*
30 Aristotle, *Meteorologika* 1.13.350b; Strabo, *Geography* 7.3.1; Jacques Desautels, "Les monts Riphées et les Hyperboréens dans le traité hippocratique *Des airs, des eaux et des lieux*," *RÉG* 74 (1971) 289–96.
31 Roller, *Ancient Geography* 102–4.
32 Braund, *Georgia* 175–6.
33 Theophanes, *FGrHist* #188.
34 Strabo, *Geography* 11.3.1–6; Braund, *Georgia* 205–9.
35 Patrokles, *FGrHist* #712.
36 Strabo, *Geography* 11.4.1–8.
37 Braund, *Georgia* 14, 46.
38 Pliny, *Natural History* 6.30; Roller, *Guide* 333.
39 Genesis 2:14, 8:4.
40 Eratosthenes, *Geography* F87.
41 Strabo, *Geography* 11.14.2.
42 Giusto Traina, "Strabone e le città dell'Armenia," in *Studi sull'xi libro dei Geographika di Strabone* (ed. Giusto Traina, Galatina 2001) 141–54.
43 Strabo, *Geography* 11.14.6.
44 Polybios 8.25; Pliny, *Natural History* 6.26; Strabo, *Geography* 11.14.15; Plutarch, *Lucullus* 25–8.
45 Ronald Syme, *Anatolica: Studies in Strabo* (ed. Anthony Birley, Oxford 1995) 58–65.
46 Strabo, *Geography* 1.15.2, 4; Roller, "Timosthenes" 56–79.
47 Strabo, *Geography* 14.6.3.
48 Strabo, *Geography* 16.2.8.
49 Pollard and Berry, *Complete Roman Legions* 136–7, 143–5.
50 Pliny, *Natural History* 12.63–5; Roller, *World of Juba* 233.
51 Hieronymos of Kardia T6 = Diodoros 19.98–100.
52 Philippe Seubert, "Délimitation et divisions de l'Arabie, d'Eratosthène à Ptolémée," *GA* 26 (2017) 23–36.
53 Pollard and Berry, *Complete Roman Legions* 156–7.
54 Hamish Cameron, *Making Mesopotamia: Geography and Empire in a Roman-Iranian Borderland* (Leiden 2018) 176–83.
55 Eratosthenes, *Geography* F87; Strabo, *Geography* 16.1.21–2.
56 Eratosthenes, *Geography* F95; Strabo, *Geography* 16.4.2.
57 Homer, *Odyssey* 3.270.
58 Strabo, *Geography* 16.1.26, 16.3.1.
59 Seubert, "Délimitation" 23–36.
60 Eratosthenes, *Geography* F52; Strabo, *Geography* 2.1.39; Michal Gawlikowski, "Thapsacus and Zeugma: The Crossing of the Euphrates in Antiquity," *Iraq* 58 (1996) 123–33.
61 Cohen, *The Hellenistic Settlements in the East From Armenia and Mesopotamia to Bactria and India* (Berkeley 2013) 129–30.
62 Pliny, *Natural History* 6.138–40.
63 Stückelberger and Graßhoff, *Klaudios Ptolemaios* 599.
64 Nahum 2:1–9.
65 Strabo, *Geography* 16.1.3.
66 Strabo, Geography 16.1.16; Pliny, *Natural History* 6.122; Cohen, *Hellenistic Settlements in the East* 100–1.
67 Herodotos 1.98–9.
68 Strabo, *Geography* 11.13.10; Roller, *Historical and Topographical Guide* 677.
69 Strabo, Geography 11.12.4–5; J. F. Standish, "The Caspian Gates," *G&R* second series 17 (1970) 17–24; Roller, *Historical and Topographical Guide* 163–4.
70 Cohen, *Hellenistic Settlements in the East* 209–10.

71 Strabo, *Geography* 15.3.4.
72 Pliny, *Natural History* 6.138–40.
73 Polybios 10.28.7; Strabo, *Geography* 11.9.1.
74 Cohen, Hellenistic Settlements in the East 204, 210–15.
75 Euripides, *Bacchants* 16.
76 Herodotos 4.44; Skylax, *FGrHist* #709. Skylax is not to be confused with the extant treatise of the mid-fourth century BC that has survived under his name: Shipley, *Pseudo-Skylax's Periplous* 4–18.
77 Juba F63–5; Pliny, *Natural History* 12.64–6.
78 Roller, *Ancient Geography* 99–102.
79 Casson, *Periplus* 157–8.
80 Strabo, *Geography* 16.3.3.
81 *Periplous of the Erythraian Sea* 27; Casson, *Periplus* 161.
82 Casson, *Periplus* 178; see also Pliny, *Natural History* 6.110.
83 Roller, *Ancient Geography* 100–2.
84 Agatharchides, *On the Erythraean Sea* (ed. Stanley M. Burstein, London 1989).
85 *Periplous of the Erythraian Sea* 30; Pliny, *Natural History* 6.153.
86 Quintus Curtius 9.10.24–8.
87 Herodotos 1.184; Ktesias of Knidos F1b.
88 Pliny, *Natural History* 6.107; Cohen, *Hellenistic Settlements in the East* 181–2.
89 Roller, *Ancient Geography* 103–4.
90 Patrokles, *FGrHist* #712.
91 Herodotos 1.205–14.
92 Strabo, *Geography* 11.7.3; 11.11.5.
93 Demodamas, *FGrHist* #428; Pliny, *Natural History* 6.49–50; John R. Gardiner-Garden, "Greek Conceptions on Inner Asian Geography and Ethnography From Ephoros to Eratosthenes," *Papers on Inner Asia* 9 (Bloomington, Ind. 1987) 44–8.
94 Vergil, *Georgics* 2.121; Lucan 10.142.
95 For examples, see W. W. Tarn, *The Greeks in Baktria and India* (revised third edition, Chicago 1997) 89, 231–2.
96 Hekataios of Miletos F291.
97 Strabo, *Geography* 2.1.14.
98 Pliny, *Natural History* 6.46–7; Cohen, *Hellenistic Settlements in the East* 245–50.
99 Cohen, *Hellenistic Settlements in the East* 220–1.
100 Ammianus Marcellinus 23.6.54.
101 Herodotos 3.93, 9.113.
102 Strabo, *Geography* 11.8.2; Tarn, *Greeks* 270–311.
103 Cohen, *Hellenistic Settlements in the East* 273, 279–82.
104 Strabo, *Geography* 11.11.2; Tarn, *Greeks* 114–15.
105 Herodotos 3.93; 7.66; Josef Markwart, "Die Sogdiana des Ptolemaios," *Orientalia* 15 (1946) 123–49.
106 Roller, *Historical and Topographical Guide* 652.
107 Arrian, *Anabasis* 4.1.3–4; Cohen, *Hellenistic Settlements in the East* 252–5.
108 Cohen, *Hellenistic Settlements in the East* 269–71.
109 J. R. Hamilton, "Alexander and the Aral," *CQ* 21 (1971) 106–11.
110 Patrokles F5; Eratosthenes, *Geography* F109; Strabo, *Geography* 11.7.3.
111 Ephoros F30a; Strabo, *Geography* 1.2.28.
112 Herodotos 4.1–142.
113 This statement does not appear in most manuscripts, but given Ptolemy's avoidance of ethnographic comments may have been excised at a late date: see Stückelberger and Graßhoff, *Klaudios Ptolemaios* 654.
114 The conventional equivalent of 26,280 stadia would be about 5000 km., but over such a large distance any conversion is unreliable.

115 *Periplous of the Erythraian Sea* 66.
116 Strabo, *Geography* 15.1.20.
117 Pliny, *Natural History* 6.88; Samuel Lieberman, "Who Were Pliny's Blue-Eyed Chinese?" *CP* 52 (1957) 174–7.
118 Under the standard conversion, this would be 7200 km.
119 Baiton, *FGrHist* #119.
120 Strabo, *Geography* 11.8.9; Cohen, *Hellenistic Settlements in the East* 260–1.
121 Aeschylus, *Prometheus Bound* 422, 717–21.
122 Eratosthenes, *Geography* F23; Strabo, Geography 11.5.5; Pliny, *Natural History* 6.30, 49.
123 Strabo, *Geography* 15.2.10.
124 Isidoros, *Parthian Stations* 19; Cohen, *Hellenistic Settlements in the East* 255–60.
125 Strabo, *Geography* 11.8.9.
126 Strabo, *Geography* 15.2.3–8.
127 Herodotos 3.97–102, 4.44.
128 Diodoros 17.93; Quintus Curtius 9.2.2–3; Arrian, *Anabasis* 5.26.
129 Megasthenes, *FGrHist* #715.
130 Strabo, *Geography* 2.5.12.
131 Augustus, *Res gestae* 31.1; Nikolaos of Damascus F100; Strabo, *Geography* 15.1.73; Orosius 6.21.
132 Casson, *Periplus* 50–93.
133 Casson, *Periplus* 228–9.
134 Eratosthenes, *Geography* F74; Strabo, *Geography* 15.1.13.
135 Klaus Karttunen, *India in Early Greek Literature* (Helsinki 1989) 101.
136 *Periplous of the Erythraian Sea* 47; Cohen, *Hellenistic Settlements in the East* 308–12.
137 Didier Marcotte, "Ptolémée ethnographe: géographie et traditions textuelles," *GA* 26 (2017) 47–60.
138 *Malabathron* is a variety of cinnamon leaf: see *Periplous of the Erythraian Sea* 56, 65; Casson, *Periplus* 220.
139 Stückelberger and Graßhoff, *Klaudios Ptolemaios* 723.
140 Plato, *Ion* 533d; Theophrastos, *On Stones* 4; *GG* 7.2.31; Earle R. Caley and John C. Richards, *Theophrastus on Stones* (Columbus, OH 1956) 67.
141 Dioskourides 5.130; Pliny, *Natural History* 36.126–30.
142 Dalby, *Food* 45–8.
143 Casson, *Periplus* 235–6.
144 Mortimer Wheeler, *Rome Beyond the Imperial Frontiers* (Harmondsworth 1955) 203.
145 *Periplous of the Erythraian Sea* 64.
146 Stückelberger and Graßhoff, *Klaudios Ptolemaios* 733.
147 Ephoros F30a.
148 Hipparchos, *Against the Geography of Eratosthenes* F4; Polybios 3.38; Strabo, *Geography* 1.1.9.
149 Eratosthenes, *Geography* F74.
150 Strabo, *Geography* 15.1.14–15.
151 Pliny, *Natural History* 6.81–91.
152 The reading may be "which was called Simoundou in former times."
153 Casson, *Periplus* 230–2; Didier Marcotte, "Le *Périple de la Mer Érythrée* et les informateurs de Ptolémée: géographie et traditions textuelles," *JA* 304 (2016) 33–46.

7 The Final Portion of the *Guide*

Introduction

With his transitional paragraph at 7.4.14, Ptolemy moved from his catalogue of toponyms to a technical discussion, including a broad description of the surface of the inhabited world (*GG* 7.5.1–16), of how to map a globe (*GG* 7.6.1–7.7.4), and how the inhabited world could be represented on 26 maps (*GG* 8.1.1–8.28.5). These concluding books of the *Geographical Guide* are a matter of map construction and the technique of plotting, and as such are of lesser interest to students of topographical geography.[1] There are issues with the material: to some extent it repeats previous information, but oddly coordinates can be slightly different from those in the earlier catalogues. It has even been suggested that some of these chapters are not the work of Ptolemy, or part of a different treatise by him, and that the manuscript tradition amalgamated the material, but this seems unlikely.[2] As is the case with Book 1, the section headings provided for the remainder of Book 7 and for Book 8 are as presented in the manuscripts.

Summary Description of the Plan of the Inhabited World

Ptolemy had recorded thousands of toponyms, along with their coordinates. He then produced a summary of the entire inhabited world, with its main surface features, immediately establishing that there were three continents, attributing this to unnamed "older sources." Continental theory had developed by the early fifth century BC: the original two were Europe and Asia, and the third continent of Libya was soon added.[3] In Ptolemy's time the existence of these three continents was no longer a matter of dispute, and after a description of the boundaries of the inhabited world, Ptolemy outlined the limits of each.

The extremities of the inhabited world are presented briefly in clockwise fashion, beginning with Sinai and Serike in the far east, and continuing around the south to the Western Ocean, and then north to Brettanike and the Northern Ocean, returning to the east. An important point is that the Indian Ocean (more properly designated as the Indian Sea) was enclosed, a fringe theory that may have originated with Hipparchos and was supported by Polybios.[4] Ptolemy's assertion of this placed the idea more in the mainstream, although oddly nothing was known about the presumed southern coast of the Indian Sea, or even where it was.

DOI: 10.4324/9781003248590-8

Within the inhabited world were two additional enclosed seas, Our Sea (the Mediterranean) and the Hyrkanian or Caspian Sea. To be sure, the Mediterranean was actually not enclosed—it had the one outlet at the Strait of Herakles—but this connection to the External Ocean was so limited that the sea was considered enclosed. Ptolemy's insistence on this may be due to a need to reject the belief that the Maiotic Lake—the modern Sea of Azov—connected at its northern extremity with the External Ocean. This was an outlier theory that survives only in the *Periplous of the Erythraian Sea*, perhaps based on the idea that the Tanais River (modern Don) was a passageway to the Ocean. There may have been more support for the idea in Ptolemy's time than is apparent from the single extant source.[5]

After defining the limits of the inhabited world Ptolemy discussed the boundaries of the individual continents, where they joined (or came close to one another), and the major bays and islands or peninsulas of each. He reiterated that the Indian Sea was enclosed, and furthermore noted that it was the largest of the three enclosed seas, but still provided no information about its southern coastline. He then viewed the continents in terms of their hydrological features, with the three enclosed seas, ten bays, and ten islands or peninsulas ranked according to size. The source may be a lost catalogic work ranking topographical features by size, but, as presented by Ptolemy, there is no data to justify the listing. The ten bays and ten islands are reasonable, since their dimensions tended to be known.

Ptolemy's next task was to provide the actual distances between the extremities of the known world. He began by reminding his readers that he was using a system that presumed a total circuit of the earth of 360 degrees, a standard established in Hellenistic times from Babylonian sources. Thus he determined that the known world extended from 16 5/12 degrees south of the equator to 63 degrees north. The former figure was at a point slightly south of Cape Prason on the east coast of Africa, reached by the explorer Dioskoros in the late first century AD (*GG* 1.9.4); the latter was the latitude of Thoule, which Pytheas of Massalia visited in the late fourth century BC.[6] This meant that the inhabited world extended for about 80 degrees north-south, and, using a calculation of 500 stadia to a degree, it meant 40,000 stadia for this distance and 180,000 stadia for the perimeter of the earth.[7]

Similar calculations were provided for the east-west extent, although more uncertain because of the difficulty in calculating longitude. The figure was determined at 119 1/2 degrees, taking into account the change in the size of a degree of longitude as one moved away from the equator. The crucial figure is that the east-west dimension of the known world (from the Blessed Islands to the Sinai metropolis) was 90,000 stadia or 180 degrees at the equator, a gross error due to the repeated east-west elongations scattered throughout the catalogues of the *Geographical Guide*.

Distance measurements, especially calculation of lengthy ones, were always problematic: the most common unit in Greek antiquity, the stadion, was highly variable.[8] The Roman mile (1.48 km.) was standardized, but Ptolemy only used it twice, in both cases quoting and rejecting Marinos (*GG* 1.15.6, 9). Regardless, all ancient measurements were inaccurate, even those with a pretense of accuracy, because they were generally converted from ancient data, especially the primeval measurements over long distances: camel days and sailing times. Yet Ptolemy, by using degrees, which limited him to fractions of 360, created a somewhat greater sense of exactness, but longitude remained a problem. Latitudes could more easily be calculated because of another factor, the length of the longest (or shortest) day. This had been used at least since Pytheas, and was a central part of the methodology of Eratosthenes.[9] Ptolemy was able to use length-of-day calculations for many of his topographical points. Nevertheless these are still highly defective, but, as with earlier scholars such as Eratosthenes and Hipparchos, his efforts were a valiant attempt to determine long distances.

The Delineation of a Ringed Sphere with the Inhabited World

Next Ptolemy described how a ringed sphere could be mapped (*GG* 7.6.1–15). By "ringed sphere" he meant a spherical depiction of the earth with latitude and longitude lines on it, which created the rings, or circles.[10] The word he used was *krikotes*, based on the standard Greek word for "ring" (*krikos*), but seemingly not used previously in this sense. This section of the *Geographical Guide* consists of technical material that belongs in the discipline of mathematics rather than geography.[11] Since Ptolemy was largely concerned with mapping, he believed it necessary to describe in detail how it would be possible to create a map on the curved surface of the spherical earth.

It had been understood since the fifth century BC that the earth was a sphere: this idea seems to have originated among Pythagorean sources.[12] This was difficult to accept: even as late as the second half of the third century BC Eratosthenes found it necessary to emphasize that his efforts presumed a spherical earth.[13] Yet it was only a matter of time before there was an attempt to create a model of it, by Krates of Mallos in the mid-second century BC. It is not known whether other globes existed by the time of Ptolemy, but he certainly had no reservations about the existence of a such a device and its utility to him.

The technical arguments advanced by Ptolemy in order to create a spherical map of the earth include determination of the appropriate size of the globe, where and how to begin the plotting of points, determination of a proper viewpoint, and how to construct the parallel (latitude) and meridian (longitude) lines. Interestingly Ptolemy would use different colors (*chromata*) for the lines, and they would have appropriate names (presumably topographical). The names of the winds would also be depicted, which could be used in determining directions. He did not indicate what toponyms would appear on the globe; the only one mentioned in this section is Syene in Upper Egypt at the summer (northern) tropic. This may be a recognition of the role that the locale played in Eratosthenes' determination of the circumference of the earth.[14] There is also passing mention of the winter (southern) tropic, necessary for a complete comprehension of the nature of the spherical earth but well to the south of any point reached by Greco-Roman explorers. In fact, the creation of a plan of the spherical earth allowed the plotting of places beyond the inhabited world; in addition to the winter tropic, Ptolemy necessarily assumed the existence of an Antarctic circle, probably a recent concept that literally meant "opposite the Bear," or at the opposite end of the cosmos from the constellation with that name.[15]

Summary Description of the Unfolding

The final section of Ptolemy's technical analysis of how to make a map of the surface of the earth concerns the issue of transferring the data on the ringed globe to a flat surface, in other words, as he expressed it, "unfolding" the representation (*GG* 7.7.1–4).[16] He had dealt with some of the issues of the relationship between a spherical and flat plan (in this case the celestial sphere) in his earlier work, *Planisphairion* (known today only in an Arabic translation).[17] In the *Guide* he used the rare word *ekpetasma*, which was the title of a work by Demokritos of Abdera from the fifth century BC.[18] Whether Demokritos' usage was in a similar context as that of Ptolemy remains uncertain, but he was, among many other disciplines, an early geographical theorist.[19] Mapping the spherical earth on a flat surface has long been an issue in cartography, and Ptolemy may have been the first to consider the matter. He was aware of the essential problem in "unfolding" the map: the failure of meridian lines to converge (as well as some anomalies to parallel lines), but he did not seem to consider this of concern, calling it "an illusion."

A concluding statement to Book 7—which has the appearance of an addendum that is not truly relevant to the mapping arguments just presented—is that the External Ocean would not be represented as flowing around the known world. It would be shown only on the borders of Libya and Europe (and, in the case of the latter, only as far north as the direction of the Thraskias, or north-northwest, wind). This is a strangely anomalous statement, since the idea of an encircling Ocean had existed since early times and was generally accepted.[20] But Ptolemy had been influenced by Hipparchos' theory of an enclosed Indian Ocean, and Hipparchos did not believe in an encircling Ocean; thus he may have been the source of Ptolemy's belief about the nature of the Ocean and one of "the more ancient writers" that he consulted at this point.[21]

The Basis for Dividing the Inhabited World into Maps

Having noted that he has presented everything that should be included in the *Geographical Guide*, Ptolemy believed that it was not necessary to catalogue all the places through which the parallels and meridians should be drawn (*GG* 8.1.1–6). He explicitly criticized his unnamed predecessors, since it had been common to create parallels through known points, a technique obviously causing difficulty. This was what Dikaiarchos of Messana had done in the late fourth century BC, who was the first to create a base parallel across the inhabited world.[22] His successors, most notably Eratosthenes, followed this practice; one suspects Marinos did the same.

Ptolemy then outlined his technique for dividing the inhabited world into local maps. The basic problem he confronted was one of scale: some maps would be crowded with toponyms (for example, Italy or Asia Minor), and others would be largely blank (such as most of Sarmatia in Asia). Ptolemy criticized his predecessors for squeezing in the margins of their maps to make everything fit, and for assuming (especially in the far east) that when they ran out of toponyms they were at the coast of the External Ocean; this was especially apparent in the little-known regions of northern Europe and Asia. Such objections gave Ptolemy yet another chance to reject the idea of an encircling Ocean. Yet he noted that this problem would be avoided by a proper selection of regional maps, and that there was no necessity to have all the maps at the same scale (*summetros*) or that they cover the same amount of territory. But he emphasized that the scale must remain constant on each individual map, and as an example provided an interesting analogy from artistic drawing.

He then returned to the issue of straight meridians and parallels. He was dismissive of this, but noted that while it could be a problem in a map of the entire known world, it would not be the case on the regional maps, since they covered smaller areas. This of course does not mean that the regional maps were lacking in positional errors, as has been frequently noted previously, but these were not due to the straight meridian and parallel lines.

As an introduction to the captions for the 26 regional maps (10 of Europe, 4 of Libya, and 12 of Asia), Ptolemy listed the material included on each of them.[23] There is the basic caption (continent and sequence number), the ratio of the parallel through its middle to the meridian, and the boundary of the map. There also is a list of the major cities and the length of the longest day (as expressed in equinoctial hours) at each, and the longitude of each city, defined in terms of equinoctial hours east or west of Alexandria.[24] Since the actual maps do not survive, it is not known whether these captions appeared on the maps themselves, or, as they do now, in a separate list.[25]

Ptolemy also stated that he would have liked to have related each locality to the fixed stars that were at the zenith in each place. By suggesting this he showed his grounding in astronomy, but this was impossible to do because of the shifting of the sphere of the fixed stars, something he had already shown in his most famous work, the *Mathematical Syntaxis*.[26] In fact, he noted that such a supplement to the captions would be superfluous, since the star globe described in the *Mathematical Syntaxis* could be used to establish the position of the stars. Thus to end the narrative of the *Geographical Guide* Ptolemy moved away from geography and returned to mathematical astronomy.

The final section of the *Geographical Guide* is the captions themselves for each of the 26 maps, with the data laid out as Ptolemy described it (*GG* 8.3.1–8.28.5). There is no conclusion to the work: the final entry is the caption to Map 12 of Asia, Taprobane, without further comment.[27]

Notes

1 This material has been discussed in detail by Berggren and Jones, *Ptolemy's Geography* 108–22, with a partial translation; see also Gerd Graßhoff et al., "Of Paths and Places: The Origin of Ptolemy's Geography," *AHES* 71 (2017) 483–503.

2 Berggren and Jones, *Ptolemy's Geography* 5.

3 Roller, *Ancient Geography* 50–1.

4 Hipparchos, *Against the Geography of Eratosthenes* F4; Polybios 3.38; Strabo, *Geography* 1.1.9.

5 *Periplous of the Erythraian Sea* 64; Casson, *Periplus* 239–41.

6 Strabo, *Geography* 1.4.2, 2.5.8; Pliny, *Natural History* 2.186–7.
7 This is quite different from Eratosthenes' 252,000 stadia (and far less accurate); Eratosthenes, *Geography* F28; *Measurement of the Earth* F1–7.
8 Diller, "Ancient Measurements" 6–9.
9 Eratosthenes, *Geography* F59–60. Pytheas actually used solar height at the solstice, relating it to that at Massalia; at some time his numbers were converted to latitudes, but under what circumstances is not known: see Roseman, *Pytheas* 42–4.
10 Otto Neugebauer, "Ptolemy's Geography, Book VII, Chapters 6 and 7," *Isis* 50 (1959) 22–9.
11 See further, Berggren and Jones, *Ptolemy's Geography* 112–16, for a translation and commentary with several useful diagrams.
12 Diogenes Laertios 8.48, 9.21.
13 Eratosthenes, *Geography* F25; Strabo, *Geography* 1.4.1.
14 Eratosthenes, *Measurement of the Earth* M6.
15 Geminos 5.16, 28.
16 Neugebauer, "Ptolemy's Geography" 22–9.
17 J. Lennart Berggren, "Ptolemy's Maps of Earth and the Heavens: A New Interpretation," *AHES* 43 (1991) 133–44.
18 Diogenes Laertios 9.47.
19 Roller, *Ancient Geography* 74–5.
20 Thomson, *History* 97–9.
21 Hipparchos, *Against the Geography of Eratosthenes* F4; Polybios 3.38; Strabo, *Geography* 1.1.9.
22 Keyser, "Geographical Work" 353–72.
23 The word translated as "caption" is *hypographein*, literally meaning "something written underneath."
24 Because the length of the day varied according to latitude, and the ancient day was divided into 12 hours regardless of its length, an hour at the equinox, which never changed in length, provided a standard measurement.
25 See Appendix 3 for a sample caption.
26 Ptolemy, *Mathematical Syntaxis* 8.3.
27 There is supplementary material in some of the manuscripts which largely repeats previous data and does not seem to have been part of Ptolemy's original work: *GG* 8.29.1–8.30.26. Stückelberger and Graßhoff, *Klaudios Ptolemaios* 909–21, rightly called it an "addendum of uncertain date."

Epilogue

Introduction

As early as the fourth century BC it was realized that it would be possible in theory to sail west from the Pillars of Herakles to India. In fact this would have been obvious since it had been believed for over a century that the earth was a sphere, a Pythagorean idea based on the essential harmony and mathematical perfection of the cosmos.[1] Aristotle seems to have been the first to suggest that a voyage west to India was possible, and that there might even be a land connection, reasoning based on the presence of elephants in both northwest Africa and India.[2] Eratosthenes thought such a sail was possible but the size of the intervening Ocean made it improbable.[3] Poseidonios had similar thoughts.[4]

Yet the distance from Iberia to India was so vast—Poseidonios thought it was 70,000 stadia (perhaps 14,000 km.)—that it began to be suggested there might be an intervening and unknown continent. Such speculations were more fantasy than the product of serious scholarship, often serving moral or ethical needs. These began with Plato's Atlantis, and the theme was further developed in the Hellenistic and Roman periods.[5] Plutarch, not long before Ptolemy, described an island 5000 stadia (about 1000 km.) in size west of the British Isles, with extensive land beyond it.[6] Slightly earlier than Plutarch, Seneca wrote:[7]

> There will come an era in the future when the Ocean will release its chaining of things and the full extent of the earth will be revealed. Tethys will disclose new worlds and Thule will not be the farthest land.

Seneca's phrase "new worlds" (*novos orbes*), reduced to the singular and becoming a proper noun, became a hypothetical toponym that inspired early modern explorers, and is still a viable one today. When Ptolemy was able to provide seemingly precise distances for the extent of the inhabited world—180 degrees or halfway around it—conditions were set for attempting the western voyage that Aristotle had assumed was possible and to discover Seneca's New World. By the fifteenth century, improvements in seamanship made the voyage plausible.

DOI: 10.4324/9781003248590-9

Ptolemy and Columbus

Of all the people the *Geographical Guide* affected, none was more profoundly energized than Christopher Columbus (1451–1506). He was a complex and contradictory personality whose life is still not fully understood, but his use of Ptolemy's treatise was a central part of the planning and execution of his voyages to the New World.

Living as he did at the very beginning of modern printing, he was able to amass a library that was astounding for the era. It included the Latin translation of the *Geographical Guide* that had been published in Rome in 1478 by Arnoldus Buckinck.[8] He was familiar with the other ancient geographical writers, including Strabo, Seneca, and Pliny. These gave him access to Eratosthenes, Nearchos, Onesikritos, and Juba II, and Ptolemy provided him with everything that was known about Marinos.[9]

Thus he had all that was available from ancient sources about the size and extent of the earth. He also knew the journal of Marco Polo (1254–1324), which seemed to indicate that the inhabited world extended farther east than even Ptolemy believed.[10] Columbus became astute in tweaking the data to his advantage, using as his starting point Seneca's statement that it was only a few days' cruise to cross the Atlantic.[11] He used the smallest possible size of a degree of latitude, probably made inaccurate conversions of units of measurements, and misread the ninth-century geographer Alfragan, who, he believed, had said that a degree of longitude was 45 miles.[12] But his primary source remained Ptolemy: in a letter of 1498 he wrote, "[V]ery little was known of any other land beyond what Ptolemy wrote."[13]

Ptolemy, of course, had determined the width of the known world at 180 degrees, so it was obvious that the remainder was the same distance. But Columbus went farther and used the figure of 225 degrees of Marinos, and this became the centerpiece of his determination of the width of the unknown world, thus giving ancient authority to his own assumptions. It was no matter that Ptolemy had vigorously rejected Marinos' figure. Further adjustments meant that Columbus believed he would only have to sail 2400 miles from the Canaries to Asia, only a quarter of the actual distance. Like anyone applying for a grant, from ancient to modern times, he knew how to put his proposal in the best possible light. But when he presented these calculations to the navigation committee of King João II of Portugal, the fallacy was recognized and the grant application was rejected. He then went to Ferdinand and Isabella of Spain.[14]

Even after he started his voyages, Ptolemy continued to be a factor in Columbus's thoughts. In 1494, when he reached what is now the Bahía de Cortés in western Cuba, where the coast turns sharply to the south, he believed that he was on the eastern side of Ptolemy's Golden Chersonesos. From there it would be easy to go around the peninsula and make the first voyage around the world. Yet he decided against the idea and turned back, due to the deterioration of his ships and low supplies.[15] But he believed that he had reached the continent of Asia and the eastern edge of Ptolemy's map of the inhabited world.

The Search for the Source of the Nile

As more was learned about the inhabited world after the time of Columbus, with the realization that it extended far beyond what Ptolemy had imagined, the *Geographical Guide* became less viable as a document for exploration. The last major use of the work came in the nineteenth century, when Victorian explorers attempted to solve one of the final great geographical mysteries for which Ptolemy could provide help: the source of the Nile. This had eluded geographers and explorers from ancient to modern times. The sources of all the other great rivers of the ancient world—the Rhone, Rhine, Danube, Tigris, and Euphrates—had been discovered by the late first century BC, but the Nile remained elusive, its upper course lost in the wilds of Africa.

Because of this, and other unusual qualities, such as its floods, the Nile was unique among the rivers of antiquity. It had been known since earliest times,[16] and by the fifth century BC Greeks were aware of its course for many months' travel upstream, but, as Herodotos pointed out, no one had any idea of its source.[17] In fact, it became a mythic construct: the reputation of the athlete Phylakidas of Aigina was linked with that of heroes known beyond the headwaters of the Nile.[18] Ptolemaic explorers went far upstream, perhaps into southern modern Sudan, but the final stages of the upper river remained elusive, although there was some hearsay knowledge about the mountains of central Africa.[19] Alternative theories developed, especially that the source was not to the south, but to the west in Mauretania: this was promoted by Juba II in his *Libyka*. Since he was king of Mauretania, such a view fit into his political agenda. Although the theory may seem unreasonable, Juba collected zoological and geographical data that seemed to support the idea, and there may have been Carthaginian antecedents.[20] But the origin of the Nile remained an elusive myth, yet one so powerful that academic controversies broke out about the river, and Julius Caesar allegedly said that he would give up the civil war if he could find the source.[21]

It was the emperor Nero who commissioned the final major search in ancient times for the headwaters of the river. His explorers went far upriver, allegedly 1996 miles, which is impossible yet indicative of a great distance. They passed through an extensive marshy region so thick that their boats became stuck, perhaps the Sudd of southern Sudan,[22] and witnessed—or perhaps heard about—a locality called the Veins of the Nile, seemingly a spring or waterfall. It is tempting to believe that they had reached the uplands of central Africa, and that the veins were the narrow channel of Murchison Falls, although this cannot be proven. What is certain, however, is that Nero's explorers went farther upstream than anyone in antiquity, probably into northern modern Uganda.

Ptolemy does not seem to have known about these explorers. But he had other significant information that set the stage for the actual discovery of the source in modern times. Sometime around AD 100 Diogenes, sailing along the East African coast, reported that he had reached the lakes that were the source of the river (*GG* 1.9.1). That information, obviously inaccurate, conflicted with a better but unattributed report that the lakes were "quite a distance inland" (*GG* 4.7.26). A further notice referred to the Selene Mountains, "from which the lakes of the Nile receive snow" (*GG* 4.8.3). On Map 4 of Libya they were placed 9 to 17 degrees of longitude inland, and 6 degrees of latitude north of the Selene Mountains. Ptolemy was aware that the upper Nile had two main branches, and his consistent tendency to have rivers originate in mountains placed their sources in the Selene Mountains: the rivers then passed through the lakes until they joined at 2 degrees north latitude.

Although the interest of Diogenes and perhaps other contemporary informants in locating the source of the Nile seems minimal, and may not have been due to autopsy but information received at Rhapta and elsewhere (hence the thought that the lakes were near the coast), the idea developed that the headwaters should not be reached by the arduous, indeed virtually impossible, route upstream, but by going inland from the East African trading posts. But this was not implemented until the nineteenth century, and even then there were still attempts to go upriver.[23]

By the middle of the nineteenth century, the mythic quality of the search for the source of the Nile, and indeed the upper river, was rampant. Much of the interest in the English-speaking world centered on the location of the Selene Mountains, now anglicized to the Mountains of the Moon. Edgar Allan Poe in his "Eldorado" and Henry David Thoreau in his *A Week on the Concord and Merrimac Rivers*, both published in 1849, turned the mountains into a paradigm; Poe, at least, connected them with the California gold rush as an elusive goal. Neither, of course, was ever anywhere close to the actual mountains.

This interest led to numerous physical attempts to find the places that Ptolemy mentioned: subsequent to the middle of the nineteenth century there were repeated attempts to find the source of the Nile and the Mountains of the Moon, which Ptolemy seemed to believe were linked. The two most important explorers were Richard Francis Burton (1821–1890) and John Hanning Speke (1827–1864), who began as colleagues but became bitter rivals. Both used Ptolemy to their own advantage and mounted several expeditions into interior Africa seeking to find the source of the river and the mountains.

This is not the place to discuss in detail these and other expeditions. Using Ptolemy as their guide, Speke and Burton set out from Zanzibar in late 1857. The following summer they separated, and in July of 1858 Speke reached the great lake that he named Victoria Nyanza, which he firmly believed was one of the two lakes mentioned by Ptolemy as existing north of the Mountains of the Moon on the upper Nile. He further confirmed his thoughts on another expedition in 1862:[24]

> I saw that Old Father Nile without any doubt rises in the Victoria N'yanza, and, as I had foretold, that lake is the great source of the holy river.

Burton, however, who was unable to make the original 1858 journey to the lake, insisted that Victoria Nyanza was not the source of the Nile, but that it was Lake Tanganyika to the south, allegedly Ptolemy's western lake. In a paper presented to the Royal Geographical Society shortly after Speke's death he wrote:[25]

> The real sources of the Nile—the "great Nile problem"—so far from being "settled for ever" by the late exploration, are thrown farther from discovery than ever before.

But Burton was greatly misled: even though Lake Tanganyika might have been one of the lakes on Ptolemy's map, it was not connected to the Nile, but was west of the African continental divide and part of the upper Congo River system.

Dominating the thoughts of both explorers was the other feature that Ptolemy showed on his map, following his view that all rivers originated in mountains. This was the Selene Mountains, the Mountains of the Moon, which Burton was given to calling, in an ingenious pun, the Lunatic Mountains, because the search for them drove people crazy.[26] Generally they have been identified with the Ruwenzori Mountains of western Uganda, on the continental divide and whose highest point is both the highest peak in Uganda (5109 m.) and the third highest peak in Africa.[27]

The discovery of the source of the Nile and, assumedly, the Selene Mountains and the lakes on Ptolemy's map, was a classic example of Victorian exploration and perhaps one of the last major uses of the data supplied by the *Geographical Guide* about remote places on the earth. There remain many places mentioned by Ptolemy that have not been identified, but the explorations of Burton and Speke, accompanied by the contemporary expeditions of others such as Henry Morton Stanley (1841–1904) and David Livingston (1813–1873) were a collective effort unlike none other to determine the location of some of the most remote places mentioned by Ptolemy.

Notes

1 Diogenes Laertios 9.21–2; Strabo, *Geography* 2.2.2.
2 Aristotle, *On the Heavens* 2.14.298a. In 1494 Columbus believed that the alligators he saw in Cuba proved that he was in India, confusing them with the Indian crocodile (M. Cary and E. H. Warmington, *The Ancient Explorers* [Baltimore 1963] 259).
3 Eratosthenes, *Geography* F33; Strabo, *Geography* 1.4.6.
4 Poseidonios F49; Strabo, *Geography* 2.3.6.
5 Plato, *Timaios* 24–6; Roller, *Through the Pillars* 52–3.
6 Plutarch, *Concerning the Face That Appears on the Globe of the Moon* 26.
7 Seneca, *Medea* 375–9.
8 Henry N. Stevens, *Ptolemy's Geography: A Brief Account of All the Printed Editions Down to 1730* (second edition, London 1908) 38.
9 Paolo Emilio Taviani, *Christopher Columbus: The Grand Design* (London 1985) 174–7, 448–9.
10 V. Frederick Rickey, "How Columbus Encountered America," *MM* 65 (1992) 223–4.
11 Seneca, *Natural Questions* 1, *Preface* 13.
12 Samuel Eliot Morison, *Christopher Columbus, Mariner* (New York 1955) 18.
13 Taviani, *Christopher Columbus* 435.
14 Rickey, "How Columbus" 223–4.
15 Morison, *Christopher Columbus* 93.
16 Homer, *Odyssey* 4.477; Hesiod, *Theogony* 335.
17 Herodotos 2.28–32.
18 Pindar, *Isthmian* 6.22–3.
19 Pliny, *Natural History* 6.191, 194.
20 Pliny, *Natural History* 5.51–4; Ammianus Marcellinus 22.15.8; Roller, *World of Juba* 192–6.
21 Strabo, *Geography* 17.1.5; Lucan 10.191–2.
22 Seneca, *Natural Questions* 4a.2.7, 6.8.3–5; Lucan 10.325–6.

23 Most notably Samuel Baker (1821–1893) and his companion Florence (1841–1916), who discovered and named Murchison Falls in 1864, but were unable to go any farther.

24 John Hanning Speke, *Journal of the Discovery of the Source of the Nile* (New York 1868) 429.

25 Richard F. Burton, "On Lake Tanganyika, Ptolemy's Western Lake-Reservoir of the Nile," *JRGS* 35 (1865) 1–15.

26 Edward Rice, *Captain Sir Richard Francis Burton* (New York 1990) 355.

27 Stückelberger and Graßhoff, *Klaudios Ptolemaios* 469.

Appendix 1

The Text of the Ivernia Section of the *Guide*

The following is a translation of the first toponymic portion of the *Geographical Guide*, describing Ivernia (Ireland) and the surrounding islands (*GG* 2.1.1–12). It reproduces the columnar nature of the manuscripts, although it is by no means certain when the heading and section divisions became attached to the text. The first figure is the longitude, and the second the latitude. Fractions have been presented in the modern form. This format and style are used consistently throughout the catalogues; significant manuscript variants appear in parentheses after the main entry. Nevertheless this shows the pattern used throughout the *Guide*, with the tendency to alternate lists of toponyms and their coordinates with descriptive narrative statements, often with ethnyms. The translation is based on the text of Stückelberger and Graßhoff, *Klaudios Ptolemaios* 142–6.

The Placement of Ivernia, a Brettanikan Island (First map of Europe).

1 Outline of the northern side, beyond which lies the Hyperboreian Ocean.
2 Cape Boreion   11   61

Cape Venniknion	12½	61⅓
the mouth of the Vidua River	13	61
the mouth of the Argita River	14½	61½
Cape Rhobogdion	16⅓	61½

3 The Venniknioi live on the western side, and the Rhobogdioi toward the east.
4 Outline of the western side, where the Dytikos Ocean lies.

After Cape Boreion, there is	11	61
the mouth of the Ravius River	11⅓	60⅔
the city of Nagnata	11¼	60¼
the mouth of the Libnios River	10½	60
the mouth of the Ausoba River	10½	59½
the mouth of the Senos River	9½	59½
the mouth of the Dur River	9⅔	58⅔

the mouth of the Iernos River	8	58
Cape Notion	7⅔	57¾

5 On the west side, after the Venniknioi, live the Erdinoi, and below them the Nagnatai (Magnatai), then the Auteinoi, then the Ganganoi, and below them the Vellaboroi.

6 Outline of the southern side, which lies along the Vergivius Ocean:

beyond Cape Notion there is	7⅔	57¾
the mouth of the Dabrona River	11¼	57
the mouth of the Birgos River	12½	57½
Hieron Cape	14	57½

7 On the side beyond the Vellaboroi live the Ivernoi, with the Vodiai beyond them and the Brigantes more to the west.

8 Outline of the eastern side, which lies along the ocean called the Ivernios:

after Cape Hieron there is	14	57½
the mouth of the Modonos River	13⅔	58⅔
the city of Manapia	13½	58⅔
the mouth of the Oboka River	13⅙	59
the city of Eblana	14	59½
the mouth of the Buvinda River	14⅔	59⅔
Cape Isamnion	15	60
the mouth of the Vinderios River	15	60¼
the mouth of the Logia River	15⅓	60⅔
After it is Cape Rhobogdion.		

9 On the east side after the Rhobogdioi live the Darinoi, with the Voluntioi below them, and then the Eblanioi, then the Kaukoi, and below them the Manapioi and then the Koriondoi above the Brigantes.

10 These cities are in the interior:

Regia	13	60⅓
Raiba	12	59¾
Laberos	13	60¼
Makolikon	11½	58⅔
another Regia	11	59½
Dounon	12½	58¾
Ivernis (Iernis)	11	58⅙

11 Lying above the island of Ivernia are the so-called Ebudai, five in number.

The westernmost is called Ebuda	15	62
east of this is another Ebuda	15⅔	62
then Rikina	17	62
then Maleos	17½	62⅙
then Epidion	18½	62

12   Farther to the east of Ivernia are these islands:

Monaoida	17⅔	61½
the island of Mona	15	57⅔
Edros (deserted)	15	59½
Limnos (deserted)	15	59

Appendix 2
Toponyms Mentioned in Book 1

The *Geographical Guide* contains thousands of toponyms and ethnyms. The most unusual are in Book 1, and a list of them follows, with the citations in the *Guide*. Further details on some of these names are in the text proper of this volume, but this gazetteer provides a brief accessible summary of their location. Names are presented with Ptolemy's orthography. References to them in the *Guide* refer only to the citations in Book 1.

Adoulikon Bay (1.15.11) is the deep Annesley Bay of the Gulf of Zula, which is the only good harbor on the west side of the Red Sea (slightly north of 15 degrees north latitude). The village of **Adoulis** (*GG* 4.7.8, 8.16.11) was in the vicinity, and by the first century AD the bay was an important port for the Axum kingdom (Casson, *Periplus* 102–6).

Adria (1.15.3, etc.) is the Adriatic Sea. It was identified by Phokaian explorers by the sixth century BC (Herodotos 1.163). Trade routes to northern Europe began at its northern end.

Agisymba (1.7–12), an African location which Julius Maternus reached after a journey of several months from the Mediterranean. It was said to be one of the farthest south points in the known world and a region of unusual fauna, especially the rhinoceros. The *Guide* is the only source to mention it; it may be in the region of Lake Chad (Berggren and Jones, *Ptolemy's Geography* 145–7).

Aithiopians (1.7–10), a generic term for the people of sub-Saharan Africa, used as such since the beginning of Greek literature (Homer, *Iliad* 1.423 etc.). All southern peoples came to be so characterized, even those in southeast Asia (*GG* 7.3.1–3).

Akamas (1.15.4), the northwest point of Cyprus, modern Cape Arnouti.

Amphipolis (1.15.7), an Athenian outpost founded in the fifth century BC on the lower Strymon River in Thrace (Thucydides 1.100 etc.; Diodoros 12.32.3) and a major archaeological site today.

Antiocheia Margiane (1.12.7), a city at Gyaur Kala (near the famous oasis of Merv, in modern Turkmenistan), that was refounded by the Seleukid king Antiochos I. It was said to have been established originally by Alexander the Great (Pliny, *Natural History* 6.46–7), although its history is contradictory and confused, as well as any exact determination of its location (Cohen, *Hellenistic Settlements in the East* 245–50).

Apokopon (Apokope) Bay (1.17.9), a locality on the east coast of Africa. The name is probably descriptive ("precipitous") rather than indigenous, referring to the nature of its surroundings.

Arabia (1.7.6 etc.), the modern Arabian Peninsula, divided in antiquity between **Arabia Eremos** (5.15.6 etc.), the deserted country in the north, **Arabia Petraia** (2.1.6 etc.), the territory around Petra and west to the borders of Egypt, and **Arabia Eudaimon** (1.7.6), Fortunate Arabia, the aromatic-producing territory in the south, known to the Mediterranean world since the fifth century BC (Euripides, *Bacchants* 16).

Arbela (1.4.2), a city in Assyria (modern Erbil in Iraq), near where Alexander the Great defeated Dareios III of Persia in 331 BC.

Areia (1.12.7 etc.), a district in western modern Afghanistan, originally a Persian satrapy (Herodotos 3.93) but in Ptolemy's day Parthian territory.

Argarikos Bay (1.13.2) is on the southern coast of India, perhaps part of the strait between it and Taprobane (Sri Lanka): see Casson, *Periplus* 226–8.

Armenia (1.15.9 etc.), the region southeast of the Black Sea, which retains its ancient name. In Ptolemy's time it was an independent kingdom, although there was sporadic Roman control.

Aromata (1.9.1 etc.), modern Cape Guardafui, the easternmost point of Africa and an important trading post. The name is Greek ("Spices"; see *Periplous of the Erythraian Sea* 30).

Assyria (1.12.5 etc.), the ancient region in northern Iraq, although the toponym was often generalized to mean various parts of Mesopotamia.

Athos (1.15.7), the easternmost of the three long peninsulas that extend into the northern Aegean from the Macedonian Chalkidike.

Azania (1.7.6), a general term for the East African coast from the outlet of the Red Sea to the vicinity of Zanzibar (a related toponym). It may have been first cited by Juba II of Mauretania in his *On Arabia* of the early first century AD (Juba F31 = Pliny, *Natural History* 6.153; see also *Periplous of the Erythraian Sea* 15–16).

Baitis (1.12.11, 1.14.9), the largest river of southern Iberia, the modern Guadalquivir.

Baktriane (1.17.5), a region of south central Asia, mostly in modern Afghanistan. It was conquered by the Persians (Herodotos 3.93) and eventually became a Greek kingdom, which lasted until late Hellenistic times. Its major city was **Baktra** (1.12.7), at modern Balkh.

Barbaria (1.17.6), a name given by Greco-Roman merchants to the east coast of Africa.

Blessed Islands see **Makaroi Nesoi**.

Brettania (1.15.6), the British Isles.

Byzantion (1.11.6; 1.12.1, 8; 1.15.8–9; 1.16.1), located on the European side of the Thracian Bosporos and established in the seventh century BC (Herodotos 4.144), was always a significant city and survives as modern Istanbul.

Carthage see **Karchedon**.

Caspian Gates see **Kaspiai Pylai**.

Chelidoniai (1.15.4), islands off the southern coast of Asia Minor, the modern Beş Adalar.

Chryse Chersonesos (1.13.9; 1.14.1, 4–8; 1.17.5), the Golden Peninsula, a trading region at the eastern extremity of the known world, beyond India. Suggestions for its location range from Burma to the Malay Peninsula and farther east (Casson, *Periplus* 235–6).

Dalmatia (1.16.1), a region and Roman province on the east coast of the Adriatic.

Dere (1.15.11), or Deire, a promontory and indigenous town on the African side of the Bab el-Mandeb at the mouth of the Red Sea (Strabo, *Geography* 16.4.4).

Ekbatana (1.12.5), at modern Hamadan in Iran, was the ancient Median capital and later a Parthian royal city.

Essina (1.17.11) was a trading post on the east coast of Africa, a few days' sail north of Rhapta. It was perhaps established in the late first century BC.

Euphrates (1.11.2; 1.12.10–11) is the famous river of eastern Asia Minor and Mesopotamia.

Gangetikos Bay (1.13.7) is the modern Bay of Bengal.

Garamaioi (1.12.5) were a population in Assyria.

Garame (1.8.5; 1.9.9; 1.10.2; 1.11.4; 1.12.2) was an oasis, city, and kingdom in the Sahara, at modern Germa in the Fezzan district of Libya.

Hekatonpylos ((1.12.5–6), or "Hundred Gates," was the Parthian capital but presumably originally a Greek foundation of uncertain origin. It is generally thought to be at modern Shahr-i Qumis in Iran, but this is far from certain (Cohen, *Hellenistic Settlements in the East* 210–15).

Hellespont (1.11.6; 1.12.1, 6–8; 1.15.7; 1.16.1), the strait from the Aegean to the Propontis, the modern Dardanelles.

Hierapolis (1.11.2; 1.12.5), a city in interior Syria (at modern Manbij), formerly called Bambyke but refounded by Seleukos I; as its name implies, it was a major religious center (Getzel M. Cohen, *The Hellenistic Settlements in Syria, the Red Sea Basin, and North Africa* [Berkeley 2006] 172–8).

Hieron Promontory (1.12.11), the Sacred Promontory (modern Cabo de São Vicente at the southwest corner of the Iberian Peninsula), and thus an important navigational and topographical point.

Himera (1.15.2), a Greek settlement on the north coast of Sicily, at modern Termini Imerese.

Hispania (1.12.11) is the Roman term for the Iberian Peninsula.

Hyrkania (1.12.6), a region on the southeastern coast of the Caspian (Hyrkanian) Sea. It included a city also named **Hyrkania**.

Hyrkanian Sea (1.12.6), an alternate name for the Caspian Sea, with the two virtually interchangeable (*GG* 7.5.4).

Imaon Mountains (1.12.7, 1.16.1), part of the modern Himalayas, although the toponym is uncertainly and variably applied.

Indike (1.7.6–8; 1.13.1; 1.16.1; 1.17.5), or India, originally limited to the basin of the Indos River, but by Ptolemy's time had expanded to include the entire subcontinent as well as much of the territory to its east.

Indos River (1.17.4), the major river of western Indike (India).

Issos (1.12.11), the bay at the northeastern corner of the Mediterranean (the modern Gulf of Alexandretta), as well as a town at the head of the bay.

Italia (1.16.1), the Italian peninsula.

Ivernia (1.11.7–8), modern Ireland.

Kaisareia Iol (1.15.2), or Caesarea, at modern Cherchel in Algeria. The town was originally the Phoenician outpost of Iol and was renamed when it became the capital of the Mauretanian allied kingdom in 25 BC; in Ptolemy's day it was the capital of the Roman province of Mauretania Caesariensis.

Kalpe (1.12.11), the prominent mountain in southern Iberia, modern Gibraltar, often considered one of the Pillars of Herakles.

Kanobos (1.15.4), or Canopus, an ancient town in Egypt, just northeast of Alexandria.

Karallis (1.12.11), modern Cagliari in Sardinia.

Karchedon (1.4.2), ancient Carthage, a Roman city in Ptolemy's time.

Kaspiai Pylai (1.12.5, 7), the Caspian Gates, a pass southeast of the Caspian Sea used by Alexander the Great and becoming an important topographical point thereafter. Its exact location is unknown, but the Sar Darreh, east of Tehran, is the most likely (Standish, "The Caspian Gates," 17–24).

Kattigara (1.11.1; 1.13.1; 1.14.1, 4–10; 1.17.5; 1.23.23), a trading post at the eastern extremity of the known world, perhaps on the coast of the South China Sea or somewhere in Indonesia (Berggren and Jones, *Ptolemy's Geography* 155–6).

Kolchikon Bay (1.13.1), on the coast of the region of Kolchoi (*Periplous of the Erythraian Sea* 59), between southern Indike (India) and Taprobane (Sri Lanka).

Komaria Promontory (1.17.3), modern Cape Comorin, the southernmost point of Indike (India).

Komedai Mountains (1.12.7), located in south central Asia on the route from Baktra to Lithinos Pyrgos (the Stone Tower), perhaps part of the modern Pamir Range.

Kory Bay (1.13.1), a variant name for Kolchikon Bay. The **Kory Promontory** is one of the capes northeast of the Komaria Promontory.

Kouroula (1.13.1, 4–5), a city on the southeastern coast of Indike (India).

Lakonike (1.12.11), a district in the south-central Peloponnesos, which includes the city of Sparta.

Leptis Megale (1.8.1, 5; 1.10.2; 1.15.2), more commonly Leptis Magna, a city on the Mediterranean coast of Africa (modern Lebda in Libya). Originally a Phoenician outpost, it became the most important regional city in the Roman period and was the point of origin for many expeditions into interior Africa.

Libye (1.8.5), or Libya, normally the Greek term for the continent of Africa, but Ptolemy in Book 1 seems to limit it to the region around Leptis Megale, an older usage.

Lilybaion (1.12.11), modern Marsala, the southwestern point of Sicily.

Limyrike (1.7.6), the southwestern region of Indike (India), the modern Malabar Coast (*Periplous of the Erythraian Sea* 51).

Lithinos Pyrgos (1.11.4, 6; 1.12.1–7; 1.17.5), the Stone Tower, an important trading and rendezvous point in south central Asia, probably somewhere in modern Tajikistan, where the toponym Tashkurgan (with the same meaning) is known.

Londinion (1.15.6), or Londinium, ancient London.

Maiotis (1.8.2), the modern Sea of Azov.

Makaroi Nesoi (1.11.1), the Blessed Islands of mythology (Hesiod, *Works and Days* 166–73); by the Roman period they were identified with the Canaries. They were just east of Ptolemy's prime meridian.

Media (1.12.5), the ancient kingdom of the northwestern Iranian plateau; by Roman times it was Parthian territory.

Megas Aigialos (1.17.10), the Great Beach, a locality on the eastern African coast north of Rhapta.

Meroë (1.7.8; 1.23.23; 1.24.4), on the upper Nile near modern Begrawiyah in Sudan, and the capital of the Kushite kingdom. Since early times (Herodotos 2.29) it had been considered one of the southernmost known places in the world, and thus was of importance to geographers.

Mesopotamia (1.12.5), the region between the Tigris and Euphrates Rivers.

Mikron Aigialos (1.17.10), the Small Beach, a locality on the eastern African coast north of Rhapta.

Mysia (1.16.1), Roman Moesia, the region immediately south of the lower Danube.

Neilos (1.9.1; 1.15.10–11; 1.17.6), the Nile.

Norikon (1.16.1), Roman Noricum, a region and province in eastern modern Austria.

Noviomagus (1.15.6), rendered as Noiomagos by Ptolemy, modern Chichester.

Okelis (1.7.4, 1.15.11), a locality and trading post on the Arabian side of the Bab el-Mandeb at the mouth of the Red Sea (*Periplous of the Erythraian Sea* 25; Casson, *Periplus* 157–8).

Opone (1.17.8), a trading post on the African coast at modern Hafun in Somalia (*Periplous of the Erythraian Sea* 13; Casson, *Periplus* 132).

Pachynos (1.12.11), modern Capo Passero, the southeastern corner of Sicily.

Palimbothra (1.12.9, 1.17.5) or Pataliputra, modern Patna on the Ganges. It was the capital of the Mauryan kingdom and retained its importance thereafter.

Paloura (1.13.5, 7), a point on the east coast of Indike (India), from which one could sail across the Bay of Bengal. It was perhaps in the vicinity of modern Masulipatam (Casson, *Periplus* 232)

Pannonia (1.16.1), a region and Roman province south of the middle Danube, largely in modern Hungary.

Paphos (1.15.4), a town in western Cyprus.

Parthia (1.12.5), the homeland of the Parthian Empire, southeast of the Caspian Sea.

Phalangis Mountain (1.17.9), on the east coast of Africa near the Zingis Promontory. The name may be Greek, a descriptive term since its multiple peaks looked like a military phalanx.

Pisa (1.15.5), the town in Tuscany.

Pontic Sea (1.16.1), the modern Black Sea. **Pontos** was a district on its south shore, generally an allied kingdom, but Ptolemy's use of the term is ambiguous and may refer to the sea itself.

Porthmos (1.12.11), the Strait, used by Ptolemy to refer to the one between the Atlantic and the Mediterranean and also called the Straits of Herakles.

Prason (1.7.2; 1.8.1; 1.9.4; 1.10.1; 1.14.3–5; 1.17.6, 12), the southernmost known point on the East African coast, perhaps modern Cabo Delgado in Mozambique. The name may be a Greek descriptive term ("seaweed": Theophrastos, *Research on Plants* 4.6.4).

Ptolemaios Theron (1.15.11), Ptolemaios of the Hunts, founded by Ptolemy II in the 260s BC as a major port and settlement for elephant hunting. By the time of the *Guide* it was in decline and under local rule. It may have been near modern Aqiq (Cohen, *Hellenistic Settlements in Syria* 341–3).

Pyrenaioi Mountains (1.15.2), the Pyrenees.

Ravenna (1.15.3), Raouenne to Ptolemy, modern Ravenna in Italy.

Rhaitia (1.16.1), Latin Raetia, a Roman province in the Alps, often Raetia et Vendelica.

Rhapta or **Rhaptos** (1.9.1, 4; 1.14.3–4; 1.17.6, 12), a cape, river, and important trading post on the East African coast, the ancestor of modern Zanzibar and Dar es Salaam. The name is Greek ("sewn"), referring to the local sewn boats (*Periplous of the Erythraian Sea* 16; Casson, *Periplus* 141–2).

Rhodes (1.11.2, 1.12.11), more properly Rhodos, the Greek island.

Sachalitis (1.17.2–3) a bay and region in the frankincense territory of southern Arabia (*Periplous of the Erythraian Sea* 29; Casson, *Periplus* 165–6).

Sada (1.13.7), a town on the eastern coast of the Bay of Bengal.

Sakai (1.16.1), a wide-ranging ethnym, linguistically similar to "Skythai," or the Skythians. Ptolemy placed the Sakai north of Indike (India); elsewhere in the *Guide* (6.12.1) they are near the Iaxartes River (modern Syr Darya).

Sarapion (1.17.11), an anchorage on the east coast of Africa, probably at modern Warsheik in Somalia (*Periplous of the Erythraian Sea* 15; Casson, *Periplus* 138–9). Its name indicates that Egyptian or Greek traders may have established a shrine to Sarapis there.

Sardo (1.12.11), modern Sardinia.

Sarmatai (1.8.2), a generic term for nomadic inhabitants in the northern part of the inhabited world, located at various times from Europe to the Caspian region.

Satala (1.15.9), an important regional center in Lesser Armenia, 60 miles south of the Black Sea, at modern Sadek.

Sebennytos (1.15.4), a city in the central part of the Egyptian Delta, and the name of the channel of the Nile just to its west.

Sera (1.11.4, 6; 1.12.1, 10, 12; 1.17.5), a city at the eastern limit of the known world and the metropolis of the Seres. It is not certainly identified but may be Luoyang (Berggren and Jones, *Ptolemy's Geography* 178).

Seres (1.11.1; 1.17.5), the Silk People. Originally they were central Asian traders in silk (Lieberman, "Who Were Pliny's Blue-Eyed Chinese?" 174–7), but by Ptolemy's day the ethnym refers to people farther east. Yet identifying them is a complex problem and not of certain resolution. See also **Sinai**.

Sikelia (1.12.11), the island of Sicily.

Simylla (1.17.3), or Semylla, a trading post in Indike (India), near modern Bombay (*Periplus of the Erythraian Sea* 53; Casson, *Periplus* 215–16).

Sinai (1.11.1; 1.14.10; 1.17.5), a people and their metropolis, probably approximately the same as the Seres, but accessed by the sea route heading east and north from Indike (India) rather than the overland one to the Seres. The name "Sinai" is linguistically related to "China" (Casson, *Periplus* 238). Some manuscripts of the *Guide* have another related form, Thinai (7.3.6, 8.27.12; see also "Thina" at *Periplous of the Erythraian Sea* 64, and also **Seres**). The **Bay of the Sinai** (1.13.1) is difficult to locate, but the most probable identification is the modern Gulf of Tonkin or elsewhere in the South China Sea.

Skythai (1.8.2), the Skythians. In Greek geographical usage, they were the widespread indigenous population in the northern quadrant of the inhabited world (Ephoros F30a = Strabo, *Geography* 1.2.28), but there were many sub-groups (Ptolemy named over a dozen), and as nomads they could not be located in a single place.

Smyrna (1.12.6), the ancient Ionian city, modern İzmir in Turkey.

Sogdianoi (1.16.1), an indigenous population of central Asia, largely located in modern Uzbekistan.

Stone Tower see **Lithinos Pyrgos**.

Strymon (1.15.7), a river of northern Greece, at times the border between Macedonia and Thrace.

Syagros Cape (1.17.3), modern Ras Fartak, a prominent headland on the southern coast of Arabia (*Periplous of the Erythraian Sea* 30; Pliny, *Natural History* 6.100; Casson, *Periplus* 166).

Syene (1.7.9, 1.9.9, 1.23–4), an Egyptian town at the First Cataract of the Nile (modern Aswan), located on the northern tropic and famous in the history of geography as the major datum point for Eratosthenes' calculation of the circumference of the earth (Kleomedes, *Elementary Theory* 1.7; Pliny, *Natural History* 6.171; Roller, *Eratosthenes' Geography* 263–7).

Tainaron (1.12.11), the southernmost point in the Peloponnesos (modern Cape Matapan).

Tamala (1.13.8–9), a town beyond the mouth of the Ganges, either on the east coast of the Bay of Bengal, or farther south near the narrows of the Chryse Chersonesos (the Golden Peninsula, probably the Malay Peninsula).

Taprobane (1.14.9), modern Sri Lanka.

Tarrakon (1.15.2), Roman Tarraco, modern Tarragona in Spain.

Tergeston (1.15.3) or Tergeste, modern Trieste.

Theainai (1.15.2) or Thena, a town on the coast of the Roman province of Africa, in southern modern Tunisia.

Thoule (1.7.1, 1.20–4) or Thule, the far northern locale discovered by Pytheas of Massalia in the late fourth century BC (Strabo, *Geography* 2.5.8), probably modern Iceland. Nevertheless its identification became almost immediately disputed, and continues to be; Ptolemy placed it farther south, probably in the Shetlands (8.3.3; Stückelberger and Graßhoff, *Klaudios Ptolemaios* 777).

Thraike (1.15.8, 1.16.1), or Thrace, the territory northeast of the Aegean.

Tigris (1.12.5), the famous river of Mesopotamia.

Tilavonton (1.15.3), or Tiliaventus River, the modern Tagliamento at the north end of the Adriatic.

Timoula (1.17.4), a variant name for Simylla, a trading post in Indike (India).

Toniki (1.17.12), a trading post at the entrance of the bay leading to Rhapta on the East African coast, perhaps the Nikon of the *Periplous of the Erythraian Sea* (15).

Trapezous (1.15.9), a town on the southeastern coast of the Black Sea, modern Trabzon.

Triakontaschoinos (1.9.9), Thirty Schoinoi, a district on the west side of the Nile near Syene. Due to the variable length of the *schoinoi* it is not possible to determine its extent, but characterizing a riverine district by its length may have been common along the upper Nile: Ptolemy also referred to one of 12 *schoinoi* south of Syene (4.5.74)

Troglodytike (1.8.1; 1.9.1–2), a toponym in the manuscripts of the *Geographical Guide* for a population located in two places: on the shore of the Red Sea toward the Nile, and farther south, in the region of Azania south of Aromata. But the term "Troglodytike" ("Land of the Cave Dwellers") is an error, perpetrated in antiquity and continuing until today, and the proper version is "Trogodytike," a territory known to the Greek world since the fifth century BC (Herodotos 4.183; G. W. Murray, "Trogodytica: The Red Sea Littoral in Ptolemaic Times," *GJ* 43 [1967] 24–33).

Zabai (1.14.1, 5–7), a city at an unknown location in the far east, probably on the Malay Peninsula or somewhere beyond.

Zingis Cape (1.17.9), a locality on the coast of East Africa several days beyond Aromata.

Appendix 3

The Caption for Map 1 of Europe

Map 1 of Europe covers the British Isles, including Ireland, Great Britain, and a number of surrounding islands. The following is a translation of the caption for that map (*GG* 8.3.1–11):

> The first map of Europe covers the Brettanic Islands along with the surrounding islands. The parallel through its middle has the ratio of nearly 11:20 to the meridian. The map has as its boundaries the Ocean everywhere, with the Germanic on the west, the Brettanic on the south as well as what is called the Vergivian, the Dytic on the west, and on the north the Hyperboreian and what is called the Douekaledonian.
>
> The island of Thoule has for its longest day 20 equinoctial hours. Its distance from Alexandria is 3 1/4 hours to the west.
>
> Important cities on the island of Ivernia: the homonymous city on the island, Ivernis, has a longest day of 18 hours. Its distance from Alexandria is 3 1/4 hours to the west. Raiba has a longest day of 18 1/2 hours. Its distance from Alexandria is 3 1/5 hours to the west.
>
> On the island of Albion: London has a longest day of 17 hours. Its distance from Alexandria is 2 2/3 hours to the west. Eborakon has a longest day of 17 5/6 hours. Its distance from Alexandria is 2 2/3 hours to the west. Katouraktonion has a longest day of 18 hours. Its distance from Alexandria is 2 2/3 hours to the west. The Pteron Stratopedon has a longest day of 18 1/2 hours. Its distance from Alexandria is 21/6 hours to the east. The island of Doumna has a longest day of 19 hours. Its distance from Alexandria is two hours to the west. The island of Vektis has a longest day of 16 2/3 hours. Its distance from Alexandria is 2 2/3 hours to the west.

Ptolemy's choice of locales is based both on their importance and the availability of data. He began with Thoule, accepted as the farthest north known place. The two Irish towns cannot be identified but their coordinates show that one (Raiba) was in the center of the island and the other (Ivernis) was in the south-central portion; thus they were probably regional capitals. The three British towns are the important places of London and Eboracum (York), as well as Cataractonium (modern Catterick, a Roman fort in North Yorkshire) and the Winged Camp (in Ptolemy's Greek,

Pteroton Stratopedon, probably a translation of Pinnata Castra), a legionary fortress or perhaps an indigenous hill fort in Scotland. The inclusion of these two sites suggests that some of Ptolemy's information came from military sources.

In addition, the two islands cited, Doumna (not certainly identified but one of the islands off the coast of Scotland), and Vektis (or Vectis, generally thought to be the Isle of Wight) bracket Great Britain and thus provide coordinates for its north-south extent. Thus the captioned locales were carefully selected to cover as wide an extent as possible with a minimum number of places.

Abbreviations

AAntHung	*Acta antiqua hungarica*
AHES	*Archive For History Of the Exact Sciences*
ANRW	*Aufstieg und Niedergang der römischen Welt*
AJP	*American Journal of Philology*
AR	*Archaeological Reports*
BA	*Barrington Atlas of the Greek and Roman World* (ed. Richard J. A. Talbert, Princeton 2000)
BAPR	*Bulletin de l'Association Pro Aventico*
BICS	*Bulletin of the Institute of Classical Studies*
BNP	*Brill's New Pauly*
C&C	*Classica et Christiana*
C&M	*Classica et Mediaevalia*
CIL	*Corpus inscriptionum latinarum*
CP	*Classical Philology*
CQ	*Classical Quarterly*
DSB	*Dictionary of Scientific Biography*
EANS	*Encyclopedia of Ancient Natural Scientists* (ed. Paul T. Keyser and Georgia L. Irby-Massie, London 2008).
FGrHist	*Fragmente der Griechischen Historiker*
G&R	*Greece and Rome*
GA	*Geographia Antiqua*
GeogAnn	*Geografiska Annaler*
GG	Ptolemy, *Geographical Guide*
GJ	*Geographical Journal*
HGSS	*History of Geo- and Space Sciences*
JA	*Journal Asiatique*
JAC	*Journal of Ancient Civilizations*
JHS	*Journal of Hellenic Studies*
JRGS	*Journal of the Royal Geographical Society*
JRS	*Journal of Roman Studies*
JSNS	*Journal of Scottish Name Studies*
MediterrAnt	*Mediterraneo Antico*
MEFR	*Mélanges de l'école française de Rome*

MM	*Mathematics Magazine*
OT	*Orbis Terrarum*
PECS	*Princeton Encyclopedia of Classical Sites*
PIR	*Prosopographia Imperii Romani*
PRIA-C	*Proceedings of the Royal Irish Academy, Section C, Archaeology*
PSAS	*Proceedings Of the Society Of Antiquaries Of Scotland*
RE	*Paulys Realencyclopädie der classichen Altertumswissenschaft (Pauly-Wissowa)*
RÉG	*Revue des études grecques*
SIG	*Sylloge inscriptionum graecarum*
TAPA	*Transactions of the American Philological Association*

Bibliography

Agatharchides. *On the Erythraean Sea* (ed. Stanley M. Burstein, London 1989).

Andrade, Nathanael. "The Voyage of Maes Titianos and the Dynamics of Social Connectivity Between the Roman Levant and Central Asia/West China," *MediterrAnt* 18 (2015) 41–74.

Arnaud, Pascal. "Marin de Tyr," in *Sources de l'histoire de Tyr* (ed. Pierre-Louis Gatier *et al.*, Beirut 2017) 87–100.

———. "Texte et carte de Marcus Agrippa: Historiographie et données textuelles," *GA* 16–17 (2007–2008) 73–126.

Aujac, Germaine. *Claude Ptolémée, astronome, astrologue, géographe: connaissance et représentation du monde habité* (Paris 1993).

———. "The 'Revolution' of Ptolemy," in *Brill's Companion to Ancient Geography: The Inhabited World in Greek and Roman Tradition* (ed. Serena Biachetti *et al.*, Leiden 2016) 313–34.

Bagrow, Leo. "The Origin of Ptolemy's *Geographia*," *GeogAnn* 27 (1945) 318–87.

Batty, Roger M. "A Tale of Two Tyrians," *Classics Ireland* 9 (2002) 1–18.

Berggren, J. Lennart. "Ptolemy's Maps of Earth and the Heavens: A New Interpretation," *AHES* 43 (1991) 133–44.

——— and Alexander Jones. *Ptolemy's Geography: An Annotated Translation of the Theoretical Chapters* (Princeton 2000).

Bernard, Alain. "Pappos of Alexandria," *EANS* 611–12.

Bertheau, Jochen. "Die Mitteleuropäischen Ortsnamen in der Geographie des Klaudios Ptolemaios," *OT* 8 (2002) 3–48.

Bianchetti, Serena. "Esplorazioni africane di età imperiale (Tolomeo, *Geogr.*, I, 8, 4)," in *l'Africa Romana: Atti dell'XI convegno di studio Cartagine, 15–18 dicembre 1994* (ed. Mustapha Khanoussi *et al.*, Ozieri 1996) 351–9.

Boardman, John. "Greek Archaeology on the Shores of the Black Sea," *AR* 9 (1962–1963) 34–41.

———. *The Greeks Overseas: Their Early Colonies and Trade* (fourth edition, London 1980).

Bondarenko, Grigory. "Goidelic Hydronyms in Ptolemy's Geography: Myth Behind the Name," in *Periphery of the Classical World in Ancient Geography and Cartography* (ed. Alexander Podossinov, Leuven 2014) 147–54.

Bowersock, G. W. "The East-West Orientation of Mediterranean Studies and the Meaning of North and South in Antiquity," in *Rethinking the Mediterranean* (ed. W. V. Harris, Oxford 2005) 167–78.

Braund, David C. *Georgia in Antiquity* (Oxford 1994).

Breeze, David J. "The Ancient Geography of Scotland," in *In the Shadow of the Brochs: The Iron Age in Scotland* (ed. Beverly Ballin Smith and Iain Banks, Stroud 2002) 10–14.

———. "Auxiliaries, Legionaries, and the Operation of Hadrian's Wall," *BICS Supplement* 81 (2003) 147–51.

Bridgman, Timothy P. *Hyperboreans: Myth and History in Celtic-Hellenic Contacts* (New York 2005).

Broderick, George. "Some Island Names in the Former 'Kingdom of the Isles': A Reappraisal," *JSNS* 7 (2013) 1–28.

Burian, Jan *et al.* "Moesi, Moesia," *BNP* 9 (2006) 115–19.

Burri, Renate. "Übersicht über die griechischen Handschriften der ptolemäischen *Geographie*," in *Klaudios Ptolemaios: Handbuch der Geographie, Ergänzungsband* (ed. Alfred Stückelberger and Florian Mittenhuber, Basel 2009) 10–20.

Burton, Richard F. "On Lake Tanganyika, Ptolemy's Western Lake-Reservoir of the Nile," *JRGS* 35 (1865) 1–15.

Caley, Earle R. and John C. Richards. *Theophrastus on Stones* (Columbus, OH 1956).

Cameron, Hamish. *Making Mesopotamia: Geography and Empire in a Roman-Iranian Borderland* (Leiden 2018).

Cary, Max. "Maës, qui et Titianus," *CQ* 6 (1956) 130–4.

Cary, Max and E. H. Warmington. *The Ancient Explorers* (Baltimore 1963).

Casson, Lionel. *The Periplus Maris Erythraei* (Princeton 1989).

Clarke, Katherine. *Between Geography and History: Hellenistic Constructions of the Roman World* (Oxford 1999).

Cohen, Getzel M. *The Hellenistic Settlements in the East From Armenia and Mesopotamia to Bactria and India* (Berkeley 2013).

———. *The Hellenistic Settlements in Europe, the Islands, and Asia Minor* (Berkeley 1995).

———. *The Hellenistic Settlements in Syria, the Red Sea Basin, and North Africa* (Berkeley 2006).

———. "*Katoikiai, katoikoi* and Macedonians," *AncSoc* 22 (1991) 41–50.

Dalby, Andrew. *Food in the Ancient World From A to Z* (London 2003).

Dana, Dan and Sorin Nemeti. "Ptolémée et la toponymie de la Dacie (I)," *C&C* 7.2 (2012) 431–7.

———. "Ptolémée et la toponymie de la Dacie (II–V)," *C&C* 9.1 (2014) 97–114.

———. "Ptolémée et la toponymie de la Dacie (VI–IX)," *C&C* 11 (2016) 67–93.

Defaux, Olivier. *The Iberian Peninsula in Ptolemy's Geography* (Berlin 2017).

Desautels, Jacques. "Les monts Riphées et les Hyperboréens dans le traité hippocratique *Des airs, des eaux et des lieux*," *RÉG* 74 (1971) 289–96.

Dicks, D. R. *The Geographical Fragments of Hipparchus* (London 1960).

Dietler, Michael. "Colonial Encounters in Iberia and the Western Mediterranean: An Exploratory Framework," in *Colonial Encounters in Ancient Iberia* (ed. Michael Dietler and Carolina Lopez-Ruiz, Chicago 2009) 3–48.

Diller, Aubrey. "The Ancient Measurements of the Earth," *Isis* 40 (1949) 6–9.

———. "The Oldest Manuscripts of Ptolemaic Maps," *TAPA* 71 (1940) 62–7.

———. "Review of Stevenson, *Geography of Claudius Ptolemy*," *Isis* 22 (1935) 533–9.

———. *The Tradition of the Minor Greek Geographers* (New York 1952).

Eichel, Marijean H. and Joan Markley Todd. "A Note on Polybius' Voyage to Africa in 146 BC," *CP* 71 (1976) 237–43.

Freeman, Philip. *Ireland and the Classical World* (Austin 2001).

Gardiner-Garden, John R. "Greek Conceptions on Inner Asian Geography and Ethnography From Ephoros to Eratosthenes," *Papers on Inner Asia* 9 (Bloomington, Ind. 1987).

Gawlikowski, Michal. "Thapsacus and Zeugma: The Crossing of the Euphrates in Antiquity," *Iraq* 58 (1996) 123–33.

Geus, Klaus. "Hellenistic Maps and Lists of Places," in *Hellenistic Astronomy: The Science in Its Contexts* (ed. Alan C. Bowen and Francesca Rothberg, Leiden 2020) 232–9.

————. "The Problem of Practical Applicability in Ptolemy's *Geography*," in *Knowledge, Text and Practice in Ancient Technical Writing* (ed. Marco Formisano and Philip Van der Eijk, Cambridge 2017) 186–99.

————. *Prosopographie der Literarisch bezeugten Karthager* (Leuven 1994).

————. "Ptolemaios über die Schulter geschaut—zu seiner Arbeitsweise in der *Geographike Hyphegesis*," in *Wahrnehmung und Erfassung geographischer Räume in der Antike* (ed. Michael Rathmann, Mainz 2007) 159–66.

————. "Wer ist Marinos von Tyros? Zur Hauptquelle des Ptolemaios in seiner *Geographie*," *GA* 26 (2017) 13–22.

———— and Florian Mittenhüber. "Die Länderkarten Africas," in *Klaudios Ptolemaios: Handbuch der Geographie, Ergänzungsband* (ed. Alfred Stückelberger and Florian Mittenhuber, Basel 2009) 282–9.

———— and Irina Tupikova. "Von der Rheinmündung in der Finnischen Golf . . . Neue Ergebnisse zur Weltkarte des Ptolemaios, zur Kenntnis der Ostsee im Altertum und zur Flottenexpedition des Tiberius im Jahre 5 n. Chr.," *GA* 22 (2013) 125–43.

Goudineau, Christian. "Tropaeum Alpium," *PECS* 936–7.

Graßhoff, Gerd *et al.* "Of Paths and Places: The Origin of Ptolemy's Geography," *AHES* 71 (2017) 483–503.

Grünzweig, Friedrich E. "Gross-Germanian," in *Klaudios Ptolemaios: Handbuch der Geographie, Ergänzungsband* (ed. Alfred Stückelberger and Florian Mittenhuber, Basel 2009) 305–11.

Halfmann, Helmut. *Itinera principum* (Stuttgart 1986).

Hamilton, J. R. "Alexander and the Aral," *CQ* 21 (1971) 106–11.

Haushalter, Arthur. "L'Ibérie de Ptolémée, entre géographie mathématique et procédés empiriques," *GA* 26 (2017) 61–73.

Heil, Mattäus and Raimund Schulz. "Who Was Maes Titianus?" *JAC* 30 (2015) 72–84.

Hewsen, Robert H. "The *Geography* of Pappus of Alexandria: A Translation of the Armenian Fragments," *Isis* 62 (1971) 186–207.

Hirt, Alfred Michael. *Imperial Mines and Quarries in the Roman World* (Oxford 2010).

Hölbl, Günther. *A History of the Ptolemaic Empire* (tr. Tina Saavedra, London 2001).

Hoppál, Krisztina. "The Roman Empire According to the Ancient Chinese Sources," *AAnt-Hung* 51 (2011) 263–306.

Huss, Werner. "Leptis Magna," *BNP* 7 (2005) 419–25.

Jones, Alexander. "Ptolemy ('Claudius Ptolemaeus')," *EANS* 706–9.

————. "Ptolemy's *Canobic Inscription* and Heliodorus' Observation Reports," *SCIAMVS* 6 (2005) 53–97.

————. "Ptolemy's Geography: Mapmaking and the Scientific Enterprise," in *Ancient Perspectives: Maps and Their Place in Mesopotamia, Egypt, Greece and Rome* (ed. Richard J. A. Talbert, Chicago 2012) 109–28.

Jones, Barri and Ian Keillar. "Marinus, Ptolemy and the Turning of Scotland," *Britannia* 27 (1996) 43–9.

Jones, G. D. B. and J. H. Little. "Coastal Settlement in Cyrenaica," *JRS* 61 (1971) 64–79.

Karttunen, Klaus. *India in Early Greek Literature* (Helsinki 1989).

Keulzer, Andreas. "Protagoras," *EANS* 700–1.

Keyser, Paul T. "The Geographical Work of Dikaiarchos," in *Dicaearchus of Messana: Text, Translation, and Discussion* (ed. William W. Fortenbaugh and Eckart Schütrumpf, New Brunswick 2001) 353–72.

Kish, George. "Mercator, Gerardus," *DSB* 9 (1974) 309–10.

Kitchell, Jr., Kenneth. *Animals in the Ancient World From A to Z* (London 2014).

Lassus, J. "Portus Magnus," *PECS* 732–3.

Leveau, Philippe. "Recherches historiques sur une région montagneuse de Maurétanie Césarienne: des Tigava Castra à la mer," *MEFR* (1977) 257–311.

Lieb, Hans. "Forum Tiberii," *BAPR* 31 (1989) 107–8.

Lieberman, Samuel. "Who Were Pliny's Blue-Eyed Chinese?" *CP* 52 (1957) 174–7.

Manconi, D. "Antas," *PECS* 58–9.

Mann, John C. and David J. Breeze. "Ptolemy, Tacitus and the Tribes of North Britain," *PSAS* 117 (1987) 85–91.

Marcotte, Didier. "Le *Périple de la Mer Érythrée* et les informateurs de Ptolémée: Géographie et Traditions Textuelles," *JA* 304 (2016) 33–46.

———. "Ptolémée ethnographe: Questions de tradition," *GA* 26 (2017) 47–60.

Marinescu, L. "Ulpia Traiana," *PECS* 946–7.

Markwart, Josef. "Die Sogdiana des Ptolemaios," *Orientalia* 15 (1946) 123–49.

Marx, Christian. "The Western Coast of Africa in Ptolemy's Geography and the Location of His Prime Meridian," *HGSS* 7 (2016) 27–52.

Mitford, Terence Bruce. "Roman Rough Cilicia," *ANRW* 2.7.2 (1980) 1230–61.

Mittenhuber, Florian. "Die Länderkarten Europas," in *Klaudios Ptolemaios: Handbuch der Geographie, Ergänzungsband* (ed. Alfred Stückelberger and Florian Mittenhuber, Basel 2009) 268–81.

Moracchini-Mazel, G. and R. Boinard. *La Corse Selon Ptolémée* (Bastia 1989).

Morison, Samuel Eliot. *Christopher Columbus, Mariner* (New York 1955).

Murray, G. W. "Trogodytica: The Red Sea Littoral in Ptolemaic Times," *GJ* 43 (1967) 24–33.

Najbjerg, Tina and Jennifer Trimble. "The Severan Marble Plan Since 1960," in *Formae Urbis Romae* (ed. Robert Meneghini and Riccardo Santangeli Valenzani, Rome 2006) 75–101.

Neugebauer, Otto. *A History of Ancient Mathematical Astronomy* (Berlin 1975).

———. "Ptolemy's Geography, Book VII, Chapters 6 and 7," *Isis* 50 (1959) 22–9.

Nobbe, C. F. A. *Claudii Ptolemaei Geographia* (Leipzig 1843–1845).

Noonan, Thomas S. "The Grain Trade of the Northern Black Sea in Antiquity," *AJP* 94 (1973) 231–42.

Phillips, E. D. "Odysseus in Italy," *JHS* 73 (1953) 53–67.

Photinos, N. C. "Marinos von Tyros," *RE Supplement* 12 (1970) 791–838.

Pingree, David. "Maximus Planudes," *DSB* 11 (1975) 18.

Polaschek, Erich. "Ptolemaios: Das Geographische Werk," *RE Supp* 10 (1965) 680–833.

Pollard, Nigel and Joanne Berry. *The Complete Roman Legions* (London 2015).

Polverini, Leandro. "Cesare e la geografia," *Semanas de estudios romanos* 14 (2005) 59–72.

Prontera, Francesco. "Geografia e corografia: Note sul lessico della cartografia antica," *Pallas* 72 (2006) 75–82.

Reichert, H. "Limios alsos," *Reallexicon der germanischen Altertumskunde* 18 (2001) 448–50.

Rice, Edward. *Captain Sir Richard Francis Burton* (New York 1990).

Rickey, V. Frederick. "How Columbus Encountered America," *MM* 65 (1992) 219–25.

Riley, Mark T. "Ptolemy's Use of his Predecessors' Data," *TAPA* 125 (1995) 221–50.

Roller, Duane W. *Ancient Geography* (London 2015).

———. *Cleopatra: A Biography* (Oxford 2010).

———. *Cleopatra's Daughter and Other Royal Women of the Augustan Era* (Oxford 2018).

———. *Empire of the Black Sea* (Oxford 2020).

———. *Eratosthenes' Geography* (Princeton 2010).

———. *A Guide to the Geography of Pliny the Elder* (Cambridge 2022).

———. *A Historical and Topographical Guide to the Geography of Strabo* (Cambridge 2018).

————. *Scholarly Kings: The Writings of Juba II of Mauretania, Archelaos of Kappadokia, Herod the Great and the Emperor Claudius* (Chicago 2004).

————. *Three Ancient Geographical Treatises in Translation* (London 2022).

————. *Through the Pillars of Herakles: Greco-Roman Exploration of the Atlantic* (New York 2006).

————. "Timosthenes of Rhodes," in *New Directions in the Study of Ancient Geography* (ed. Duane W. Roller, University Park 2019) 56–79.

————. *The World of Juba II and Cleopatra Selene: Royal Scholarship on Rome's African Frontier* (New York 2003).

Roseman, Christina Horst. *Pytheas of Massalia: On the Ocean* (Chicago 1994).

Schön, Franz. "Limes V: Danube," *BNP* 7 (2005) 574–83.

Seubert, Philippe. "Délimitation et divisions de l'Arabie, d'Eratosthène à Ptolémée," *GA* 26 (2017) 23–36.

Shcheglov, S. I. "The Accuracy of Ancient Cartography Reassessed: The Longitude Error in Ptolemy's Map," *Isis* 107 (2016) 687–706.

————. "The Length of Coastlines in Ptolemy's *Geography* and in Ancient *Periploi*," *HGSS* 9 (2018) 9–24.

Shipley, Graham. *Pseudo-Skylax's Periplous* (Exeter 2011).

Shütte, Gudmund. "A Ptolemaic Riddle Solved," *C&M* 13 (1952) 236–84.

Sidebotham, Steven E. *Berenike and the Ancient Maritime Spice Route* (Berkeley 2011).

Sidoli, Nathan. "Heron's *Dioptra* 35 and Analemma Methods: An Astronomical Determination of the Distance Between Two Cities," *Centaurus* 47 (2005) 236–58.

Speke, John Hanning. *Journal of the Discovery of the Source of the Nile* (New York 1868).

Standish, J. F. "The Caspian Gates," *G&R Second Series* 17 (1970) 17–24.

Stevens, Henry N. *Ptolemy's Geography: A Brief Account of All the Printed Editions Down to 1730* (second edition, London 1908).

Stevenson, Edward Luther. *Claudius Ptolemy, the Geography* (New York 1932).

Stiehle, R. "Der Geograph Artemidoros von Ephesos," *Philologus* 11 (1856) 193–244.

Stückelberger, Alfred, "Zu den Quellen der *Geographie*," in *Klaudios Ptolemaios: Handbuch der Geographie, Ergänzungsband* (ed. Alfred Stückelberger and Florian Mittenhuber, Basel 2009) 122–33.

————. "Zu Sprache und Stil der *Geographie* der Ptolemaios," in *Klaudios Ptolemaios: Handbuch der Geographie, Ergänzungsband* (ed. Alfred Stückelberger and Florian Mittenhuber, Basel 2009) 432–9.

———— and Gerd Graßhoff. *Klaudios Ptolemaios: Handbuch der Geographie* (Basel 2006).

Sulkimirski, T. *The Sarmatians* (New York 1970).

Sullivan, Richard D. *Near Eastern Royalty and Rome, 100–30 BC* (Toronto 1990).

Svennung, J. *Skandinavien bei Plinius und Ptolemaios* (Uppsala 1974).

Syme, Ronald. *Anatolica: Studies in Strabo* (ed. Anthony Birley, Oxford 1995).

Talbert, Richard J. A. "*Urbs Roma* to *Orbis Romanus*: Roman Mapping on the Grand Scale," in *Ancient Perspectives: Maps and Their Place in Mesopotamia, Egypt, Greece and Rome* (ed. Richard J. A. Talbert, Chicago 2012) 163–91.

Tarn, W. W. *The Greeks in Baktria and India* (revised third edition, Chicago 1997).

Taviani, Paolo Emilio. *Christopher Columbus: The Grand Design* (London 1985).

Thomson, J. Oliver. *History of Ancient Geography* (Cambridge 1948).

Thoresen, Lisbet. "Archaeogemmology and Ancient Literary Sources on Gems and Their Origins," in *Gemstones in the First Millenium AD: Mines, Trade, Workshops and Symbolism* (ed. Alexandra Hilgner *et al.*, Mainz 2017) 155–217.

Tierney, J. J. "The Greek Geographic Tradition and Ptolemy's Evidence for Irish Geography," *PRIA-C* 76 (1976) 257–65.

Toomer, G. J. "Ptolemy," *DSB* 11 (1975) 186–206.

Traina, Giusto. "Strabone e le città dell'Armenia," in *Studi sull'xi libro dei Geographika di Strabone* (ed. Giusto Traina, Galatina 2001) 141–54.

Tsetskhladze, Gocha R. "A Survey of the Major Urban Settlements in the Kimmerian Bosporos (with a Discussion of their Status as *Poleis*)," in *Yet More Studies in the Ancient Greek Polis* (ed. Thomas Heine Nielsen, Stuttgart 1997) 93–81.

Tupikova, Irina et al. *Travelling Along the Silk Road: A New Interpretation of Ptolemy's Coordinates* (n.p. 2014).

Wheeler, Mortimer. *Rome Beyond the Imperial Frontiers* (Harmondsworth 1955).

Wieber, Reinhard. "Marinos von Tyros in der Arabischen Überlieferung," in *Historische Interpretationen: Gerold Walser zum 75. Geburtstag* (ed. Marlis Weinmann-Walser, Stuttgart 1995) 161–90.

Winkler, Gerhard and Florian Mittenhuber. "Die Länderkarten Asiens," in *Klaudios Ptolemaios: Handbuch der Geographie, Ergänzungsband* (ed. Alfred Stückelberger and Florian Mittenhuber, Basel 2009) 290–304.

Zaninović, M. "Scardona," *PECS* 812–13.

List of passages cited

Italicized numbers are citations in ancient texts; romanized numbers are pages in this volume.

Greek and Latin Literary Sources

Aeschylus; *Persians 865*, 174n12;
  Prometheus Bound 422, 177n121;
  717–21, 177n121
Agathemeros *1*, 174n7; *1.1*, 5n8; *2*, 25n46,
  25n47, 25n62, 45n17, 45n28; *5*,
  45n28
Airs, Waters, and Places 13, 67n1, 93n14,
  174n21
Ammianus Marcellinus *22.8.10*, 26n74;
  22.15.8, 188n20; *23.6.54*,
  176n100
Apollonios of Rhodes, *Argonautika 4.1384*,
  117n34
Appian; *Civil War 2.90*, 6n14; *4.54*,
  117n21; *Illyrike 3–5*, 93n49
Aristotle; *On the Heavens 2.14.298a*,
  25n58, 188n2; *Meteorologika
  1.13.350b*, 69n58, 93n24, 175n30;
  2.5.362b, 25n62
Arrian; *Anabasis 1.1-8*, 93n47; *1.1-13*,
  93n30; *3.2.3-4*, 117n41; *3.30.6-9*,
  93n25; *4.1.3-4*, 176n107; *5.26*,
  177n128; *Periplous of the Euxine
  Sea 11.4*, 174n16
Athenaios *2.44d*, 117n27; *3.83b*, 118n54;
  5.201c, 24n22, 45n22
Augustus, *Res gestae 26*, 69n47, 69n60;
  31.1, 177n131
Avienus *110–12*, 68n7; *111*, 68n13; *112–19*,
  68n17; *117*, 68n7; *383*, 68n8;
  412, 68n8

Caesar, Julius: *Gallic War 1.1*, 69n49;
  4.20-36, 68n21; *5.7-23*, 68n21;
  5.13, 68n7; *6.24*, 69n62
Cassiodorus, *Institutiones 1.25*, 26n77

Dikaiarchos of Messene *F122-5*, 25n47;
  F124, 25n49; *F125*, 117n13
Dio Cassius *48.45.1-3*, 117n9; *51.23.2*,
  69n46; *53.12*, 94n61; *55.1*, 69n65;
  60.9, 117n12; *68.6-14*, 93n42
Diodoros of Sicily *5.11*, 93n12; *5.32.5*,
  69n46; *5.38*, 69n45; *12.32.3*, 193;
  17.93, 177n128; *19.98-100*, 175n51
Diogenes Laertios *2.1*, 174n7; *2.2*, 5n8,
  6n12; *5.51*, 5n10; *8.48*, 5n2,
  183n12; *9.21*, 5n2, 183n12, 188n1;
  9.47, 183n18
Dioskourides *5.130*, 177n141

Ephoros *F30*, 45n20, 93n51, 118n57,
  176n111, 177n147
Eratosthenes ; *Geography F1*, 44n1;
  F23, 177n122; *F25*, 183n13;
  F28, 183n7; *60*, 183n9; *F60*,
  25n56; *F66*, 69n48; *F68*, 25n44;
  F74, 177n134, 177n149; *F87*,
  175n40, 175n55; *F95*, 175n56;
  F106, 117n13; *F109*, 176n110;
  F117, 94n58; *F148*, 93n48; *F150*,
  69n62; *Measurement of the Earth
  M1-7*, 183n7; *M1-8*, 44n9; *M6*,
  183n14
Eudoxos of Knidos *F276a*, 25n46
Euripides; *Andromache 795*, 94n58;
  Bacchants 16, 176n75, 194

Florus *2.30*, 69n65

Geminos *5.16*, 183n15; *5.28*, 183n15

Hanno of Carthage *1*, 117n24; *6–7*, 117n11;
  8, 118n58

Hekataios of Miletos (*FGrHist* #1) *F191-2*,
174n26; *F291*, 176n96

Herodotos; Book 1: *6*, 174n12; *46*, 117n41;
75, 174n12; *98–9*, 175n67; *163*,
193; *173*, 174n8; *184*, 176n87;
196, 69n80; *205–14*, 17n91; Book
2: *4*, 117n43; *15*, 23n4; *28-32*,
188n17; *29*, 45n27, 197; *29–30*,
118n61; *30–1*, 117n26; *32*, 117n41,
118n51; *34*, 25n45, 25n50, 45n42;
55, 117n41; *178*, 117n42; Book 3:
93, 176n101, 176n105, 194; *97-
102*, 117n127; *115*, 69n43; Book
4: *1–142*, 176n112; *8*, 117n15; *13*,
68n12; *22*, 93n33; *32*, 175n29;
42, 118n74; *44*, 176n76, 177n127;
45, 67n1, 93n14, 118n50, 174n21;
49, 69n73, 93n49; *87–140*, 93n46;
110–17, 93n16; *116–17*, 174n20;
144, 194; *150–8*, 117n30; *172*,
117n39; *174*, 118n55; *183*, 117n39,
118n55, 200 ; Book 5: *49–50*, 5n9,
174n7; Book 6: *36*, 94n60; Book 7:
66, 176n105; *89*, 24n34; Book 9:
113, 176n101

Hesiod; *Theogony 335*, 188n16; *339*,
69n75; *Works and Days 166–73*,
197

Hieronymos of Kardia (*FGrHist* #154) *T6*,
175n51

Hipparchos of Nikaia; *Against the
Geography of Eratosthenes F4*,
67n3, 118n73, 177n148, 182n4,
183n21; *F11*, 5n7, 44n11; *F11-15*,
5n5; *F12*, 6n11; *F12-15*, 25n51;
F45, 24n19; *Commentary on the
Phenomena of Eudoxos and Aratos
1.7.11*, 25n55

Homer; *Iliad 1.423*, 118n56, 193; *2.844-
5*, 94n54; *4.228*, 23n2; *10.252-3*,
25n54; *10.434-5*, 94n54; *Odyssey
3.270*, 175n57, *4.477*, 188n16;
4.85, 116n12; *9.82, 104*, 117n25;
11.107, 93n10

Horace, *Ode 1.4*, 69n42

Isidoros of Charax, *Parthian Stations 19*,
177n124

Juba II of Mauretania (*FGrHist* #275) *F6*,
118n54; *F31*, 194; *F43*, 25n59,
117n14; *F43-4*, 25n59, 118n60;
F63-5, 176n77

King Nikomedes Periplous 196, 117n24

Kleomedes, *Elementary Theory 1.7*, 199

Kosmas Indikopleustas, 177, *26n78;* 182,
26n78

Ktesias of Knidos (*FGrHist* #688) *F1b*,
176n87

Livy *30.39*, 93n6; *39.51*, 174n6; *43.1*,
69n80; *43.5*, 69n76

Lucan *9.355*, 117n35; *10.142*, 176n94;
10.188-192, 6n14; *10.191-2*,
118n62, 188n21; *10.268-331*, 6n14;
10.325-6, 188n22

Markianos of Herakleia *2.2*, 26n76

Nikolaos of Damascus (*FGrHist* #90)
F100, 177n131

Olympiodoros, *Commentary on Plato's
Phaidon 17.72b*, 23n5

Orosius *6.21*, 177n131

Ovid, *Epistulae ex Ponto 1.2.77*, 93n34

Patrokles (*FGrHist* #712) *F5*, 176n110

Pausanias *10.17.2*, 93n8

Periplous of the Erythraian Sea 1–12,
118n69; *12*, 118n66; *13*, 197;
13–18, 118n67; *15*, 118n77, 198,
200; *15–16*, 45n24, 194; *16*, 198;
25, 197; *27*, 176n81; *29*, 198; *30*,
176n30, 194, 199; *32*, 45n37; *47*,
177n136; *51*, 196; *53*, 198; *56*,
24n21, 177n138; *59*, 196; *63–5*,
45n34; *64*, 45n44, 174n3, 177n145,
182n5, 199; *65*, 177n138; *66*,
174n2, 177n115

Pindar, *Isthmian 6.22-3*, 188n18

Plato; *Ion 533d*, 177n140; *Timaios 24–6*,
188n5

Pliny the Elder, *Natural History;* Book 2:
167, 69n60, 69n66; *180*, 44n13;
186–7, 183n6; Book 3: *4*, 117n13;
7, 69n41; *81*, 93n3; *92*, 93n12;
141, 69n81; Book 4: *55*, 94n62;
59, 94n63; *83*, 174n24; *95*, 24n20,
45n32, 93n19; *96*, 6n16, 69n68;
102, 68n16; *103*, 25n67, 68n15,
69n68; *104*, 68n37; *111*, 69n42;
120, 117n15; Book 5: *2*, 117n10;
8–11, 118n53; *14–15*, 117n12; *27*,
117n22; *31*, 117n35; *36*, 45n21;
49, 117n44; *51–4*, 118n63, 188n20;

73, 25n63; *76*, 24n36; Book 6: *19*,
93n21; *26*, 175n44; *30*, 175n38,
177n122; *46–7*, 176n98, 193; *49,
93n25, 174n4; 49-50*, 176n93;
81–91,25n68, 177n151; *84–5*,
24n26; *88*, 45n35, 177n117; *100*,
199; *107, 176n88; 110*, 176n82;
122, 175n66; *138–40*, 175n62,
176n72; *153*, 176n85, 194; *171*,
199; *177*, 67n2; *181–7*, 45n25; *191*,
188n19; *194*, 188n19; *199*, 118n53;
201, 117n14; *201-5*, 25n59;
212-20, 25n56; Book 7: *197*,
69n44; Book 9: *63*, 69n74; Book
12: *53*, 45n29; *63–5*, 175n50; *64*,
5n4; *64–6*, 176n77; Book 18: *22–3*,
118n53; Book 19: *38–45*, 117n32;
Book 22: *100–6*, 117n32; Book 36:
126–30, 177n141; Book 37: *33*,
24n20, 45n32, 93n20; *36*, 24n20,
45n32, 93n20

Pliny the Younger, *Letters 10.179*, 174n5

Plutarch; *Alexander 26–7*, 117n41;
Antonius 63, 93n38; *Concerning
the Face that Appears on the Globe
of the Moon 26*, 188n6; *Lucullus
25–8*, 175n44; *On the Obsolescence
of Oracles 18.419e*, 68n23; *Solon
26.1*, 23n4

Polybios *1.6*, 174n9; *10.28.7*, 176n73;
12.2a.4, 6n25; 3.38, 67n3,182n4,
183n21; *8.25*, 175n44; *31.18.9*,
117n28; *34.1.5*, 44n2; *34.5.7-6.14*,
25n53; *34.11.8*, 6n25; *34.15.7*,
6n13; *34.15.7-9*, 117n7, 118n53;
34.29.4, 6n25

Pomponius Mela *1.30*, 117n17; *2.120*,
93n12;, *2.122*, 93n4; *3.13*, 69n42;
3.24, 69n74; *3.33*, 69n56; *3.54*,
6n16, 68n35, 69n68

Poseidonios of Apameia *F49*, 118n46,
188n4

Pseudo-Skylax *81*, 174n16; *108*, 117n38;
109–10, 117n22

Ptolemy of Alexandria; *Geographical
Guide:* (Book 1: *1–9*, 27; *1.2*, 2;
1.7, 12; *1.9*, 115; *2.1-3.33*, 48; *2.1-
8*, 28; *2.2*, 9; *2.1-12*, 25; *3.1-33*, 49;
3.4, 117; *4.1-2*, 29; *4.2*, 5, 10, 44n6,
194, 196; *5.1.2*, 30; *6.1*, 12, 13; *6.1-
4*, 30; *6.2*, 12; *7*, 12; *7.1*, 199; *7.2*,
148; *7.4*, 5, 24n19, 197; *7.6*, 10,
114, 194, 196; *7.6-8*, 195; *7.8*, 197;

7–9, 12; *7.9*, 199; *7.10*, 193; *7.12*,
193; *8–9*, 115; *8.1*, 196, 198, 200;
8.1-7, 32; *8.2*, 197, 198, 199; *8.5*,
11, 195, 196; *9.1*, 10, 186, 194, 197,
198; *9.1-2*, 200; *9.4*, 10, 33, 116,
179, 198; *9.9*, 145, 199, 200; *10.1*,
198; *10.2*, 11, 195, 196; *11–12*, 119,
157; *11.1*, 196, 197, 198; *11.1-8*,
36; *11.2*, 195, 198; *11.4*, 36, 195,
196, 198; *11.6*, 194, 195, 196, 198;
11.7-8, 196; *11.8*, 13, 24n19, 36, 61,
93n20; *12.1*, 194, 195; *12.1-7*, 196;
12.2, 195; *12.3*, 36; *12.5*, 194, 195,
196, 197; *12.5-6*, 195, 199; *12.6*,
195, 199; *12.7*, 193, 194, 195, 196;
12.8, 36, 194; *12.9*, 197; *12.10*,
198; *12.10-11*, 195; *12.11*, 194,
195, 196, 197, 198, 199; *12.12*,
198; *12.13*, 12; *13.1*, 195, 196, 199;
13.1-9, 38; *13.2*, 194; *13.4-5*, 196;
13.5, 197; *13.7*, 195, 197, 198;
13.8-9, 199; *13.9*, 195; *14.1*, 168,
195, 196, 200; *14.1-3*, 10; *14.1-10*,
39; *14.2-3*,170; *14.3*, 10; *14.3-5*,
198; *14.4*, 10; *14.4-8*, 195; *14.4-10*,
196; *14.5-7*, 200; *14.9*, 194, 199;
14.10, 198; *15–16*, 12; *15.1-16.1*,
40; *15.2*, 24n17, 40, 195, 196, 198,
199; *15.3*, 12, 193, 198, 199; *15.4*,
24n17, 40, 193, 195, 196, 197, 198;
15.5, 197; *15.6*, 5, 12, 40, 179,
194, 197; *15.7*, 193, 194, 195, 199;
15.8, 199; *15.8-9*, 194; *15.9*, 5, 40,
179, 194, 198, 200; *15.10-11*, 197;
15.11, 195, 198; *16*, 40; *16.1*, 194,
195, 196, 197, 198, 199; *17*, 9;
17.1-2, 41; *17.2-3*, 198; *17.3*, 196,
198, 199; *17.4*, 196, 199; *17.5*, 194,
195, 196, 197, 198; *17.6*, 35, 116,
194, 197, 198; *17.8*, 197; *17.9*, 194,
197, 200; *17.10*, 197; *17.11*, 195,
198; *17.12*, 198, 200; *18.1-24-2*,
42; *19.1-3*, 42; *19–20*, 12; *20.1-7*,
43; *20.4*, 199; *21.1*, 42; *22.1-6*, 19;
23–4, 199; *23.1-23*, 43; *23.23*, 196,
197; *24.1-33*, 44; *24.4*, 197); (Book
2: *1.1-10*, 46; *1.1-11*, 18; *1.1-12*,
48, 190; *1.2*, 67; *1.6*, 194; *2.1*, 19;
2.1-10, 20; *2.1-12*, 25n16; *2.1-3.33*,
48, 49; *3.4*, 50; *3.6*, 50; *3.12*, 60;
3.13, 51; *3.17*, 51; *3.19*, 51; *3.27-8*,
51; *3.30*, 51; *3.31*, 53; *3.32*, 18, 51;
3.33, 53; *3.38*, 12; *4.1-6*, 78, 53;

4.5, 54; *4.15*, 54; *4.16*, 54, 99; *5.10*, 54; *6.3*, 54; *6.11*, 54, 57; *6.76-8*, 54; *7.1-10.21*, 56; *7.9*, 59; *8.17*, 59; *9.4*, 57; *9.5*, 57; *9.15*, 59; *9.16*, 59; *9.17*, 59; *9.20*, 59; *10.2*, 54; *10.3*, 57; *10.15*, 59; *11.1-35*, 59; *11.2*, 61; *11.4*, 60, 61; *11.5*, 64; *11.5-7*, 60; *11.7*, 60, 62, 79; *11.10*, 60; *11.11*, 61; *11.27*, 24n44, 64, 28n27; *11.28*, 61; *11.31-2*, 61; *11.33*, 62; *11.35*, 61; *12.1-16.4*, 62; *12.5*, 62, 64; *31.1-4*, 64; *14.1-15.8*, 65; *14.3*, 65; *14.6*, 65; *14.7*, 62; *15.1-8*, 65; *15.2*, 65; *15.4*, 65; *15.5*, 62; *16.1*, 62, 85; *16.1-4*, 66; *16.6*, 85; *16.5*, 62; *16.6*, 62; *16.7*, 62); (Book 3: *1.1*, 57; *1.1-80*, 70; *1.2*, 72; *1.4*, 72; *1.1072*; *1.24*, 70; *1.52*, 72; *1.67*, 159; *1.75*, 72; *1.78-9*, 72; *1.80*, 72; *2.1-7*, 72; *2.1.35*, 99; *3.1-8*, 74; *4.1-17*, 75; *4.7*, 77; *4.10*, 77; *4.16*, 77; *5.1*, 78; *5.1-31*, 77; *5.2*, 60, 78; *5.3*, 78; *5.6*, 80; *5.10*, 78, 79; *5.13*, 81; *5.15*, 80; *5.19*, 78; *5.22*, 78; *5.26*, 80, 131; *5.31*, 78; *5.50*, 79; *6.1*, 78; *6.1-6*, 81; *7.1-2*, 83; *8.1-10*, 84; *9.1-6*, 85; *10.1-17*, 86; *10.2-4*, 86; *10.10*, 24; *11.1*, 83; *11.1-14*, 88; *11.2*, 83; *12.1-4*, 88; *13.1-17.11*, 91; *15.1*, 91; *15.14*, 91; *17.1-11*, 92); (Book 4: *1.7*, 97; *1.10*, 98; *3.1-16*, 97; *2.26*, 100; *3.1-47*, 95, 102; *3.10*, 103; *3.13*, 103; *3.47*, 77; *4.1-14*, 106; *4.1-77*, 106; *4.4*, 106; *4.8*, 107; *4.10*, 107; *4.11*, 106; *5.12*, 110; *5.13*, 106; *5.26*, 108; *5.30*, 108; *5.33*, 108; *5.53*, 109; *5.54*, 109, 10; *5.74*, 109; *6.1-34*, 110; *6.33*, 99, 113; *6.34*, 6, 99, 113; *7.1-41*, 113; *7.8*, 193; *7.10*, 34, 114; *7.12*, 115; *7.20-4*, 114; *7.26*, 86, 114; *7.29*, 114; *7.31*, 114; *7.34*, 114; *7.40*, 34; *8.1*, 22; *8.1-7*, 112, 115; *8.2*, 18; *8.3*, 11, 114, 186; *8.5*, 33; *8.6*, 116; *8.7*, 18*)*; *(*Book 5: *1.1-7.12*, 121*; 1.1-15*, 121; *2.1-34*, 122; *2.9*, 92; *2.16-17*, 123; *2.28*, 89; *3.1-9*, 123; *4.10*, 124; *4.1-12*, 124; *4.8*, 159; *5.1-10*, 125; *6.1-18*, 118; *6.12-18*, 126; *7.1-12*, 127, 135; *8.1-7*, 127; *8.5*, 127; *9.1-32*, 77, 129; *9.2-5*, 81, *10.1-6*, 133; *10.1-13.22*, 131; *10.2*, 127; *11.1-3*, 134; *12.1-8*, 134;

13.1-22, 136; *14.1-7*, 138; *15.1-27*, 138; *15.6*, 194; *15.21*, 138; *16.1-10*, 140; *16.2*, 110; *17.1-7*, 140; *18.1-7*, 141; *19.1-7*, 142; *20.1-8*, 143); (Book 6: *1.1-7*, 145; *2.1-18*, 145; *2.6*, 145; *3.1-6*, 147; *3.2*, 146; *4.1-8*, 147; *5.1.4*, 147; *6.1-2*, 148; *7.1.47*, 148; *7.7*, 45n18; *7.43-5*, 115; *8.1-16*, 151; *9.1-80*, 154; *10.1-4*, 154; *11.1-9*, 155; *12.1*, 163; *12.1-6*, 155; *12.1-11*, 37; *13.1-3*, 156; *14.1- 14*, 157; *15.1-4*, 157, 159; *16.1-8*, 160; *17.1-8*, 162; *18.1-5*, 162; *19.1-5*, 163; *20.1-5*, 163; *21.1-6*, 163); (Book 7: *1.1-96*, 165; *1.15*, 38; *1.46*, 168; *1.73*, 168; *2.1*, 118n72; *2.1-31*, 168; *2.12*, 166; *2.15- 17*, 169; *2.25*, 166; *2.29*, 39; *2.31*, 177; *3.1-3*, 119, 193; *3.5*, 16; *3.6*, 39, 47, 199; *4.1*, 19, 169, 173; *4.1-13*, 20, 173; *4.14*, 174, 178; *5*, 18; *5.1-16*, 178; *5.4*, 195; *5.11*, 68n6; *5.1-16*, 178; *5.2*, 118n73; *6.1-7.4*, 178; *6.1.15*, 180; *7.1-4*, 181; *7.4*, 9); (Book 8: *1.1-6*, 30, 181*; 1.1-7*, 32; *1.1-28.5*, 178; *1.4*, 118n73; *1.5*, 44n4*; 2.3*, 7; *3.1-28.5*, 182; *3.1.11*, 201; *3.3*, 199; *16.11*, 193; *24.2*, 22; *27.2*, 199; *29.1-30.26*, 183; *Mathematical Syntaxis 2.13*, 23n8*; 5.1*, 44n8; *8.3*, 45n28, 183n26; *9.7*, 23n3; *11.5*, 23n3; *Tetrabiblios 2.2-3*, 23n7)

Pytheas (ed. Roseman) *F2*, 68n32; *F2-3*, 68n20; *F3*, 68n26; *T1-2*, 68n20; *T5*, 68n20; *T8*, 68n20; *T21*, 69n57; *T23*, 69n68; *T25*, 69n57

Quintus Curtius *9.2.2-3*, 177n128; *9.10.24-8*, 176n86

Sallust, *Jugurtha 17.3*, 117n3
Seneca; *Medea 375–9*, 188n7*; Natural Questions 1.Preface 13*, 188n11; *4a.2.7*, 118n64, 188n22; *6.7.2-3*, 118n64; *6.8.3-5*, 45n25, 188n22
Strabo, *Geography; Book 1: 1.1*, 44n1*; 1.9*, 67n3, 177n148, 182n4, 183n21; *1.12*, 5n7, 44n11; *2.17*, 117n25; *2.25*, 67n2; *2.27*, 69n46; *2.28*, 118n57, 176n111, 199; *3.4*, 44n15, 117n41; *4.1*, 183n13; *4.2*, 183n6; *4.4*, 68n7, 68n8; *4.6*, 188n3; *15.2*,

175n46; *15.4, 175n46;* Book 2: *1-3,*
25n48; *1.13,* 68n8; *1.14.* 176n97;
1.14-15, 45n26; *1.20,* 25n44; *1.39,*
175n60; *2.2,* 188n1; *3.4,* 118n46;
3.6, 188n4; *3.8,* 117n3; *4.1,* 68n18,
93n13; *4.1-3,* 25n49; *4.2-3,* 25n53;
5.7, 93n17; *5.8,* 45n50, 183n6;
5.10, 6n12, 25n65, 45n47, 118n80;
5.12, 118n46, 117n130; *5.38-41,*
25n56; Book 3: *1.8,* 117n9; *4.20,*
68n39; *5.4,* 117n15; *5.5,* 117n13;
Book 4: *5.4,* 25n67; *6.3,* 69n46;
Book 5: *2.6,* 93n3; Book 6: *1.2,*
69n53; *1.15,* 45n37; *2.1,* 93n10;
3.9, 92n2; Book 7: *1.3,* 69n63;
1.5, 69n63, 69n74; *2.4,* 93n32;
3.1, 93n24, 175n30; *3.13,* 69n74,
69n75; *3.15,* 93n48; *3.16,* 93n53;
3.17, 93n17; *4.5,* 93n15; *7.4,* 6n26;
F21, 94n60; Book 8: *1.3,* 44n5 *;*
Book 11: *2.2,* 93n22; *2.3,* 174n25;
2.10, 174n23; *3.1-6,* 172n34; *4.1-*
8, 174n27, 175n36; *5.5,* 177n122;
7.3, 176n92, 176n110; *8.2,*
176n102; *8.9,* 177n120, 177n125*;*
9.1, 176n73; *11.2,* 176n104; *11.5,*
173n92; *12.4-5,* 175n69; *13.10,*
175n68; *14.2,* 175n41; *14.6,*
175n43; *14.15,* 175n44; Book 12:
1.4, 174n15; *8.18,* 174n11; Book
14: *6.3,* 175n47; Book 15: *1.13,*
177n134; *1.20,* 177n116; *1.14-15,*
25n68, 177n150; *1.73,* 177n131;
2.3-8, 177n126; *2.10,* 177n123;
3.4, 176n71; Book 16: *1.3,* 175n65;
1.16, 175n66; *1.21-2,* 175n55; *1.26,*
175n58; *2.8,* 175n48; *3.1,* 175n58;
3.3, 176n80; *3.4,* 24n34; *4.2,*
175n55; *4.4,* 195; Book 17: *1.4-5,*
117n40; *1.5,* 188n21; *1.17,* 23n4;
1.24, 45n29; *1.43,* 117n41; *3.3,*

117n8; *3.18,* 117n22; *3.19,* 45n21;
3.20, 117n35; *3.22,* 117n29; *3.23,*
117n39; *3.25,* 94n61
Suda, "*Ptolemaios [2]*", 23n6
Suetonius; *Augustus 21,* 69n72; *Divine*
Julius 52.1, 6n14, 117n5
Synkellos *631,* 94n57

Tacitus; *Annals 4.73,* 24n14, 68n27;
12.30, 93n35; *Agricola 10,* 68n36;
24, 68n9; *29–38,* 68n22, 68n27;
Germania 39, 69n64; *46,* 69n69;
Histories 3.5, 93n35
Theophanes of Mytilene (*FGrHist* #188)
F5, 174n27; *F6,* 174n27
Theophrastos; *On Stones 4,* 177n140;
Research on Plants 4.6.4, 118n72,
198
Thucydides *1.10,* 193*; 2.96,* 93n49; *6.2,*
93n10
Timaios of Tauromenion (*FGrHist* #566)
F13, 69n53
Timosthenes of Rhodes (ed. Roller) *F11,*
24n17; *F29,* 24n17

Valerius Maximus *8.7 ext. 2,* 69n53
Velleius *2.39,* 69n72; *2.96,* 69n77; *2.116,*
69n77, 69n81
Vergil, *Georgics 1.30,* 68n33; *2.121,*
176n94
Vitruvius *2.9.13,* 117n3; *8.3.24,* 117n33

Biblical Sources

Acts 13:9, 24n29; *28:5,* 69n51
Genesis 2:14, 175n39; *8:4,* 175n39
Nahum 2:1-9, 175n64

Epigraphic Sources

CIL 1.2.21, 6n24
SIG 407, 24n34; *762,* 93n37

Index

Variant spellings may be shown after the main entries. Modern names are often listed in parentheses, but not all are shown. See also the List of Passages Cited for references to authors quoted in the text.

Abnoba Mountains 60
Abos, Mt. 135–6
Abyle 98
Achaia 14, 91
Achilles, Island of 88
Achilles, Sanctuary of 130
Actium 84, 91, 97
Adana 129
Adoula Alps 57, 62
Adoulikon Bay 193
Adoulis 193
Adria 193
Adriatic Sea 4, 62, 64, 66–7, 70, 72, 91–2,
    193, 195, 199; Northern 49, 61;
    Upper 41
Aegean Sea 80, 83, 88–9, 119, 121, 194–5,
    199; culture of 88, 124; islands in
    70, 89, 91–2, 123; northeastern 89;
    size of 81; western 70
Aelia Capitolina 140
Africa 12, 14, 21, 46, 95, 107, 112–16,
    121, 186, 188, 193, 196; central
    10–11, 104, 186; eastern 9–11, 18,
    32–5, 39–41, 110, 114–15, 179,
    187, 194–5, 197–200; Horn of 40,
    113, 151; northern 9, 33, 75, 97;
    northwestern 3, 97, 99, 184; Roman
    province of 95, 99–100, 102–4, 106,
    110, 199; southern 12, 22, 46; sub-
    Saharan 21, 193; western 16, 112
African Sea 74
Agatharchides of Knidos 151
Agbatana (Ekbatana) 146
Agisymba 12, 31, 33, 35, 37, 44, 115, 193
Agricola, Cn. Julius 48–51

Agrippa II of Judaea 140
Agrippa, M. Vipsanius 4; map of 9, 19
Agrippinensis (Köln) 57
Aigaion, Mt. 92
Aigimios (Zembra) 104
Ailantic Gulf 141
Ainos (Inn) River 62
Aiolian Islands 77
Aithale (Aithalia) 72
Aithiopia 14, 108, 171, 193; Below Egypt
    95, 110, 113–15; Fish Eating 171;
    Inner or Interior 95, 112, 115–16
Aithiopian Mtns. 113
Aitne (Etna), Mt. 77
Akamas 193
Akila 150
Alabastrinos Mtn. 110
Albana 134
Albania 130, 133–5, 146
Albanian Gates 135
Albanos River 134
Albion 48–53, 201
Albis (Elbe) River 61
Alexander, Altars of 80, 131
Alexander, Columns of 131
Alexander the Great 7, 21, 28, 30, 36–7, 80,
    83, 116, 131, 140, 146, 160, 162–5,
    173, 193–4, 196; and Arabia 150–1;
    and Baktria 155–6; Camp of 108;
    and Egypt 106, 108; and India 21,
    165–8; and Istros 83, 86–7; and
    Karmania 151–2; manipulation
    of topography by 80, 163; and
    Margiane 154; and Mesopotamia
    141, 143, 145; and Parthia 148

Alexander, Island of 147
Alexandria Eschate 156
Alexandria in Arachosia 11, 163
Alexandria in Areia 162
Alexandria-on-the-Issos 139
Alexandria in Karmania 152
Alexandria in Margiane 154
Alexandria-next-to-Egypt 7–9, 20, 24, 29,
    42, 108–9, 182, 196, 201
Alexandria Oxeiane 156
Alexandros, explorer 10, 39, 168–71, 174
Alfragan 185
Alokiai Islands 61
Aloutas (Olt) River 84–5
Alps 21, 48, 60, 62, 64, 70, 98, 198;
    Maritime 56–7; Transylvanian 84
Amanos Gates 138
Amanos Mountains 125, 138
Amardokaia 143
Amardos River 146
Amaseia 126
Amathous 138
Amazons 78
amber 10, 21, 60, 64
Ammon (Siwa) 108
Amphipolis 91, 193
Ampsaga River 95, 99–100, 102, 104
Anaphe 92
Anas (Guadiana) River 53
Anatolia 37, 121–7
Anaximandros of Miletos 3
Anaximenes of Miletos 80
Ancyra 124
Andros 92
Andu 11
Annius Plocamus 11, 173
Antas 75
Anthedon 106, 110
Anthropophagoi 169
Anti-Lebanon Mtns. 139
Antiocheia in Galatia 124
Antiocheia Margiane 154, 193
Antiocheia in Pamphylia 125
Antioch-on-the-Orontes 11, 139
Antiochos I, Seleukid king 154, 193
Antitauros Mtns. 126–7
Antivestaian Cape 50
Antoninus, Roman emperor 8
Antonius, Marcus 84, 97, 106, 125
Apameia, in Media 148
Aphrodite, Island of 107
Aphrodite Sanctuary, in Pyrenees 54, 57
Aphroditopolis 109
Apokopon (Apokope) Bay 194

Apollonopolis 109
Appennines (Apennines) 70
Aprositos 16
Apsoros, or Apsyrta (Cres) 67
Apsorros River 127
Apsyrtides 67
Aquae Augustae 59
Aquae Calidae (Bath) 51
Aquae Calidae, in Mauretania 100, 103
Aquae Sextiae 59
Aquincum (Budapest) 65, 79, 81
Aquitania 56–7, 59
Arabia 1, 9, 21, 32, 109, 138, 166, 197–9;
    Bay of 47; Eremos 136, 141–3,
    194; Eudaimon 41, 141–2, 148–51,
    194; Petraia 136, 140–1, 194
Arabian Gulf 106, 109
Arabian Peninsula 1, 3, 21, 31, 121, 140,
    142, 148, 194
Arabon (Raab or Rába) River 65
Arachosia 11, 36, 160, 163–6
Arachotos 163
Arakia 147
Aral Sea 154, 156, 162
Ararat, Mt. 135–6
Aravene 127
Araxes (Aras) River 133, 136
Arba (Rab) 67
Arbela (Erbil) 29, 145, 194
Arbis (city) 165
Arbis River 163, 165
Archelaos I of Cappadocia 125–6
Areia 145, 148, 160, 162–3, 194
Areia Lake 162
Areios River 162
Argarikos Bay 194
Argentorate (Strasbourg) 59
Argita River 190
Argonauts 89, 133
Arikamedu 166
Aristagoras of Miletos 3
Aristoboulos of Lesser Armenia 127
Aristophylians 163
Arkadia 91
Armaktika 134
Armenia 29, 126–7, 133–4, 194; Greater
    133, 135–42, 145; Lesser 121, 125,
    127, 135, 198
Arnus (Arno) River 72
Aromata 34, 40, 114–15, 194, 200
aromatics 1, 114–15, 140, 150, 194
Arsamosata 136
Arsessa, Lake 136
Arsinoë, Red Sea port 109–10

Artanissa 134
Artaxata 136
Artaxias I of Armenia 136
Artemidoros of Ephesos 4
Artemis, Oracle of 150
Artemita 143
Artoarta 163
Asia 1, 119–78, 182, 185, 193–4, 196–9
Asia Minor 11, 20, 32, 56, 89, 92, 119, 121, 135, 159, 181, 195
Askatankas 157
Askibourgion Mountains 60–1
Aspabata 159
Aspendos 125
Asphaltitis Lake (Dead Sea) 140
Asseria (Podgradje) 67
Assyria, Assyrians 126, 129, 131, 135, 142–3, 145–8, 152, 194–5
Astaboras River 114
Astapous River 114
astrolabos 28–9
astronomy 2, 7–8, 20, 28, 31, 34, 38, 182
Aternos (Pescara) River 72
Athena 115
Athens 91; Lyceum at 3
Athos, Mt. 194
Atlantic Ocean 2–3, 21–2, 46, 56, 95–9, 106, 110, 112, 115, 185, 197
Atlantis 184
Atlas Mountains 97–100, 106, 113
Augila 108
Augustan Period 3, 11, 36, 57, 60, 62, 165
Augusta Raurikon (Augst) 57
Augusta Triberon (Trier) 57
Augusta Vindelicum (Augsburg) 60, 62, 64
Augustus (emperor) 53
Ausoba River 190
Auteinoi 191
Auzakia 159
avians 169
Axiopolis (Hinog) 86
Azania 32, 194, 200

Babylon, in Babylonia 141–4
Babylon, in Egypt 109
Babylonia 136, 140–5, 147, 179
Baetica (Baitike) 53–4, 98
Bagoon Mtns. 162
Bagradas (Mejerda) River, in Africa 104
Bagradas River, in Persis 147–8
Baikolikon Mtns. 107
Bainakos Lake (Lago di Garda) 70
Baithana 166
Baitis (Guadalquiver) River 53, 194

Baiton 28, 162
Bakchos 115
Baker, Samuel and Florence 189
Baktra 155, 194, 196
Baktria 152, 154–6, 162
Baktriane 194
Baktros River 155
Baliaric Islands 54
Balkans 14
Baltic Sea 3, 10, 21, 57, 59–61, 64, 70, 77–80, 129
Barbaria 194
Barbarian Gulf 115
Barke 106
barley 170
Batanaia 138
Bautisos River 160
Bears (constellations) 10, 31, 180
Bebioi (Biblioi) Mountains 65
Belgica 56–7
Bepyrrhon Mtns. 169
Berenike, in Kyrenaika 106–7
Berenike, Red Sea ports 110
beryllos 173
Besadians 169
Besyngitians 169
Birgos River 191
Bithynia 121–2, 124
Black Forest 59–60, 62
Black Sea 5, 14, 21–2, 41, 70, 80–6, 88–9, 124–6, 130–3, 163, 194, 197; coast of 2–3, 21, 87–9, 135, 200; lands north of 33, 48, 129, 156; northern 2, 77–9, 121–2, 124; western 64, 83; *see also* Euxeinos Pontos; Pontic Sea
Blessed Islands *see* Makarioi (Makaroi) Nesoi
Bocchus II 97, 100
Bogudes II 97
Boiotia 91
Bonna (Bonn) 59
Borsippa 143
Borysthenes (Dneiper) River 79–81, 83, 87
Borysthenes Island (Berezan) 87–8
Bosporanian kingdom 81, 130, 133
Bosporos, Kimmerian 78–81, 129–30
Bosporos, Thracian 83, 88–9, 122, 194
Bostra (Bosra) 141
Bouiaimon 61
Boukephala (Boukephalos Alexandria) 168
Bracheia Sea 115
Bragodurum 64
Brettania (Brettanic) Islands 49, 201

Brettania (Brettanike) 50, 178, 190, 194
Brettanic Ocean 49, 56, 201
Brigands, Land of the 169
Brigantes 191
Brigantium (Bregenz) 62, 64
Brigetio (Szöny) 65
Brisoanas River 147
British Isles 2–3, 21, 48, 50–1, 54, 56, 184,
    194, 201
Brundisium (Brindisi) 4, 72
Bruttium 74
Burned Laodikeia 124
Burton, Richard Francis 187–8
Buvinda River 191
Buzara Mtn. 104
Byke Lake 79, 81
Byrebistas 84–5
Byzantion 37, 194

Caesar, Altars of 80
Caesarea Maritima 138, 140
Caesarea in Mauretania 112, 196; *see also*
    Kaisareia
Caesarea Panias (Banias) 138
Cambodunum (Kempten) 62
camel caravans 1, 8, 108, 140–1, 150, 180
Campania 74
Canary Islands 113, 171
cannibals 116, 169–70
Cappadocia 121, 124–7, 133, 135–6, 138
Carnuntum 65
Carthage 98, 102–3, 196; defeat of 2–3, 95,
    97, 102; location of 29, 112
Caspian Gates 37, 135, 146, 196
Caspian Mtns. 133
Caspian Sea 22, 80, 121, 130–6, 145–8,
    151–9, 162, 171, 179, 195–8
Castra Regina (Regensburg) 60
Cataractonium 201
Caucasus Mtns. 3, 99, 130–6, 163
Caucasus near India 162–3, 165
Chaboras River 142
Chandragupta 165
Charax 147
Charindas River 146
Chaurana 160
Chelidoniai 195
Chelidonophagoi 152
Chesinos River 78
China, Chinese 11, 22, 119, 157, 160,
    171, 199
Chios 92, 123
Choathras Mtns. 145
chorographia 27–8

Chorseas River 138–40
Chouthoi 148
Chronos River 78
Chryse (Chrysos) Chersonnesos 38, 166,
    168, 195, 199; *see also* Golden
    Peninsula
Chrysoras River 139
Chrysoun Mountain (Monte Cinto) 74
Cilicia 121, 125, 127–9, 138
cinnamon 114
Claudius, emperor 4, 14, 51, 85, 88, 173
Claudius Nero, Ti. 75
Claudivium 64
Colchis 127, 133; *see also* Kolchis
colors 39, 169, 180
Columbus, Christopher 16, 185–6, 188
continental theory 22, 46, 178
Corinth 91
Cornelius Balbus, L. 33
Cornelius Scipio Aemilianus, P. 3
Cretan Sea 92
Crete 70, 89, 91–2
Cuba 185, 188
Cyprus 121, 136, 138, 193, 197
Cyrenaica 95, 102, 104, 106–8, 110, 112
Cyrus the Great 152, 154

Dabrona River 191
Dacia 79, 83–5, 87
Daix (Ural) River 157
Dalmatia 66, 89, 195
Damascus 139
Danube River 14–15, 41, 48, 62–7, 78–87,
    186, 197; *see also* Istros
Daqin 11
Dareios III of Persia 194
Dareios I of Persia 86–7, 156, 160
Dargomanes River 162
Darinoi 191
Darnis (Derna) 106–7, 109
Darouernon (Duroverum, Canterbury) 50
Dauaba 159
degrees 16, 180
Dekebalos 84
Demeter 115
Demetrios of Tarsos 50, 68
Demodamas of Miletos 119, 152
Demokritos of Abdera 31, 181
Deva (Chester) 51
Dia 92
diaperama 38
Dikaiarchos of Messana 15, 36, 181
Dindymos Mt. 124
Diodoros of Samos 10, 30, 32

Diodotos I of Baktria 155
Diogenes, East African explorer 10, 12, 34, 115–16, 186–7
Diokaisareia 129
Diomedeiai Islands (Isole di Tremiti) 72
Dioskourias 127, 133
Dioskourides, Arabian island 151
Dioskouroi, Island of *see* Dioskourides
Dioskouros, East African explorer 115–16
Diour Mountains 98
Domitian, emperor 33
Doumna 53, 201–2
Dounon 191
Dourios (Duero) River 53
Drakontios 104
Drangiane 160–5
Dreinos River 66
Drilon (Drin) River 66
Drusus, brother of Tiberius 60–1, 64
Drusus, Monument of 61
Durostorum (Silistra) 87
Dur River 190
dye, purple 13, 99
Dytikos Ocean 49, 190

earth, spherical 1, 43, 171–2, 180–1, 184
earthquakes 124
Eblana 48, 191
Eblanioi 191
Eboracum (Eborakon, York) 50–1, 201
Ebudai Islands 191
Echinos 106
eclipses 2, 29
Edoudai Islands 49
Edros 192
Egypt, Egyptians 14, 36, 95, 105–10, 113–14, 116, 166, 180, 196, 199; borders of 106, 136, 194; Delta of 198; location of 15, 106; traders in 198
Egyptian Sea 104, 108–9
Elamites 56, 147
Elana 141
Elantic Gulf 141
Elephant Eaters 114
Elephant Mountain 114
elephants 110, 115, 173, 184, 198
Eleutheros River 139
emeralds 110
Emoda Mountains 159–60
Emona (Ljubjana) 62, 64–6
Emporikos Kolpos 97
England 50
English Channel 56

Epeiros 91
Ephesos 122
Epidion 191
Erasmus 23
Eratosthenes of Kyrene 1–2, 18, 27–31, 44, 57–61, 87, 143, 165, 180–1, 185, 199
Erdinoi 191
Erytheia, places so named 99
Essina 41, 145
Etruscans 53, 72
Eukratides I of Baktria 155
Eukratidia 155
Eulaios River 146–7
Euphrates River 11, 36–8, 121, 125, 127, 133, 135–43, 147, 157, 186, 195, 197
Europe 53, 60, 119, 190, 198, 201; coasts of 21, 102; continent of 47, 110, 129, 178, 181; eastern 70, 77–92; northern 11, 14, 21, 37, 61, 182, 193; western 4, 14, 124
Europos 146
Eurymedon River 125
Euxeinos Pontos 70, 121; *see also* Black Sea
Exapolis 130
External Ocean 22, 112–13, 119, 152, 159, 179, 181

Ferdinand of Spain 185
Finnoi (Phinnoi) 61–2
Fish Eaters 116, 150
Flaviopolis 122
Flavius Arrianus, L. 14
Flexum 65
Forma Urbis Romae 4
Fortunate Arabia *see* Arabia
frankincense 1, 9, 150, 198

Gabreta Forest 60
Gadanopydres 148
Gades (Cádiz) 16, 54, 99
Gaetulians 99
Gaia 104
Gaius Caligula, emperor 97
Galaktophagoi 159
Galata 104
Galatia, Galatians 56, 121–2, 124–5, 159
Galilee 140; Sea of 139, 140
Gallia 46, 57
Galloi 124
Ganganoi 191
Gangara 134

Ganges River 37–8, 165–6, 168–70, 174, 197, 199
Gangetic Gulf 168
Gangetikos Bay 195
Gan (Kan) Ying 11
Garama (Garame) 11, 33, 35, 195
Garamaioi 195
Garamantes 113
Garganos Mtn. 70, 72
Gaul 3, 56, 124
Gaulos (Gozo) 104
Gaza 1, 140, 143, 150
Gedrosia 151, 160, 163–6
Gennesarit (Sea of Galilee) 139–40
geographia 27–8
Gerasa (Jerash) 141
Germania Megale (Greater Germany) 59–61, 79
Germania Superior 60
Germanic Ocean 50, 59, 61, 201
Germanikopolis 124
Gerrha 150
Gerrhos River 134
ginger 173
Glaukon 109
Gods, Islands of the (Theoi Nesoi) 54–6
gold 151, 169–70, 173
Golden Land 168–9
Golden Peninsula or Chersonesos *see* Chryse (Chrysos) Chersonnesos
Gordyaia, Mt. 135–6
Gorgos River 145
Gorya River 162
Graupius, Mt. 51
Great Gulf 168–70
grid system 1, 15, 42, 143

Hadrian, emperor 41, 140
Hadrian's Wall 50–1
Haimos Mountains 85–8
Halys River 124, 126
Hannibal of Carthage 122
Hanno of Carthage, explorer 22, 97–8, 112–13
Harmozan, Cape 152
Hebudes Islands 49
Hedaphtha (Gidaphtha) 103
Hekatonpylos 148, 195
Helios Harbor 173
Helios Spring 108
Hellas, Central Greek region 91
Hellenes 8
Hellespont 37, 83, 89, 195
Heptadelphoi 98

Hera 104
Hera, island 16, 113
Herakleian stone 170
Herakles 104; Altars of 145; Pillars of 15–16, 72, 98, 184, 196; Sanctuary of 72; Sands of 107; Straits of 46, 179, 197
Hera, Sanctuary of, on Corsica 75
Hera, Temple of, in Iberia 54
Herculaneum 30
Herkynian Forest 60–1, 77
Hermaia, Cape 102
Hermos River 123
Hermoupolis 109
Herod the Great 140
Heron of Alexandria 24
Heroonpolis 46, 106, 110
Hesperides, Gardens of 106
Hesperou Keras 113
Hibernia 1, 48
Hierapolis, in Syria 195
Hierasos River 84, 87
Hierni 49
Hieron Cape, in Ivernia 191
Hieron Mouth, of Istros 86
Hieron Promontory, in Iberian Peninsula 195
Hieropolis in Syria 36
Hierosolymna *see* Jerusalem
Himalayas 3, 15, 41, 156–7, 159–60, 195
Himera 195
Himilko of Carthage 49
Hipparchos of Nikaia 2, 5, 10, 21–2, 28–32, 46, 116, 178, 180–1
Hippika Mtns. 131
hippopotamos 107
Hippos, in Arabia 150
Hippos (star) 32
Hispania 53–4, 56, 195
Hispania Ulterior 54
Hittites 126, 129
Homer 3–4, 9, 21, 30
honey 173
Horrea, in Upper Moisia 86
hyakinthos 173
Hydata Augousta 59
Hydata, in Dacia 85
Hydatae Therma, in Africa 103
Hydata Lesitana (Sorgenti di Benetutti) 75
Hydata Sextia 59
Hydata Therma (Bath) 51
Hydata Therma, in Mauretania 100
Hydra (in Kyrenaika) 106
Hydras 104

hydrology 129, 143, 154, 166, 179
Hypanis (Bug) River 81
Hyperboreian Mtns. 131
Hyperboreian Ocean 19, 48–9, 190, 201
Hyperboreians 48
Hyrkania 145, 148, 152, 154, 195
Hyrkanian Sea 133, 154, 179, 195

Iabadiou 39, 170–1
Iasonion, Mt. 146
Iasonion in Parthia 155
Iaxartes River 80, 154, 156, 198
Iazyges, Wandering 83–4
Iberia (in Caucasus) 130, 133–5, 146
Iberia, Iberian peninsula 3, 18, 46, 53–4,
    75, 98–9, 184, 194–6
Iberos (Ebro) River 53
Ichthyophagoi 150
Ida, Mt. 123
Idios Asia 121–2
Idumaea 140
Iernos River 191
Ieron 49
Ikarian Sea 92, 121
Illyris (Illyria) 66–7, 70, 83, 85, 89, 91
Ilva (Elba) 72
Imaon Mtns. 41, 152, 156–7, 159–60, 165,
    168–9, 195
Imaos Mountains 15
India (Indike) 16, 32, 121, 145, 152, 159–60,
    162–3, 194–200and Alexander
    the Great 21; beyond the Ganges
    168–73; routes to 3, 9–10, 16, 21,
    34, 37–8, 41, 150–1, 162, 165, 168,
    184; within the Ganges 165–8; trade
    with 110, 119, 156, 160
Indian Ocean 3, 22, 39, 44, 95, 110,
    112–16, 148, 150, 160–5, 168, 170;
    enclosed nature of 39, 44, 116,
    178–9, 181
Indians 119, 171
Indian Sea 178
Indike *see* India
Indos River 10, 32, 151, 162, 165–8, 173,
    195–6
Ionia 122
Iordanos (Jordan) River 138
Iourassos (Jura) Mountain 57
Ireland 10, 20–1, 37, 46, 48–9, 54, 190,
    196, 201
Iris River 126
Irykai 83
Isabella of Spain 185
Isamnion, Cape 191

Isar (Isère) River 57
Isatichai 148
Isidoros of Charax 11, 36, 165
Iska (Exeter) 51
Issa (Vis) 67
Issos Bay or Gulf 121, 196
Istrianos River 81
Istros River 15, 62, 64, 78–9, 83–4, 86–7;
    see also Danube
Italia (Italy) 62, 66, 70–4, 196
Iuliopolis 122
Iulium Carnicum (Zoglio) 64
Ivernia 10, 13, 20, 46–9, 190–2, 196, 201
Ivernios Ocean 49, 191
Ivernis (Iernis) 191, 201
Ivernoi 191

Jason, mythological hero 133, 146, 155,
    163
Java 22, 39, 170–1
Jerusalem 140
João II of Portugal 185
Jordan River *see* Iordanos
Juba I of Numidia 102
Juba II of Mauretania 3–4, 14, 16, 97–103,
    150, 185–6, 194
Judaea 110, 136, 140–1
Julius Caesar, C. 3, 50, 54–7, 60–1, 85, 97,
    102–3, 114, 186
Julius Marinus Caecilius Simplex, L.
    13–14
Julius Marinus, L. 13
Julius Maternus 11–13, 32, 115, 193
Juno, Promontory of, in Iberia 54
Juno, Sanctuary of, on Sardinia 75

Kabeira 126
Kaikos River 123
Kainopolis 106
Kaisareia (Caesarea) Iol 100, 140, 196
Kaisareia in Cappadocia 126
Kaisareia of Straton 140
Kaledonian Forest 50
Kalpe, Mt. 196
Kambysos River, in Media 146
Kameloboskoi 152
Kanaria 16
Kane 150
Kanobos (Canopus) in Egypt 7–9, 196;
    Wings of 8
Kanobos (Canopus, star) 32
Kantion (Kent) 50
Kappadokia *see* Cappadocia
Kapraria (Capraia) 72

Kaprea (Capri) 72
Kapros River 145
Karallis (Cagliari) 196
Karambis, Cape 124
Karbones 78–9
Karchedon *see* Carthage
Karia 92, 122
Karmania 145, 147–8, 151–2, 160
Karmania Eremos 148
Karmanian Mtn. 152
Karmelos (Carmel) Mt. 139
Karpathian Sea 121
Karpatos (Carpathian) Mountains 80
Karrhai (Harran) 142, 155
Kasios River 134
Kasperia 16
Kaspiai Pylai *see* Caspian Gates
Kassios Mt. 139
Kassiotis 139
Kassiterides 56
Katabathmos 106, 108
Katakeukamene 124
Kataonia 127
Katarrhaktes River 125
Kattigara 12, 36, 39, 41, 43, 170–1, 196
Kaudos 92
Kaukoi 191
Kaystros River 123
Kelainai 125
Kelainai, Ridge of 124–5
Keltogalatia 56–9, 64
Kelts 53, 56, 124, 171
Kentouria 16
Kephallenia 91
Keraunia Mtns. 131
Kerkina (Kerkenna) 104
Kerkyra 91
Kerne 99, 113
Kestros River 125
Ketaion, Cape 173
Kiabros (Tsibrica) River 85–6
Kilikia 125, 127; *see also* Cilicia
Kilikian Channel 121
Kilikia Pedias 127
Kilikia Tracheia 127
Kimbrian Chersonesos 61
Kimmerian Bosporos 78–80, 129–30
Kinnanomophoros 114
Kinyps River 103–4
Kirrhadia 169
Kirta (Constantine) 103
Kisterna 103
Klaudiopolis in Galatia 124
Klaudiopolis in Mysia 122

Klaudios Ptolemaios *see* Ptolemy of
  Alexandria
Kleomenes of Sparta 3
Kleopatra Selene 97, 100
Kleopatra VII 3, 7, 84, 97, 106, 129
klima 15
Klimax, town in Marmarike 109
Klimax Mtn. 150
Koile Syria 139
Kolchikon Bay 196
Kolchis 130, 133–4, 155
kolchoi (mussel) 170
Kolchoi (region) 196
Kolenton (Colentum, or Murter) 67
Komana in Cappadocia 126
Komana in Pontos 126
Komaria Promontory 196
Komedai Mountains 137, 195–6
Korankalians 169
Korax Mtns. 131
Korax River 129, 133
Koriondoi 191
Korkyra Melaina (Korčula) 67
Kory Bay 196
Kory, Cape or Promontory 38, 166,
  173, 196
Kossyra (Pantelleria) 104
Kotes Promontory 97–8
Kounos 53
Kourikta (Krk) 67
Kouroula 196
Kourta 65
Kratas, Mt. 77
Krates of Mallos 3, 19, 43, 116, 180
Kroton (Crotone) 72
Ktesiphon 145
Kyaneai Islands 89
Kydnos River in Cilicia 129
Kydnos River in Pontos 126
Kyklades 92
Kyrenaika *see* Cyrenaica
Kyrene 103, 106
Kyrnos (Corsica) 70, 72–5
Kyros River, in Caucasus region 134–6
Kyros River, in Media 146

Laberos 191
Labotas River 139
Laganikoi, Caves of 107
Laia 107
Lakonike 196
Laodikeia in Galatia 124
Laodikeia, places so named in Syria 139
Laodikene 139

Lapethos River 138
Larios Lake (Lago di Como) 70
Lathon River 107
latitude 12, 14–16, 18, 29, 31–5, 46, 170,
　180, 185, 190
Latium 4, 74
Laviansene 127
Lebanon Mtns. 139
Legio I Adiutrix 65
Legio I Italica 87
Legio I Minervia 59
Legio II Adiutrix 65
Legio II Augusta 51
Legio III Augusta 103
Legio III Cyrenaica 141
Legio III Gallica 139
Legio IV Flavia Felix 85
Legio V Macedonia 87
Legio VI Victrix 51
Legio VII Claudia 85
Legio VIII Augusta 59
Legio X Gemina 65
Legio XI Claudia 13–14 87
Legio XIV Gemina 65
Legio XVI Flavia 139
Legio XX Valeria Victrix 51
Legio XXII Primigenia 59
Legio XXX Ulpia 59
Lemnos 91
Leptis Magna (Leptis Megale) 33, 35,
　103–4, 115, 196
Leptis Mikra 103
Lesbos 123
Letoa 92
Leukas 91
Leukon 109
Leukopetra, Cape (Punta di Pellaro) 70
Leukos Island (Zmeinij) 88
Levant 11, 14, 37, 104, 106, 110, 121, 129,
　139–40, 142
Libanotophoroi 150
Libnios River 190
Liburnia 66–7
Libya (Africa) 44, 46–7, 95, 107, 119, 171,
　178, 181–2, 186; Inner or Interior
　98, 100, 110, 112–13
Libya Lake 104
Libyan Mountains 109, 114
Libye *see* Libya
Libyphoenicians 103
Libyssa 122
Licinius Crassus, M., the elder 142
Licinius Crassus, M., the younger 85
Liger (Loire) River 56–7

Ligystike (Liguria) 70
Ligystikian (Ligurian) Sea 74
Likias (Lech) River 64
Lilybaion (Marsala) 75, 196
Limia River 54
Limnos 192
Limyrike 196
Lipari Islands 77
Lissos (Lezha) 62, 66
Lithinos Pyrgos (Stone Tower) 36, 157, 196
Livingston, David 88
Lix (Lixos), city in Mauretania 98
Lixos River 98
Logia River 191
Londinion (London) 12, 40, 197, 201
Londobris 54
longitude 2, 12, 14–19, 29, 31, 36, 39–40,
　112–13, 173, 179–82, 190
Long Wall (Makron Teichos) 89
Lotoa 91
Lotophagitis 103–4
Lotophagoi 103
Louppia 61
Lugdunensis 56–7
Lugdunum 56
Luna (Carrara) 60, 72
Luna Forest 60
Luoyang 160, 198
Lusitania 18, 53–4
Lydia 122–3
Lykia (Lycia) 14, 121, 123, 125
Lykian Sea 121
Lykomedes Lake 108
Lykos River, in Assyria 145
Lykos River, on Cyprus 138
Lykos River, in Pontos 126–7
Lysimacheia 89
Lysimachos 89

Macedonia, Macedonians 7, 11, 37, 48, 66,
　83, 85, 88, 91, 150, 168, 194, 199
Maes Titianos 11–13, 36–7, 119, 157, 160
Magdalensberg 64
Magna Graecia 72
Magnatai 191
Magnesia at Sipylos 123
magnetic stone 170
Mago of Carthage 104
Maiandron Mtns. 169
Maiandros River 123
Mainoumena Mtns. 74
Maiotic Lake or Sea 46, 77–80, 129–31, 179
Makarioi (Makaroi) Nesoi 14, 99, 113,
　194, 197

Makolikon 191
Malabar Coast 10, 32, 196
malabathron 169
Malay Peninsula 10, 38, 169–71, 195,
    199–200
Maloua River 97, 99
Manapia 49, 191
Manapioi 191
Maniolai 170
mapping 2–3, 20, 180–1
Marakanda (Samarkand) 155–6
Marcius Turbo, Q. 84
Marcus Aurelius, emperor 8
Margiane 152, 154–5, 162
Margos (Marghab) River, in Margiane
    154–5
Margos (Morava) River 66
Marianon Mountain 54
Marinos of Tyre 9–16, 18, 20, 30–42, 50–1,
    59, 84, 97–8, 113, 170, 179, 181,
    185
Maritime Alps 56–7
Marmarike 107–8
Marobodos 61
Masikytos Mtns. 123
Masion Mtns. 142
Massagetians 154, 157
Massalia (Marseille) 21, 56, 183
mathematics 2, 7–8, 20, 28, 180, 184
Matiane Lake 146
Mauretania 97–8, 102, 186, 196;
    Caesariensis 95, 97, 99–101, 110,
    196; Tingitana 95, 97–9, 110
Mauryans 37, 163, 165, 168, 197
Maxeras River 146, 148, 154
Mazaka 126
measurements 15, 16, 28–9, 179–80;
    astronomical 2; and distance 1–2, 5,
    15, 179; length of day 183; units of
    4–5, 185
Medaba 141
Medea 133, 146, 155
Media, Medians 133, 135, 145–8, 152, 195,
    197
Mediterranean 1, 36, 38, 54, 59, 102, 106,
    131, 154, 166; coast of 56–7, 95,
    97, 99, 123, 136, 141; eastern 12,
    21, 121, 138; enclosed nature of
    179; narrows of 104; shape of 15,
    19, 31; western 2, 21, 46; world of
    3, 5, 11, 33, 40, 49, 78–80, 112–15,
    126, 130, 134, 170, 173, 194
Medoura 168
Mega Akroterion 169

Megale Hellas 72
Megalos Kolpos 168
Megas Aigialos 197
Megas Harbor 113
Megasthenes 165, 173
Melana Mtns. 141
Melas, rivers so named 125–7
Melibokon Mountains 60
Melite (Malta) 104
Melite (Mljet) 67
Melitene 127
Melos 92
Menelaos, city 109
Meninx 103
Menouthias 116
merchants 3, 9–13, 22, 34, 37–8, 48, 114,
    130, 166, 194
meridians 14–19, 29, 42–4, 46, 66, 113,
    171, 180–2, 197, 201
Meroë 35, 43–4, 113–14, 197
mesembria 15
Mesembria (Nesebur) 88
Mesopotamia 21, 121, 125, 131, 133–8,
    141–6, 155, 162, 194–5, 197, 199
Messenia 91
Metapontion 72
Mikron Aigialos 197
miles 4–5, 32, 35, 40, 179
Miletos 122–3
milia see miles
Minios (Mi˜no) River 53
Misynos 104
Mithridates I of Pontos 126
Mithridates, Land of 131
Mithridates VI of Pontos 3, 122, 124–5
Modomastike 148
Modonos River 191
Moesia 66, 197
moira 16
Moisia Inferior (Lower) 70, 83–4, 86–8
Moisia, Upper 83, 85–6, 88–9
Mokontiakon (Mainz) 59
Molibodes Island 75
Mona 192
Monaoida 49, 192
Mosaios River 147
Mourimene 127
Murchison Falls 186, 189
Mykale, Mt. 123
Mykonos 92
Myrmex 107
myrrh 114, 150
Mysia 66, 148, 197
Mysoi 86

Naarsares River 143
Nagnata 48, 190
Nagnatai 191
Narbonensis 56–7, 59
Narona (Vid) 66
Naron (Neretva) River 66
Naukratis 109
Neapolis, in Kyrenaika 103, 106
Neapolis (Naples) 72
Nearchos 151, 185
Nedinon (Nadin) 66
Neilos *see* Nile
Nero, emperor 34, 114, 186
Nessos (Nestos) River 83, 88–9, 91
New World 184–5
Nigeir River 112
Niger River 98, 112
Nigritian Aithiopians 113
Nikaia (Nice) 70, 72
Nikomedeia 23, 122
Nikomedes IV of Bithynia 122
Nikopolis 91
Nile 9, 21, 35, 46, 86, 107–10, 113, 197–200; Delta of 106, 108; First Cataract of 108–9, 113–14, 199; and Julius Caesar 3; source of 10–11, 34, 40, 114–16, 119, 186–8; Veins of 186
Nineveh (Ninos) 145
Niphates Mtns. 133
Niphates River 145
Nisaia 155
Nisibis in Areia 162
Nisibis (Nusaybin) in Mesopotamia 142
nomoi 109
Noricum (Norikon) 64–6, 197
Northern (Boreios) Cape 48
Notion, Cape 191
Novae 87
Noviomagus (Chichester) 12, 40, 50, 197
Numidia 97, 102–3

Oboka River 191
Ocean 50, 119, 179, 181–2, 184, 201
Octavian (Augustus) 98
Odrangidan Aithiopians 113
Odysseus 1, 103
Odysseus Cape 77
Oichardes River 159
oikoumene 2, 25
Oiskos (Gigen) 87
Okelis 31, 150, 197
Olbe 129
Oligas Mt. 124

Olympiodoros of Alexandria 8
Olympos, Mt., on Cyprus 138
Olympos, Mt., in Mysia 122
Onesikritos 151, 185
Ophiodes River 113
Opone 197
Oppidium Novum 100
Orbelos Mtn. 83, 85, 88
Orbisene 127
Orcades (Orkneys) 51
Orchoe (Warka) 141, 143
Orion (constellation) 32
Orkas Cape 50
Orkynian Forest 60–1
Orodes II of Parthia 155
Orontes (el-Asi) River 138–9
Orontes Mtn. 146
Orrea 51
Orsene 127
Ortospana 163
Osteodes 77
Ouelpa (Velpa) Mtn. 107
Ouenedikos (Venedicus) Gulf 78
Oustika (Ustica) 77
Oxeiane Lake 156
Oxos River 37, 154–7, 162

Pachynos (Passero) Cape 75, 197
Pados (Po) River 70, 86
Paina 99
Pakonia 77
Paktolos River 123
Paktye 89
Palai Simoundou 173
Palestina Judaea 110, 136, 139–41
Palimbothra 37, 41, 168, 197
Pallas Lake 104
Palmyra 138
Paloura 38, 197
Pamphylia 121, 123–5, 127
Pamphylian Sea 121
Pandion 168
Pannonia 65–7, 79, 83–4, 197
Panon 41
Pantikapaion (Kerch) 81, 130
Paphlagonia 124
Paphos, Old and New 138, 197
parallels 8, 15, 29, 36–7, 41–4, 180–2
Parchoathras Mtns. 146, 148
Parmenides of Elea 1, 116
Paropanisadians 160, 162–3, 165–6
Paropanisos Mtns. 155, 162–3
Parsiana 163
Parsis 165

Parthenope 72
Parthia, Parthians 8, 11, 14, 142–9, 154–5, 160, 162–5, 194–5, 197
Parthyene 148
Paryadres, Mt. 135–6
Passalians 169
Pataliputra 165, 197
Patrokles, Seleukid explorer 152, 154, 156–7
Paul of Tarsos 11
pearls 151
Pediaios River 138
Peloponnesos 13, 91–2, 196, 199
Peloros Cape 75
Peparethos 91–2
Pergamon, Pergamenes 122
Perge 125
periploi 20, 72–4, 91
Periplous of the Erythraian Sea 3–4, 41–2, 119, 165–6, 169, 179
Persia, Persians 3, 8, 21, 88, 135, 145–7, 151–2, 154–5, 163, 165, 194
Persian Gulf 13, 136, 142–3, 145, 147–8, 150–1
Persian Mtns. 152
Persopolis (Persepolis) 147
Pessinous 124
Petra 140–3, 150, 194
Peutinger Map 4
Phalangis, Mt. 41, 197
Phanagoreia 130
Pharan, Cape 141
Pharia (Hvar) 67
Phasis, city 133
Phasis (Rioni) River 133
Philadelphia (Amman) 138
Philainos, Altars of 102
Philemon 10, 13, 30, 37, 48, 61, 78
Phoenicians 13, 53, 74, 88, 97, 103, 116, 138, 196
Phokaia, Phokaians 21, 54, 56, 193
Phorbia, Cape 92
Phoros Tiberiou 59
Phylakidas of Aigina 186
Piada 160
Piera Mtns. 139
Pindos Mtns. 91
Pinnata Castra 51, 202
Pisa, in Italy 197
Planoudes, Maximos 23
Pliny the Elder 4–5, 9, 14, 19
Ploubion 75
Pluvalia 16
Poe, Edgar Allan 187

Poetovio (Ptuj) 65
Poinina Lake 70
Polaris (star) 10, 31
pole, celestial 10, 24, 31–2; terrestrial 15, 18, 43, 107, 116
Polo, Marco 185
Polybios Island 151
Polybios of Megalopolis 3, 5, 14
Pomentine Plain 4
Pompeii 30
Pompeiopolis 124
Pompeius, Cn. (Pompey the Great) 3, 134–5
Pomponius Mela 4–5, 14
Pontia 104
Pontic Mtns. 135
Pontic Sea 197
Populonium 72, 74
Porthmos 197
Portus Magnus 100
Poseidion, Cape, in Arabia 150
Praetoria Augusta, in Dacia 85
Praetorium 65
Prason, Cape 18, 22, 31, 33–5, 39, 43, 112, 116, 179, 198
Prettanike 49
Proikonesos (Marmara) 89
Prokyon (star) 32
Prometheus 163
Propontis 83, 88, 122, 195
Prousias 122
Prousias I of Bithynia 122
Pseudostomos 86
Pteron Cape 86
Pteron Stratopedon 201
Ptolemaios, use of name 7
Ptolemaios Theron 198
Ptolemies 114, 127
Ptolemy Apion 106
Ptolemy II 9, 33, 40, 138, 198
Ptolemy of Mauretania 97
Pyramos River 127, 129
Pyrenaioi *see* Pyrenees
Pyrenees 53–4, 56, 198
Pyrrhaian Aithiopians 143
Pyrrhon Pedion 98
Pythagoreans 1, 180, 184
Pytheas of Massalia 21, 51, 183, 199
Pythodoris of Pontos 126–7

Raetia and Vindelica 62–4, 198
Raiba 191
Raitiaria (Arçar) 86
Ravenna 12, 40, 198

Ravius River 190
Red Sea 10, 34, 46, 95, 106, 110, 112–15, 141, 148–51, 166, 193–7, 200
Regia (Rhegia), cities so named 49, 191
Renaissance exploration 16, 22, 72, 116, 185
Rhabon River 84
Rhagai 146
Rhaitia *see* Raetia and Vindelica
Rhaphaneia 139
Rhapton, Cape 43, 116
Rha (Volga) River 86, 130–1, 157, 159
Rhegianon (Kozlodui) 87
Rhegia *see* Regia
Rhenos (Rhine) River 46, 56–7, 59–60, 62, 64
rhinoceroses 11–12, 33, 115, 193
Rhipaia Mountains 80, 131
Rhobogdioi 48, 190–1
Rhobogdion Cape 48, 190–1
Rhodanus (Rhone) River 21, 56–7, 86, 186
Rhodes 36–7, 43–4, 123, 198
Rhodian Sea 121
Rhogomanis River 147
Rhoimetalkes III 88
Rhosikos Promontory 139
Rhoxolanoi 78
Rhyndakos River 122
rice 173
Rikina 191
ringed sphere 180
Romans 3–5, 9, 10, 14, 21, 33, 95, 106, 122, 133, 136, 150, 194–9; in Asia Minor 122, 124, 126, 129; in eastern Europe 62, 64, 66, 83, 88, 91; in Egypt 114; in western Mediterranean 56, 72, 74, 96, 102
Rome 3–4, 24, 33, 38, 61, 84, 102, 124, 173, 185
Root Eaters 114
Roubon River 78
Roxane, Baktrian princess 155
Royal River 141–3

Sabadibai 170
Sabana 169
Sabbatha (Sabata) 150
Sachalitis 198
Sacred Cape of Ivernia 48
Sacred Cape of Lusitania 18
Sada 38, 198
Sahara Desert 1–2, 21, 95, 104, 112, 195
Sakai 198
Salamis, on Cyprus 138

Sala River 97
Salike, Salians 173
Salinae 85
Salinon (Adony) 65
Salonae (Solin) 66
Samaria 140
Samnitians 159
Samos 123
Samosata 133, 138–40
Samothrake 89
Sangarios River 122
Saokoras River 142
Sarapion, in east Africa 41, 198
Sarata 16
Sardinia *see* Sardo
Sardis 122–3
Sardo 74–5, 198
Sardoan Sea 74, 99
Sardopater 75
Sardos 75
Saripha Mtns. 160
Sarmatai 198
Sarmatia 84, 87, 134–5; in Asia 129–31, 152, 157, 181; in Europe 77–81, 83–4
Sarmatian Mountains 60, 79, 81
Sarmatian Pass 131
Sarmizegetusa 84–5
Sauromatai 78
Savaria (Szombathely) 65
Savos (Sava) River 62
Saxonoi Islands 61
Scandinavia 3, 21–2, 53, 61
schoinos 36
Scotland 50–4, 202
seamanship 1, 13, 184
Sebaste, Greek version of Augustus 59, 103
Sebasteia, on Black Sea 126
Sebastopolis 126
Sebastos *see* Augustus
Sebennytos 198
Sekoana (Seine) River 56–7
Selene Mountains (Mountains of the Moon) 11, 22, 107, 116, 186–8
Seleukeia in Pamphylia 125
Seleukeia-on-the-Tigris 142
Seleukids 11, 125, 127, 133–5, 138–9, 145–8, 152–5, 162–3, 193
Seleukos I 146, 195
Semanos Forest 60
Semiramis, Mt. 152
Senos River 190
Septimius Flaccus 11, 13, 32, 115
Sera 36–8, 160, 198

Seres 11, 41, 198–9
Serike 159–60, 168, 171, 178
Seros River 170
Sestius, Altars of 54, 56
Sestius Quirinalis Albinianus, L. 54
Shetlands 51, 199
Sicily (Sikelia) 15, 21, 32, 74–7, 102, 104, 195–8
Sigeion, Cape 119
silk 154, 157, 160, 198
Silk People 11, 21, 119, 157, 198; *see also* Seres
Silk Road 154
Silphiophoros 106
silphium 106–7
silver 170, 173
Silver, Land of 38, 169
Sinai, far eastern region 16, 22, 36, 39–41, 46, 119, 168, 170–1, 178–9, 198–9; Gulf or Bay of 171, 199; Peninsula 46, 142
Sinai, Mt. 141
Singaras Mtn. 142
Singas River 139
Singidunum (Belgrade) 85–6
Sinians 168
Sinibra (Erzurum) 121, 125, 133
Sinope 124
Sipphara (Sippar) 142
Sipylos, Mt. 123
Sirenousai (Li Galli) 72
Siriptolemaios 166
Sirius (Dog star) 32
Skandia 61–2
Skandiai 62
Skardon, Mt. 62, 66, 83, 85
Skardona (Skradin) 67
Skiathos 92
skiotheres 28
Skitis 53
Skopelos 92
Skordiskos Mtn. 126
Skorpiophoros 162
Skoupoi (Skopje) 86
Skylax of Karyanda 148, 176
Skythai *see* Skythia, Skythians
Skythian Issedon 159–60
Skythia, Skythians 33, 87, 130, 152, 156–60, 168, 171, 198–9
Smyrna 37, 123, 199
Smyrnaphoroi 150
Smyrnophoros region 114
Soanas River 130, 133–4
Sogdianians 41, 152, 155–6, 199

Soita 160
Sokandas River 154
Sokrates Island 151
Solva (Estergom) 65
Soudeta Mountains 60
Sounion, Cape (on Paros) 92
Sousa 147
Sousiane 143, 145–7
Sparta 91, 196
Speke, John Hanning 187–9
sphere, celestial and terrestrial concept of 42–4, 180–2, 184
Sporades 91–2
Stanley, Henry Morton 188
stars 8–10, 28, 31–2, 182
Stone Tower *see* Lithinos Pyrgos
Strabo of Amaseia 4–5, 9–10, 30, 57, 165, 185
strategiai 88, 126
Straton I of Sidon 140
Straton River 146
Strymon River 83, 193, 199
Sub-Saharan Africa 21, 193
Suebos River 61
Suetonius Paulinus, C. 98–9
Syagros Cape 199
Syene 29, 35, 43–4, 108–9, 180, 199–200
Symplegades 89
Syria 13–14, 36, 110, 121, 136, 138–42, 145, 195
Syriac language 23
Syria-Palestina 140
Syros 91

Tagos (Tajo) River 53
Tainaron 199
Takoraians 169
Tamala 38, 199
Tamerians 169
Tamesa (Thames) River 53
Tanagra, in Persis 147
Tanais River 21, 46, 77–81, 129–31, 179
Tanais, town 129–30
Tanatis 53
Taprobane 11, 14, 19–21, 38, 68, 121, 166, 169, 173–4, 182, 194, 196, 199
Taras (Taranto) 72
Tarentum 4
Tarracina 4
Tarraconensis (Tarrakonesia) 53–4
Tarrakon (Tarraco, Tarragona) 53, 199
Tasgaetium 64
Tauric Chersonesos 77, 79–81
Tauros Mtns. 133

Taurounon (Zemum) 62, 65–7, 83
Taxiana 147
Tektosagai, in Galatia 124
Tektosagians, in Skythia 159
Tenedos 122
Tenos 92
Tergeste (Trieste) 12, 40, 199
Tethys 184
Tetios River 138
Thammes Mtn. 104
Thasos 89
Theainai 199
Theganousa (Venetiko) 91
Theon Ochema 112
Theophanes of Mytilene 131, 134–5
Theophilos, explorer 10, 34, 115
Theophrastos of Eresos 3
Thera 92
Therasia 92
Thermodon River 126
Thessaly 91
Thiagola (Thiagos) Marsh 86
Thina (Thinai) 119, 170–1, 199
Thoreau, Henry David 187
Thospitis, Lake 136
Thrace 83, 85, 88–9, 193, 199
Thracian Bosporos 83, 88, 194
Thracian Chersonesos 88–9
Thraike *see* Thrace
Thraskias wind 181
Thrinakia (Trinakia) 75
Tiberius, emperor 7, 57, 60–1, 64
Tibiskos River 83–4
tigers 115–16, 173
Tigranes II of Armenia 135, 136
Tigranokerta 136
Tigris River 133, 136, 141–3, 145, 147,
     186, 197, 199
Tiladians 169
Tilavonton River 199
Timagenes Island 151
Timoula 199
Tingis (Tangier) 98
Tin Islands *see* Kassiterides
Titos (Krka) River 67
Tmolos, Mt. 123
Tolistobogoi 124
Tomis 83
Toniki 41, 200
topographia 28
Tourountos River 78
Tragourion (Trogir) 67
Traiana 72
Traianos Limen 72

Trajan, Column of 84
Trajan, emperor 14, 79, 85, 103, 126,
     141, 143
Trajan River 110
Trapezous 200
Triakontaschoinos 200
Triballians 87
Triglyphon (Trilingon) 169
trigonometry 2
Trikornioi 86
Triton Lake and River 104
Troad 122
Troesmis (Igliţsa) 87
Troglodytai (Trogodytai) 87
Troglodytike (Trogodytike) 34, 200
Trokmoi 124
Tropaea Augusta (Tropaia Sebastou) 72
Tropics 31, 35, 98, 114, 180, 199
Tyrangeitian Sarmatians 87
Tyras (Dniester) River 83–4, 87–8
Tyros (Tyre), in Phoenicia 13
Tyros (Tyre), on Persian Gulf 13
Tyrrhenikian Sea 74

Ulpianon (Ulpianum) 85
Ulpia Traiana Sarmizegetusa 85
Ultima Thule 51
Urgo (Gorgona) 72
Utica (Ityke) 102

Varus (Var) River 57
Vectis (Vektis) 53, 201–2
Vedra River 50
Vellaboroi 191
Venniknioi 48, 190–1
Vergivios Ocean 49, 191, 201
Vespasian, emperor 85, 140
Vetera (Xanten) 59
Vetus Salina *see* Salinon
Via Appia 4
Via Latina 4
Via Salaria 4
Victoria, in Mauretania 100
Victoria, in Scotland 51
Victoria Nyanza, Lake 34, 40, 187
Vidua River 190
Viminacium (Kostolac) 85
Vinderios River 191
Vindobona (Vienna) 65
Virunum 64
Vistula River 48, 59–60, 77–8, 80
Vodiai 191
Vologaesus I 143
Vologaisia 143

Volubilis 98–9
Voluntioi 191
vulcanism 124

Wales 50
Winged Camp 51
Wings of Kanobos 8

Xanthos 123
Xanthos River 123
Xinjiang 37

Zabai 39, 169–71, 200
Zagros Gates 146
Zagros Mtn. 145–6
Zanzibar 10, 116, 187, 194, 198
Zariaspa 155
Zeugma, in Dacia 85
Zeus, Cape 173
Zilia 98
Zingis Cape or Promontory 41, 197, 200
zodiac 15, 31
zones (terrestrial) 31, 116

For Product Safety Concerns and Information please contact our EU
representative GPSR@taylorandfrancis.com
Taylor & Francis Verlag GmbH, Kaufingerstraße 24, 80331 München, Germany

www.ingramcontent.com/pod-product-compliance
Lightning Source LLC
Chambersburg PA
CBHW060254220326
41598CB00027B/4092

9 781032 164427